DRUG DEVELOPMENT FOR CANCER AND DIABETES

A Path to 2030

DRUG DEVELOPMENT FOR CANCER AND DIABETES

A Path to 2030

Edited by

K. Saravanan, PhD
Chukwuebuka Egbuna, MSc
Horne Iona Averal, PhD
S. Kannan, PhD
S. Elavarasi, PhD
Bir Bahadur, PhD

Apple Academic Press Inc.
4164 Lakeshore Road
Burlington ON L7L 1A4, Canada

Apple Academic Press Inc.
1265 Goldenrod Circle NE
Palm Bay, Florida 32905, USA

© 2021 by Apple Academic Press, Inc.

Exclusive worldwide distribution by CRC Press, a member of Taylor & Francis Group

No claim to original U.S. Government works

International Standard Book Number-13: 978-1-77188-860-8 (Hardcover)
International Standard Book Number-13: 978-0-42933-049-0 (eBook)

All rights reserved. No part of this work may be reprinted or reproduced or utilized in any form or by any electric, mechanical or other means, now known or hereafter invented, including photocopying and recording, or in any information storage or retrieval system, without permission in writing from the publisher or its distributor, except in the case of brief excerpts or quotations for use in reviews or critical articles.

This book contains information obtained from authentic and highly regarded sources. Reprinted material is quoted with permission and sources are indicated. Copyright for individual articles remains with the authors as indicated. A wide variety of references are listed. Reasonable efforts have been made to publish reliable data and information, but the authors, editors, and the publisher cannot assume responsibility for the validity of all materials or the consequences of their use. The authors, editors, and the publisher have attempted to trace the copyright holders of all material reproduced in this publication and apologize to copyright holders if permission to publish in this form has not been obtained. If any copyright material has not been acknowledged, please write and let us know so we may rectify in any future reprint.

Trademark Notice: Registered trademark of products or corporate names are used only for explanation and identification without intent to infringe.

Library and Archives Canada Cataloguing in Publication

Title: Drug development for cancer and diabetes : a path to 2030 / edited by K. Saravanan, PhD [and 5 others]
Names: Saravanan, K. (Professor of zoology), editor.
Description: Includes bibliographical references and index.
Identifiers: Canadiana (print) 20200220551 | Canadiana (ebook) 20200220799 | ISBN 9781771888608 (hardcover) | ISBN 9780429330490 (ebook)
Subjects: LCSH: Antineoplastic agents—Development. | LCSH: Herbs—Therapeutic use. | LCSH: Medicinal plants. | LCSH: Cancer—Treatment. | LCSH: Diabetes—Treatment.
Classification: LCC RC271.H47 D78 2020 | DDC 616.99/4061—dc23

Library of Congress Cataloging-in-Publication Data

Names: Saravanan, K. (Professor of zoology), editor.
Title: Drug development for cancer and diabetes : a path to 2030 / edited by, K. Saravanan, Chukwuebuka Egbuna, Horne Iona Averal, S. Kannan, S. Elavarasi, Bir Bahadur.
Description: Burlington, ON ; Palm Bay, Florida : Apple Academic Press, [2020] | Includes bibliographical references and index. | Summary: "Drug Development for Cancer and Diabetes: A Path to 2030 focuses on new developments in the discovery of drugs for the treatment of cancer and diabetes. This new book presents important recent advances, emerging trends, and novel innovations for these two world-leading diseases. It is structured in two parts. Part I consists of chapters that explores cancer cell cycle checkpoints, cancer biomarkers, essential drug properties, stages in anti-cancer drug development, and various bioactives that are effective against cancer. Part II presents a complete coverage of important advances and research discoveries in antidiabetic drugs. It details the potentials of medicinal plants and phytocompounds as sources for antidiabetic drugs and highlights the ethnobotanical basis of the compounds. With chapters written by professionals in the field of phytochemistry and pharmacology from key institutions around the world, this volume provides a wide coverage and will be useful especially for scientists in pharmaceutical field. Key features: Presents novel discoveries and mechanisms for the treatment of cancer and diabetes Details the role of plants for the treatment and management of cancer and diabetes Discusses the role of phytocompounds as ligands for cancer and diabetic targets Reviews plants and the potential of phytochemicals as antidiabetic and anticancer drugs Explores the green biosynthesis of nanoparticles and their efficiency in the treatment of diabetes and cancer"-- Provided by publisher.
Identifiers: LCCN 2020015809 (print) | LCCN 2020015810 (ebook) | ISBN 9781771888608 (hardcover) | ISBN 9780429330490 (ebook)
Subjects: MESH: Antineoplastic Agents, Phytogenic | Neoplasms--drug therapy | Antihypertensive Agents | Diabetes Mellitus--drug therapy | Phytotherapy--methods | Drug Discovery
Classification: LCC RS431.A64 (print) | LCC RS431.A64 (ebook) | NLM QZ 267 | DDC 615.7/98--dc23
LC record available at https://lccn.loc.gov/2020015809
LC ebook record available at https://lccn.loc.gov/2020015810

Apple Academic Press also publishes its books in a variety of electronic formats. Some content that appears in print may not be available in electronic format. For information about Apple Academic Press products, visit our website at **www.appleacademicpress.com** and the CRC Press website at **www.crcpress.com**

About the Editors

K. Saravanan, PhD
Associate Dean of Academic Affairs and Assistant Professor of Zoology, Nehru Memorial College, Puthanampatti – 621 007, India, Tel.: +919443757052, E-mail: kaliyaperumalsaravanan72@gmail.com

K. Saravanan, PhD, is Associate Dean of Academic Affairs and Assistant Professor of Zoology, Nehru Memorial College (Autonomous), an affiliated institution of Bharathidasan University, Tiruchirappalli, Tamilnadu, India. He has been engaged in teaching and research for over 15 years. He has guided 6 Ph.D and 38 M.Phil scholars and at present several scholars are working under his guidance in the field of environment and natural drug research. He has published research papers, book chapters, and a book. He is currently editing two books with Elsevier and Astral Publishers. He is a reviewer for various journals, including the *Indian Journal of Experimental Biology, Journal of Threatened Taxa,* and *Journal of Medicinal Plant Research.* He is an editorial member of the *Indian Journal of Natural Science* and his university's college magazine, *The Rose.* He has edited conference abstracts and proceedings. Dr. Saravanan is a recipient of Har Gobind Khorana Best Scientist Award given by the Bose Science Society, Pudukottai, Tamilandu, India. He has completed one major research project and organized national and international conferences as well as several state-level workshops with financial assistance of various funding agencies. His fields of interest are natural drug development, rodent pest management, and ecology. He obtained his BSc and PhD degrees in zoology and he obtained his postgraduate degree in wildlife biology and post-MSc degree in bioinformatics.

Chukwuebuka Egbuna, MSc
*Research Biochemist, Department of Biochemistry,
Faculty of Natural Sciences, Chukwuemeka Odumegwu Ojukwu University,
Anambra State – 431124, Nigeria, Tel.: +2347039618485,
E-mail: egbuna.cg@coou.edu.ng*

Chukwuebuka Egbuna is a chartered chemist, a chemical analyst, and an academic researcher. He is a member of the Institute of Chartered Chemists of Nigeria (ICCON), the Nigerian Society of Biochemistry and Molecular Biology (NSBMB), the Society of Quality Assurance (SQA) (USA), and the Royal Society of Chemistry (RSC) (United Kingdom). He has been engaged in a number of roles at New Divine Favor Pharmaceutical Industry Limited, Akuzor Nkpor, Anambra State, Nigeria, and Chukwuemeka Odumegwu Ojukwu University (COOU), Nigeria. He has collaboratively worked and published quite a number of research articles in the area of phytochemistry and its applications. Mr. Egbuna has edited a 3-volume set book of phytochemistry with Apple Academic Press (*Phytochemistry, Volume 1: Fundamentals, Modern Techniques, and Applications; Phytochemistry, Volume 2: Pharmacognosy, Nanomedicine, and Contemporary Issues;* and *Phytochemistry, Volume 3: Marine Sources, Industrial Applications and Recent Advances*). He is also working on other books with Elsevier and Springer Nature. He is a reviewer and editorial board member of various journals, with his most recent role as a website administrator for the *Tropical Journal of Applied Natural Sciences* (TJANS), a journal of the faculty of Natural Sciences, COOU. His primary research interests are in phytochemistry, food and medicinal chemistry, analytical chemistry, and nutrition and toxicology. He obtained his BSc and MSc degrees in biochemistry at Chukwuemeka Odumegwu Ojukwu University.

Horne Iona Averal, PhD
*Dean of Science, Associate Professor of Zoology, Holy Cross College,
Trichy, Tamil Nadu, India, Tel.: +919443644858,
E-mail: profionahorne@gmail.com*

Horne Iona Averal, PhD, is an Associate Professor at Holy Cross College, Trichy, India. She has been honored with the Har Gobind Khorana Best Scientist Award and the Fellow of Bose Science Society Award. Also, she is a recipient of many other awards and honors, including a gold medal for best research paper from an international symposium. She has been engaged with many academic roles in her home institution and other institutions and universities. She is a member of academic bodies at 21 different colleges. Dr. Averal is a life member of the Bose Science Society and the Society of Reproductive and Comparative Endocrinology. She is an active reviewer of the *British Journal of Medicine and Medical Research*. She has undertaken major and minor research projects relevant to anti-repellency and polycystic ovary. Dr. Averal has published four nucleotide sequences in the NCBI-GENBANK and has developed many e-lessons in the area of metabolic disorders, genetic disorders, and gene therapy. She obtained her BSc and MSc degrees in zoology at Holy Cross College and her MPhil and PhD degrees in zoology at Bharathidasan University.

S. Kannan, PhD
Professor and Head, Department of Zoology, Periyar University, Salem, Tel.: 09952891773, E-mail: sk_protein@periyaruniversity.ac.in

Soundarapandian Kannan, PhD, is a specialist in cancer biology and nanomedicine. He has been engaged in a number of roles at national and international academic and scientific bodies. He has collaboratively worked and published a large number of research articles in the area of nanomedicines related to cancer. Dr. Kannan has published four sets of books relevant to molecular biology and cancer therapy and is a reviewer in six different journals. His recent studies have demonstrated the presence of a target for the development of therapeutic agents for lung and breast cancers, namely cytosolic phospholipase A2α (cPLA2α) and cycloxiginase. Secondly, he has performed target validation for cPLA2α and COX2 with a view to the development of highly specific, tumor-directed anticancer drug vehicles such as enzyme inhibitors and enzyme-targeted prodrugs, namely Pyrrolidine-2 and Doxorubicin loaded antibody dependent nanoparticles, etc. He has also formulated multifunctional magnetic polymers nano-combinations, which are gaining wide recognition in cancer nanotheranostics due to their safety and their potential in delivering targeted functions. M/S Sigma Aldrich has provided citations for his methodologies to extract products, such as sandal wood oil/natural oil, Fucoidan, R-Phycoerythrin, and application of 1-aminoanthracene. Professor Kannan obtained his BSc and MSc degrees in zoology from Madurai Kamaraj University and his PhD degree in zoology and PhD in biotechnology at Bharathidasan University, Tiruchirappalli, India.

S. Elavarasi, PhD
Assistant Professor, PG and Research Department of Zoology,
Holy Cross College (Autonomous), Tiruchirappalli, Tamil Nadu, India,
Tel.: +919003991615, E-mail: elavarasi888@gmail.com

S. Elavarasi, PhD, is an Assistant Professor at Holy Cross College, an affiliated institution of Bharathidasan University, Tiruchirappalli, Tamilnadu, India. She has been engaged in teaching and research for more than eight years. She was awarded a NTS-PhD fellowship–2010 by the Central Institute Indian Languages, Mysore, Karnataka, India, for pursuing her PhD degree. She has also awarded National Post Doctoral Fellowship–2016 by SERB-DST, New Delhi, India. She has collaboratively worked and published 25 research articles in various fields of life sciences. She also published two book chapters and a book. She has edited a conference souvenir. Her primary research interests are herbal technology and drug designing. She obtained her BSc, MSc, MPhil, PhD degrees in zoology, and PGDCA (PG Diploma in Computer Application) at Nehru Memorial College, Puthanampatti, Tiruchirappalli, India.

Bir Bahadur, PhD
Professor (Rtd.), Chairman and Head of the Department,
Dean of the Faculty of Science, Kakatiya University, Warangal, India

Bir Bahadur, PhD, was formerly Professor, Chairman, and Head of the Department, as well as Dean of the Faculty of Science at Kakatiya University in Warangal, India, and has also taught at Osmania University in Hyderabad, India. During his long academic career, he was honored with the Best Teacher Award by the Andhra Pradesh State Government for mentoring thousands of graduate and postgraduate students, most of whom went on to occupy high positions at various universities and research organizations in India and abroad. Dr. Bahadur has been the recipient of many awards and honors, including the Prof. Vishwambar Puri Gold Medal from the Indian Botanical Society for his original research contributions in various aspects of plant sciences. He is listed as an Eminent Botanist of India and is a BharathJyoti Awardee for his sustained academic and research career at New Delhi, India. Long active in his field, he is a fellow and member of over dozens professional bodies in India and abroad. He is a Fellow of the Linnean Society (London); Chartered Biologist & Fellow of the Institute of Biology (London); Member of the New York Academy of Sciences; Recipient of the Royal Society Bursary and Honorary Fellow of Birmingham University (UK); among others. He was an Independent Director at Sri Biotech Laboratories India, Ltd., Hyderabad, India, for several years. Professor Bahadur has published over 250 research papers in national and international journals as well as reviews, book chapters, and several books. He has been an invited speaker at over 120 national and international conferences and has served as Chief Editor of *Proceedings of Andhra Pradesh Akademi of Sciences* and the *Journal of Palynology*. A member of editorial boards of several national and international journals, he has made significant contributions in several areas of plant biology, especially in heteromorphic incompatibility, genetics, mutagenesis, morphogenesis, plant tissue culture and biotechnology, plant asymmetry and handedness, and the application of SEM pollen and seeds in relation to systematic and medicinal plants. He has supervised many PhD theses and has taught genetics, biotechnology, plant molecular biology, plant reproduction, and related subjects for over 45 years and has accumulated research experience in these areas for 50 years.

Contents

Contributors .... *xv*
Abbreviations .... *xix*
Foreword .... *xxv*
Preface .... *xxvii*

PART I: CANCER AND LEADS FOR CANCER TREATMENT 1

1. **Biomarkers for Screening of Premalignant Lesions and Therapeutic Surveillance** 3
   Chukwuebuka Egbuna

2. **Kinase Targets in Cancer Drug Discovery** 21
   S. Usha

3. **Essential Drug Properties and Stages in Anti-Cancer Drug Development** 29
   Andrew G. Mtewa, Duncan Sesaazi, Amanjotannu, and Serawit Deyno

4. **Plant Constituent as Anti-Cancer Drugs** 43
   T. Veni, T. Pushpanathan, and G. Vimala

5. **Anticancer Properties of Some Effective Phytochemicals** 59
   Murat Alan and Behnaz Aslanipour

6. **Therapeutic Aspects of Some Extracts and Purified Cardiac Glycosides Obtained From *Nerium oleander* L.** 79
   Behnaz Aslanipour and Murat Alan

7. **Challenging Compounds of Some Medicinal Plants Against Cancer** 95
   N. Shaista Jabeen, L. Jagapriya, B. Senthilkumar, and K. Devi

8. **Medicinal Plants as Therapeutic Agents in the Treatment of Cancer** 103
   Jaya Vikaskurhekar

9. **Cytotoxic, Apoptosis Inducing Effects and Anti-Cancerous Drug Candidature of Jasmonates** 117
   Parth Thakor, Ramalingam B. Subramanian, Sampark S. Thakkar, and Vasudev R. Thakkar

10. **Apoptosis Activity of 1,2,-Benzene Dicarboxylic Acid Isolated from *Andrographis paniculata* on KB, SiHa and IMR Cancer Cell Line** 129
    P. Krishnamoorthy and D. Kalaiselvan

11. Anticancer Drug Discovery From the Indian Spice Box 141
   M. V. N. L. Chaitanya, Santhivardhan Chinni, P. Ramalingam, and Y. Padmanabha Reddy

12. Phytochemical Investigation and Evaluation of Anti-Breast
   Cancer Activity of Chloroform Extract of *Tagetes erecta* 153
   R. Arul Priya, K. Saravanan, and Chukwuebuka Egbuna

13. Isolation, Identification and *In Silico* Evaluation of
   Anticancer Activity of Flavone from *Pisum sativum* 167
   R. Arul Priya, K. Saravanan, and P. Karuppannan

14. Functional Lead Phytochemicals from the Rutaceae and
   Zingeiberaceae Plants for Development of Anticancer Drugs 177
   K. Saravanan, B. Umarani, V. Balan, P. Premalatha, and Bir Bahadur

15. *In Vitro* Anticancer Activity of *Biophytum sensitivum* on Liver Cancer
   Lines (HEPG2) 191
   M. P. Santhi, K. Saravanan, and P. Karuppannan

16. Saffron and Its Active Ingredients: A Natural Product with
   Potent Anticancer Property 199
   Ashutosh Gupta and Abhay K. Pandey

PART II: ANTIDIABETIC DRUG DISCOVERY 217

17. Effects of *Pterocarpus marsupium* in the Management of
   Type 2 Diabetes Mellitus 219
   Priscilla Suresh

18. Antidiabetic and Antihyperlipidemic Activities of *Cyathea nilgiriensis*
   (Holttum) on STZ Induced Diabetic Rats 233
   S. Elavarasi, G. Revathi, K. Saravanan, and Horne Iona Averal

19. Antidiabetic Activity of Drug Loaded Chitosan Nanoparticle 249
   G. Revathi, S. Elavarasi, and K. Saravanan

20. Antidiabetic Activity of Silver Nanoparticles Biosynthesized Using
   Ventilago maderaspatana Leaf Extract 263
   P. Karuppannan, K. Saravanan, and Horne Iona Averal

21. Hypoglycemic Activity of *Biophytum sensitivum* Whole Plant
   Extracts on Alloxan Induced Diabetic Rats 275
   C. Renuka, K. Saravanan, and P. Karuppannan

22. Antidiabetic Activity of *Ventilago maderaspatana* Leaf Extracts on
   STZ Induced Diabetic Rats 283
   P. Karuppannan, K. Saravanan, and P. Premalatha

Contents

23. **Isolation, Identification, and Molecular Docking of Antidiabetic Compounds of *Cyatheia nilgiriensis* (Holttum)** 293

    S. Elavarasi, G. Revathi, and K. Saravanan

24. **Antidiabetic Potentials of Major Indian Spices: A Review** 305

    Pratibha, Hema Lohani, S. Zafar Haider, and Nirpendra K. Chauhan

25. **Aloe Species: Chemical Composition and Therapeutic Uses in Diabetic Treatment** 343

    Amit Kumar Singh, Ramesh Kumar, B. Umarani, and Abhay K. Pandey

Index *349*

Contributors

Murat Alan
Department of Obstetrics and Gynecology, Izmir Tepecik Training and Research Hospital, Tepecik – 35120, Izmir, Turkey

Amanjotannu
Department of Pharmacy and Pharmacology, Mbarara University of Science and Technology, Mbarara, Uganda

Behnaz Aslanipour
Department of Bioengineering, Faculty of Engineering, Ege University, Izmir, Turkey,
E-mail: Behnaz_Aslanipour@yahoo.com

Horne Iona Averal
P.G. and Research Department of Zoology, Holy Cross College (Autonomous), Trichy, Tiruchirappalli, Tamil Nadu, India

V. Balan
P.G. and Research Department of Zoology, Nehru Memorial College (Autonomous), Puthanampatti – 621 007, Tiruchirappalli, Tamil Nadu, India

Bir Bahadur
Department of Botany, Kakatiya University, Warangal, India

M. V. N. L. Chaitanya
Department of Pharmaceutical Quality Assurance, Phytomedicine and Phytochemistry Division, Raghavendra Institute of Pharmaceutical Education and Research (Autonomous), Saigram, Anantapuramu – 515721, Andhra Pradesh, India,
E-mails: chaitanya.phyto@gmail.com, drchaitanya@riper.ac.in

Nirpendra K. Chauhan
Center for Aromatic Plants (CAP), Industrial Estate Selaqui – 248011, Dehradun, Uttarakhand, India

Santhivardhan Chinni
Department of Pharmacology, Pharmacology and Toxicology Division, Raghavendra Institute of Pharmaceutical Education and Research (Autonomous), Saigram, Anantapuramu – 515721, Andhra Pradesh, India

K. Devi
Department of Zoology, Thiruvalluvar University, Vellore, Tamil Nadu India

Serawit Deyno
Department of Pharmacy and Pharmacology, Mbarara University of Science and Technology, Mbarara, Uganda, Department of Pharmacology, School of Medicine, College of Medicine and Health Sciences, Hawassa University, Hawassa, Ethiopia, Pharm-BioTechnology and Traditional Medicine,
A World Bank-Africa Center of Excellence (ACE II), Mbarara University of Science and Technology, Mbarara, Uganda

Chukwuebuka Egbuna
Department of Biochemistry, Faculty of Natural Sciences, Chukwuemeka Odumegwu Ojukwu University, Anambra State – 431124, Nigeria, Tel.: +2347039618485,
E-mails: egbuna.cg@coou.edu.ng, egbunachukwuebuka@gmail.com

S. Elavarasi
P.G. and Research Department of Zoology, Holy Cross College (Autonomous), Trichy, Tiruchirappalli,
Tamil Nadu, India, E-mail: elavarasi888@gmail.com

Ashutosh Gupta
Department of Biochemistry, University of Allahabad, Allahabad – 211002, India

S. Zafar Haider
Center for Aromatic Plants (CAP), Industrial Estate Selaqui – 248011, Dehradun, Uttarakhand, India,
Tel.: +91-9450743795, E-mails: zafarhrdi@gmail.com, zafarhaider.1@rediffmail.com

N. Shaista Jabeen
Department of Zoology, Dhanabagyam Krishnaswamy Mudaliar College for Women (Autonomous),
Vellore – 632 001, Tamil Nadu, India

L. Jagapriya
Department of Zoology, Dhanabagyam Krishnaswamy Mudaliar College for Women (Autonomous),
Vellore – 632 001, Tamil Nadu, India

D. Kalaiselvan
Department of Zoology, Periyar E.V.R College (Autonomous), Thiruchirappalli – 620 023,
Tamil Nadu, India

P. Karuppannan
P.G. and Research Department of Zoology, Nehru Memorial College (Autonomous), Puthanampatti,
Trichy District, Holy Cross College (Autonomous), Trichy, Tiruchirappalli, Tamil Nadu, India,
Tel.: 9790298098, E-mail: vikramprabhu88@gmail.com

P. Krishnamoorthy
Department of Zoology, Periyar E.V.R College (Autonomous), Thiruchirappalli – 620 023,
Tamil Nadu, India, E-mail: pkmoorthy68@rediffmail.com

Ramesh Kumar
Department of Biochemistry, University of Allahabad, Allahabad – 211002, Uttar Pradesh, India

Hema Lohani
Center for Aromatic Plants (CAP), Industrial Estate Selaqui – 248011, Dehradun, Uttarakhand, India

Andrew G. Mtewa
Chemistry Section, Department of Applied Science, Malawi Institute of Technology,
Malawi University of Science and Technology, Thyolo, Malawi, Department of Pharmacy and
Pharmacology, Mbarara University of Science and Technology, Mbarara, Uganda,
Pharm-BioTechnology and Traditional Medicine, A World Bank-Africa Center of Excellence
(ACE II), Mbarara University of Science and Technology, Mbarara, Uganda,
E-mails: andrewmtewa@yahoo.com, amtewa@must.ac.mw, ORCID: 0000–0003-2618-7451

Abhay K. Pandey
Department of Biochemistry, University of Allahabad, Allahabad – 211002, Uttar Pradesh, India,
E-mail: akpandey23@redifmail.com

Pratibha
Center for Aromatic Plants (CAP), Industrial Estate Selaqui – 248011, Dehradun, Uttarakhand, India

P. Premalatha
P.G. and Research Department of Zoology, Nehru Memorial College (Autonomous),
Puthanampatti – 621 007, Trichy District, Tiruchirappalli, Tamil Nadu, India

Contributors

R. Arul Priya
P.G. and Research Department of Zoology, Nehru Memorial College (Autonomous), Puthanampatti – 621 007, Trichy District, Tiruchirappalli, Tamil Nadu, India, E-mail: arulpriyaphd@gmail.com

T. Pushpanathan
Department of Zoology, St. Xavier's College (Autonomous), Palayamkottai – 627 002, Tamil Nadu, India

P. Ramalingam
RERDS-CPR, Raghavendra Institute of Pharmaceutical Education and Research (Autonomous), Saigram, Anantapuramu – 515721, Andhra Pradesh, India

Y. Padmanabha Reddy
Department of Pharmaceutical Analysis, Raghavendra Institute of Pharmaceutical Education and Research (Autonomous), Saigram, Anantapuramu – 515721, Andhra Pradesh, India

C. Renuka
P.G. and Research Department of Zoology, Nehru Memorial College (Autonomous), Puthanampatti, Trichy District, Pin – 621 007, Tamil Nadu, India, E-mail: renukaphd@gamil.com

G. Revathi
P.G. and Research Department of Zoology, Nehru Memorial College (Autonomous), Puthanampatti, Trichy District, Tiruchirappalli, Tamil Nadu, India, E-mail: revathiphdz@gmail.com, ORCID: https/orcid.org/

Sampark S. Thakkar
Department of Advanced Organic Chemistry, P.D. Patel Institute of Applied Sciences, Charusat, Changa – 388421, Gujarat, India

M. P. Santhi
P.G. and Research Department of Zoology, Nehru Memorial College (Autonomous), Puthanampatti, Trichy District, Pin – 621 007, Tamil Nadu, India

K. Saravanan
P.G. and Research Department of Zoology, Nehru Memorial College (Autonomous), Puthanampatti – 621 007, Trichy District, Tiruchirappalli, Tamil Nadu, India, E-mail: kaliyaperumalsaravanan72@gmail.com, ORCID: https/orcid.org/

B. Senthilkumar
Department of Zoology, Thiruvalluvar University, Vellore, Tamil Nadu India

Duncan Sesaazi
Department of Pharmacy and Pharmacology, Mbarara University of Science and Technology, Mbarara, Uganda, Pharm-BioTechnology and Traditional Medicine, A World Bank-Africa Center of Excellence (ACE II), Mbarara University of Science and Technology, Mbarara, Uganda

Amit Kumar Singh
Department of Biochemistry, University of Allahabad, Allahabad – 211002, Uttar Pradesh, India

Ramalingam B. Subramanian
P.G. Department of Biosciences, Sardar Patel Maidan, Bakrol-Vadtal Road, Satellite Campus, Bakrol, Sardar Patel University, Vallabh Vidyanagar, Gujarat, India

Priscilla Suresh
P.G. and Research Department of Zoology, Bishop Heber College, Tiruchirappalli – 620 017, India, Tel.: 9789164989, E-mail: priscisf@gmail.com

Vasudev R. Thakkar
P.G. Department of Biosciences, Sardar Patel Maidan, Bakrol-Vadtal Road, Satellite Campus, Bakrol, Sardar Patel University, Vallabh Vidyanagar, Gujarat, India, E-mail: vasuthakkar@gmail.com

Parth Thakor
P.G. Department of Biosciences, Sardar Patel Maidan, Bakrol-Vadtal Road, Satellite Campus, Bakrol, Sardar Patel University, Vallabh Vidyanagar, Gujarat, India, E-mail: parth7218@gmail.com

B. Umarani
P.G. and Research Department of Zoology, Nehru Memorial College (Autonomous), Puthanampatti – 621 007, Tiruchirappalli, Tamil Nadu, India, E-mail: umamiraa86@gmail.com

S. Usha
Department of Bioinformatics, Bharathiar University, Coimbatore – 641 046, Tamil Nadu, India, E-mail: usha@buc.edu.in

T. Veni
Department of Zoology, Ayya Nadar Janaki Ammal College (Autonomous), Sivakasi – 626 124, Tamil Nadu, India, Tel.: 099431 26685, E-mail: venientomology@gmail.com

Jaya Vikaskurhekar
Department of Microbiology, Dr. Patangrao Kadam Mahavidyalaya, Sangli – 416416, Maharashtra, India

G. Vimala
Aringnar Anna Government Arts College for Women, Walajapet – 632 513, Tamil Nadu, India

Abbreviations

ACF	aberrant crypt foci
ADA	American Diabetes Association
AFP	alpha-fetoprotein
AgNPs	silver nanoparticles
ALP	alkaline phosphatase
ANOVA	analysis of variance
AO	acridine orange
ATF3	activating transcription factor 3
ATP	adenosine tri-phosphate
AuNPs	gold nanoparticles
BBB	blood-brain barrier
BDNF	brain-derived neurotrophic factor
CA-4	combretastatin A-4
CCl_4	carbon tetrachloride
CDKs	cyclin-dependent kinases
CEA	carcinoembryonic antigen
CIN	cervical intraepithelial neoplasia
CJ	cis-jasmone
CLIA	chemiluminescence immune assay
CLL	chronic lymphocytic leukemia
CML	chronic myelogenous leukemia
CNPs	chitosan nanoparticles
COX-2	cyclooxygenase-2
CPCSEA	Committee for the Permission and Control of Supervision on Experimental Animals
CS	chitosan
CSCs	cancer stem cells
DADS	diallyl disulfide
DCIS	ductal carcinoma *in situ*
DEN	diethylnitrosamine
DLA	Dalton's lymphoma ascites
DMSO	dimethyl sulfoxide
DNA	deoxyribonucleic acid
DRE	digital rectal examination
EAC	Ehrlich ascites carcinoma

EB	ethidium bromide
EGCG	epigallocatechin-3-gallate
EGFR	epidermal growth factor receptor
ELISA	enzyme-linked immunosorbent assay
EMT	epithelial-mesenchymal trans-differentiation
FAK	focal adhesion kinase
FDA	Food and Drug Administration
FE-SEM	field emission scanning electron microscope
FGF-2	fibroblast growth factor-2
FMLP	formylpeptide
FOB	fecal occult blood
FOBT	fecal occult blood test
GC-MS	gas chromatography-mass spectrometry
GI	gastrointestinal
GK	glucokinase
GLUT	glucose transporter
GOD	glucose oxidase method
GOD/POD	glucose oxidase-peroxidase
GPI	glycosylphosphatidyl inositol
GPx	glutathione peroxidase
GSE	grape seed extract
GSH	glutathione
GST	glutathione S-transferase
HBA	hydrogen bond acceptors
HbA$_1$C	glycosylated hemoglobin
HBD	hydrogen bond donors
HCC	Hepatocellular carcinoma
hCG	human chorionic gonadotropin
HDAC	histone deacetylase
HDL	high-density lipoproteins
HDNC	hydroxy-5,7-dimethoxy-2-naphthalene-carboxaldehyde
HIA	human intestinal absorbance
HOMA	homeostasis model assessment
HPV	human papillomavirus
Hsp90	heat shock protein 90
HTLV-1	human T-cell leukemia virus type-1
HVA	homovanillic acid
IC	inhibition concentration
IGF-II	insulin-like growth factor 2

Abbreviations

IL-1	interleukin-1
IP	intraperitoneal method
IV	intravenous
JA	jasmonic acid
KLK3	kallikrein-3
KSHV	Kaposi sarcoma-associated herpesvirus
LDH	lactate dehydrogenase enzyme
LDHA	lactate dehydrogenase-A
LDL	low-density lipid
LPO	lipid peroxidation
MCA	methylcholanthrene
MCF-7	Michigan cancer foundation-7
MDA	malondialdehyde
MDS	molecular design suit
MF	molecular formula
MJ	methyl jasmonate
MNC	mononuclear cell
MS	molecular structure
MSG	monosodium glutamate
MTT	methyl thiazol tetrazolium
MW	molecular weight
NCCS	National Center for Cell Science
NCI	National Cancer Institute
NF-κB	nuclear factor kappa-light-chain-enhancer from activated B chain
NGF	nerve growth factor
NIDDM	non-insulin dependent diabetic
NIST	National Institute of Standards and Technology
NPs	nanoparticles
Nrf2	nuclear factor erythroid 2-related factor 2
ODC	ornithine decarboxylase
OS	osteosarcoma
OVCAR-3	human ovarian carcinoma
PA	peak area
PD	pharmacodynamics
PDB	protein data bank
PEPCK	phosphoenolpyruvate carboxykinase
PK	pharmacokinetics
PMNs	polymorphonuclear cells

PPARs	peroxisome proliferator-activated receptors
PPB	plasma protein binding
PSA	polar surface area
PSA	prostate-specific antigen
PTH	parathyroid hormone
RB355	retinoblastoma cancer cells
RGNF	Rajiv Gandhi National Fellowship
RLUs	relative light units
RotB	rotatable bonds
RT	retention time
S.E	standard error of mean
SAMC	S-allylmercaptocysteine
SARs	structural-activity-relationships
SC	subcutaneous
SCC	skin cancer center
SD	standard deviation
SEM	scanning electron microscopy
SIL	squamous intraepithelial lesion
SMKIs	small-molecule kinase inhibitors
SNK	student-Newman Keuls
SOD	superoxide dismutase
SPSS	statistical package for the social sciences
STZ	streptozotocin
TCC	transitional cell carcinoma
TEM	transmission electron microscope
TG	triglyceride
TNF	tumor necrosis factor
TNFR	TNF receptor
TNF-α	tumor necrosis factor-α
Topo I	topoisomerase I
TPA	tetradecanoylphorbol-13-acetate
TPC	thyroid cancer cell
TPP	tripolyphosphate
TRADD	TNF receptor-associated death domain
TRAIL	TNF-related apoptosis-inducing ligand
UGC	University Grant Commissions
VCR	vincristine
VEGF	vascular endothelial growth factor
VIA	visual inspection with acetic acid

VLDL	very low-density lipoprotein
VMA	vanillylmandelic acid
WHO	World Health Organization
XRD	x-ray diffraction

Foreword

Drug discovery and development has evolved significantly over the last century. This book extrapolates the recent advancements, emerging trends, and novel innovations and speculates about the future perspective in the discovery of drugs for the treatment of cancer and diabetes. As a complete package for the readers, this book provides a valuable window on the revolutionary sector of drug discovery. Part I and Part II of the book together provides a comprehensive content to understand the theoretical aspect as well as practical advances and research discoveries. The novel approach of depicting the role of plants and the potential of phytochemicals as antidiabetic and anticancer drugs is a key feature.

I appreciate the endeavor of authors for providing the readers with a broader perspective of the discipline, together with the challenges, thus, enhancing the enthusiasm, tenacity, and dedication to develop new methods and provide new solutions to keep up with the ever-changing requirements in the field of drug development. I wish all the best for grand success and wide distribution of this book.

—Dr. P. Padmanabhan
Deputy Director, Lee Kong Chian School of Medicine,
Nanyang Technological University,
Singapore

Preface

Plants have been very essential to human life from the beginning of human civilization, and records are available for using plants since about 5000 years ago. Several civilizations have contributed to the indigenous system of medicines. The active principles are identified, which provide a lead for the development of several lifesaving drugs that are used in day-to-day life. Now, there is a belief that allopathic drugs cause serious side effects on the human body; in contrast, herbal medicines work better and provide long-lasting healing effects and without any side effects. During the past few decades, people in industrial countries became dramatically interested in traditional complementary and alternative medicines (herbal medicines). Many pharmaceutical companies are evincing much interest in the production and marketing of herbal medicines. Plant medicines have great demand in developed as well as developing countries for primary health care because of their vast biological activities, higher safety margins, and lesser costs. Researchers make more effort for the development of phytopharmaceuticals against severe metabolic syndromes including cancer and diabetes. Several plants and their secondary metabolites have been screened for medicinal purpose, but only a very few have reached the clinical level. They must be explored for their cancer and diabetes potential.

Based on this above rationale, the book *Drug Development for Cancer and Diabetes: A Path to 2030* presents information on review/research communications received from eminent scientists, providing recent and current state-of-the-art data on therapeutic properties, actions, and uses of medicinal plants in the combating of cancer and diabetes. It is hoped that our present book will attract a wide readership of phytochemists, pharmacologists, medicinal professionals of cancer and diabetes, and other biologists to facilitate further research.

Chapters included in this book are written by professionals from key institutions around the world. Most of the authors are experts in the field of phytochemistry and pharmacology. Ideas in this book are carefully arranged and integrated sequentially with consistency and continuity. The insights are grouped into two parts. Part I covers the fundamentals of cancer and phytochemical leads to cancer treatment. Part II of this book presents the phytochemical drug research for the development of the diabetic drug. Each chapter has a short abstract that explains the scientific basis of the chapter and what readers might expect. The literatures in the chapters ensure that all the information is authentic and verifiable.

Chapter 1 written by Egbuna explains the biomarkers for screening of premalignant lesions and therapeutic surveillance. Chapter 2 deals with the cancer drug discovery related to kinase targets. Usha structurally presents the various kinase targets and the mechanisms involved. Chapter 3, written by Andrew et al., is an overview of the drug properties and stages in anti-cancer drug development. Veni et al. present the plant-derived natural products as leads to anticancer drugs in Chapter 4, while authors Murat Alan and Behnaz Aslanipour discuss anticancer properties of some effective phytochemicals in Chapter 5. Chapter 6 details the therapeutic aspects of some extracts and purified cardiac glycosides obtained from *Nerium oleander*. Chapter 7 by Shaista Jabeen et al. is an overview of challenging compounds of some medicinal plants against cancer. Jaya Vikaskurhekar in Chapter 8 describes the potent medicinal herbs and their therapeutic agent- for treatment of cancer. Parth et al. in Chapter 9 provide an overview of cytotoxic, apoptosis-inducing effects and anti-cancerous drug candidature of *Jasmonates*. In Chapter 10, Krishnamoorthy, and Kalaiselvan present their findings of apoptosis activity of 1,2,-benzene dicarboxylic acid isolated from *Andrographis paniculata* on KB, SiHA, and IMR cancer cell lines. Chaitanya et al. in Chapter 11 describe anticancer drug discovery from the Indian spice box. Arul Priya et al. in Chapter 12 explain their result of anti-breast cancer effect of *Tagetes erecta* plant extract. Arul Priya and Saravanan describe the method of isolation, identification, and evaluation of anticancer activity of flavonoids from *Pisum sativum* in Chapter 13. Saravanan et al. in Chapter 14 present the functional lead compounds from the Rutaceae and Zingiferacea plants for development of anticancer drugs. Santhi et al. in Chapter 15 present mechanism of in vitro anticancer activity of *Biophytum sensitivum* on liver cancer lines (HEPG2). Gupta and Pandey in Chapter 16 describe the role of saffron in cancer treatment. In Part II, Chapter 17 by Priscilla Suresh is explained the effects of *Pterocarpus marsupium* (Gammalu) in the management of type 2 diabetes mellitus. Elavarasi et al. describe antidiabetic activity of *Cyathea nilgiriensis* (Holttum) on STZ induced diabetic rats in Chapter 18. Revathi et al. in Chapter 19 explain their results on antidiabetic activity of drug-loaded chitosan nanoparticle. In Chapter 20, Karuppannan et al. present their findings of the synthesis of silver nanoparticles using *Ventilago maderaspatana* leaf extracts and their role in diabetic treatment. Renuka et al. in Chapter 21 emphasize hypoglycemic activity of *Biophytum sensitivum* whole plant extracts on alloxan induced diabetic rats. Karuppannan et al. in Chapter 22 present diabetic research findings on *Ventilago maderaspatana* leaf extracts on STZ induced diabetic rats. The authors present photomicrographs of the diabetic and extract-treated pancreas. Chapter 23 by Elavarasi et al. describe the mechanism of a compound isolated from *Cyathea nilgiriensis* and their *in silico* antidiabetic activities. Pratibha et al. in Chapter 24 present a review of antidiabetic potentials of major Indian spices. In Chapter 25, Singh et al. explain the chemical composition and therapeutic application of Aloe species related with diabetes.

In a nutshell, the present book provides a wide scope and would be useful especially for scientists in the pharmaceutical field. We sincerely thank the contributors of this book for their constant support. We also extend our sincere appreciation to our family for their support and patience during the editorial process of this book. Our gratitude also goes to Apple Academic Press for their guidance from the onset of this book project.

We recommend this book to everyone who is interested in cancer and diabetic research areas. We welcome reviews, suggestions, and areas that will need improvement with an open heart.

—**Editors**

PART I
Cancer and Leads for Cancer Treatment

CHAPTER 1

Biomarkers for Screening of Premalignant Lesions and Therapeutic Surveillance

CHUKWUEBUKA EGBUNA

Department of Biochemistry, Faculty of Natural Sciences, Chukwuemeka Odumegwu Ojukwu University, Anambra State – 431124, Nigeria, Tel.: +2347039618485, E-mails: egbuna.cg@coou.edu.ng, egbunachukwuebuka@gmail.com

ABSTRACT

Biomarker, a portmanteau of "biological marker" is a naturally occurring molecule, gene, or characteristic capable of serving as a measurable indicator for some biological state or disease condition. Cancer biomarkers over time have gained lots of recognition for their usefulness in the early detection of cancer in symptomatic and asymptomatic patients, surveillance of patients with a high risk of cancer recurrence and as surrogate endpoint markers for primary prevention strategies. Biomarkers are also useful for the evaluation of potential drug therapies and drug developmental processes including drug developmental studies, safety measurements, proof of concept studies, and molecular profiling. This chapter provides an overview of the role of biomarkers in cancer research, screening, and drug development. It went further to detail the various cancer biomarkers utilized for screening, diagnosis, and prognosis.

1.1 INTRODUCTION

Cancer is a group of diseases that affects the human body. It manifests through the proliferation of previously healthy cells into tumor cells by some causative factors. Different types of cancers numbering over 200 are now known and classified based on the type of cell that is initially affected. For instance, cancers arising from epithelial cells are called carcinoma, those from connective tissue are called sarcoma, and those from hematopoietic (blood-forming) are called lymphoma and leukemia while those from pluripotent cells and immature "precursor" cells/embryonic tissue are respectively known as germ cell tumor and blastoma. The transformation of healthy cells can give rise to tumor cells which eventually becomes cancer through a multistage process usually from a pre-cancerous lesion to a malignant tumor (WHO, 2018a). In the disease condition, old cells do not die, instead grows out of control, forming new abnormal cells. These abnormal cells often result in a tumor although some kinds of cancers do not form tumors, e.g., leukemia. The tumor can grow and obstruct the function of the nervous, digestive, and circulatory systems, with the possibility of releasing hormones that can alter the body functions. Tumors that do not spread and with limited growth are called benign tumors while those capable of spreading are called a malignant tumor. There are six hallmarks criterion for malignant tumor formation. They include cell growth and division, continuous growth and division even when given contrary signals, circumventing programmed cell death, uncontrolled number of cell divisions, promoting blood vessel construction, metastases formation and invasion of tissue (Hanahan and Weinberg, 2000). The potential signs and symptoms of cancer include unusual bleeding or discharge (blood in the urine, bleeding between periods, blood during cough and vomit), lump in the breast, coughing, chest pain and breathlessness, changes in bowel or bladder habits, indigestion or difficulty in swallowing, a sore that does not heal, obvious changes is warts or moles, and unexplained weight loss.

In the United States alone, cancer is notoriously the second leading cause of death after heart disease which has retained the number one spot for the past 15 years (CDC, 2017). One-half of men and one-third of women have a lifetime cancer risk. In 2015, about 90.5 million people worldwide had cancer which resulted in 8.8 million deaths (15.7% of deaths) (GBD, 2016). Yearly, about 14.1 million new cases occur excluding skin cancer. According to WHO (2018a), the most prevalent type of cancers is lung cancer which caused 1.69 million deaths, liver cancer which caused 788,000 deaths, the colorectal cancer which led to 774,000 deaths, stomach cancer which caused 754,000 deaths and the breast cancer which caused 571,000 deaths. Approximately, 70% of these deaths occur in the low- and middle-income countries. This may partly be because of the lack of early-stage screening facilities or poor diagnostic conditions. In 2017, it was found that only 26% of low-income countries have pathology services available in the public sector compared to more than 90% for high-income countries (WHO, 2018a). Globally, about one in six deaths are caused by cancer. A major fraction, 90–95% of cases of cancer, are as a result of genetic mutations which is influenced by environmental factors (not genetically inherited, e.g., lifestyle, behavioral, and economic factors), while the remaining 5–10% cases of cancer are due to inherited genetics (Anand et al., 2008). From records, it was found that one-third of deaths from cancer are linked to five leading behavioral and dietary risks. These are high body mass index, lack of physical activity or exercise, low consumption of fruits and vegetables, tobacco, and alcohol use. These factors culminate between a person's genetic factors and three categories of external agents, including: physical carcinogens, such as radiations (ultraviolet, ionizing radiation, etc.); chemical carcinogens, e.g., asbestos, benzidine, aristolochic acids, mists from strong inorganic acids, e.g., sulfuric acid, formaldehyde, aflatoxin (a food contaminant), tobacco smoke components, arsenic; biological carcinogens, e.g., infections of certain viruses (e.g., human papillomavirus (HPV), Hepatitis B and C viruses, Kaposi sarcoma-associated herpesvirus (KSHV) and human T-cell leukemia/lymphoma virus type-1 (HTLV-1), bacteria (e.g., *Helicobacter pylori*), or parasites (e.g., *Opisthorchis viverrini*, *Schistosoma haematobium*). Up to 25% of cancer cases result from hepatitis and HPV infections in low- and middle-income countries (Plummer et al., 2016).

According to the American Cancer Society (2018), the increasing cases of deaths from cancer can be prevented by systematic approach by improved lifestyle through proper dieting and physical activity, reduction of tobacco use, good dieting, physical activities, and the use of tobacco and application of established screening tests. It can reduce death from cancer. In 2010, the total annual economic cost of cancer was estimated to be US$ 1.16 trillion (Stewart et al., 2014). However, there appear some prospects from the advent of new techniques for cancer research in molecular biology and biochemistry together with the declaration of "War on Cancer."

1.1.1 EARLY DETECTION OF CANCER

The destructive effects of cancer can be reduced if cases are detected and treated early. The early detection of cancer may lead to decreased morbidity with improved survival, and in some situations treatment may require only surgery if identified early enough (Schiffman et al., 2015). The two basic components of early detection are early diagnosis and screening. In early detection, education plays a vital role in creating awareness for possible warning signs of cancer, among physicians, nurses, and other health care providers as well as among the general public. Some early signs to look up for are lumps, sores that fail to heal, abnormal bleeding, persistent indigestion, and chronic hoarseness (WHO, 2018b). The risk for overdiagnosis remains a major concern in screening, whereby lesions of no clinical consequence may be detected and thus create difficult management decisions for the clinician and patient. For instance, a 2013 Cochrane review indicated the use of mammography in breast cancer screening had no effect in reducing mortality probably because of overdiagnosis and overtreatment (Gotzsche and Jorgensen, 2013).

1.1.1.1 EARLY DIAGNOSIS

The early diagnosis of cancer is a crucial step taken when one is a risk or when there is suggestive evidence. The early diagnosis is particularly relevant to cancers of the breast, cervix, mouth, larynx, colon, and rectum, and skin (WHO, 2018b). Although, each diagnostic procedure has limitations, and improved methods are required. Persons susceptible are expected to undergo blood tests, X-rays or CT scans and endoscopy. Also, tissue diagnosis from biopsy[1] can show the type of proliferating cell, the histological grade, abnormalities in genes, or other characteristics useful for prognosis or treatment.

1.1.1.2 EARLY DETECTION OF BREAST CANCER

There are many tests available for early detection of breast cancer. They include clinical breast exams, breast self-exams, mammography[2], magnetic resonance imaging, ultrasound, computerized tomography, positron emission tomography, and biopsy Maurer Foundation (2018). Although some of these techniques are very expensive, time consuming and not suitable for young women, there is need for coordinated effort for the development of biosensors utilizing different biomarkers for early breast cancer detection (discussed in the next section). Mammography has proven to be a useful

---

[1] Biopsy is a medical practice which involves the examination of tissue removed from a living body to discover the presence, cause, or extent of a disease.
[2] A mammogram is simply an X-ray of the breast.

screening test for breast cancer because it can detect breast cancer up to two years before the tumor can be felt. According to Hellquist et al. (2015) and Onega et al. (2016), although mammography has proven helpful, it does not provide more information of eventual disease outcome. Again, it is less effective in detecting very small tumors especially those < 1 mm, about 100,000 cells. There are some risk factors that should be noted when diagnosing cancer as noted by Cancer Care (2016). They are:

- **Age:** Over 80% of breast cancer is seen in women older than 50 years.
- **Genetic Factors:** Genetic mutation affecting women on the BRCA1 and BRCA2 genes are at a higher risk of developing cancer.
- **Personal History of Breast Cancer:** A woman with breast cancer in one breast has higher risk of developing cancer on the other breast.
- **Family History of Breast Cancer:** Women with family or relative developing breast cancer are at a higher risk.
- **Childbearing and Menstrual History:** Women who menstruate at an early stage (<12 years of age), those who goes through menopause lately (>55 years of age) and women who never gave birth to a child are at high risk of developing breast cancer.

1.1.1.3 *EARLY DETECTION OF PROSTATE CANCER*

Prostate cancer is one of the most prevalent form of cancer in men that occurs in the prostate—a small walnut-shaped exocrine gland in men that produces the seminal fluid which is essential for the nourishment and transportation of sperm through the journey to fuse with a female ovum. Prostate-specific antigen (PSA), a protein produced by the prostate, enables the semen maintain its liquid state. An abnormal increase of PSA in the blood usually is taken as signs of prostate cancer (Nordqvist, 2017a). Detection also involves digital rectal examination (DRE) in which a clinician places his hands gently to feel the size of the prostate. Usually, it is recommended a biopsy is done when one is at risk. Prostate cancer is characterized by difficulty commencing and maintaining urination in men, frequent urges to urinate including at night, decreased force in the stream of urine, painful urination, semen containing blood, pelvic area discomfort, bone pain, and erectile dysfunction. The early detection of prostate cancer may help increase the chances of successful treatment before it spreads. Many risk factors have been identified and they include old age, race (black men are at most risk), family history of prostate cancer or breast cancer gene (BRCA1 or BRCA2), and obesity. Although prostate cancer can't be inherited, a man can inherit genes that can increase the risk.

1.1.1.4 *EARLY DETECTION OF OVARIAN CANCER*

As the name implies, ovarian cancer results from cancerous growth that originates from the ovary. According to Nordqvist (2017b), ovarian cancer is the fifth most common cause of cancer deaths in women and the tenth most common cancer among women in the United States. The witnessing of unusual pain in the lower abdomen, pelvis, back, heartburn, signs of rapidly feeling full when eating, frequent urination are early symptoms of ovarian cancer. Others are constipation or changes in bowel habit and painful sexual intercourse. Some risk factors associated with ovarian cancer are reproductive history and older age, birth control pills, infertility or fertility treatment, breast cancer or family history of a BRCA1 or BRCA2 gene mutation. Other risk factors are hormone therapy, obesity, and overweight, gynecologic surgery, and endometriosis. Unlike breast, cervical or colon cancer, there's no effective screening test to detect the disease aside the routine pelvic examination in which a certified health care professional examines the ovaries and uterus while noting the size, shape, and consistency through the help of laparoscopy[3]. Other screening options available are imaging technique (transvaginal sonography, MRI, or a CT scan), CA125 marker, two-stage strategies (combines CA125 and imaging techniques), colonoscopy, abdominal fluid aspiration, biopsy, risk of ovarian cancer algorithm, and serum biomarkers (HE4, mesothelin, prostasin, kallikriens, osteopontin, and lysophosphatidic acid).

1.1.1.5 *EARLY DETECTION OF CERVICAL CANCER*

Cervical cancer is a type of cancer that arises from the cervix. Symptoms may include abnormal vaginal bleeding, feeling of pain in the pelvis, or painful sexual intercourse. Risk factors include infection by HPV infection, a sexually transmitted deoxyribonucleic acid (DNA) virus which causes more than 90% of cases. Other risk factors are a compromised immune system,

---

[3]A laparoscope is a thin viewing tube with a camera at the end, which is inserted through a small incision in the lower abdomen.

excessive smoking, early sex or multiple sexual partner, and long-term use of birth control pills. Cervical biopsy is usually requested by a pathologist for the examination of intraepithelial neoplasia which is a precursor to cervical cancer. The most common screening test is Papanicolaou test (Pap test).

1.1.1.6 EARLY DETECTION OF BOWEL CANCER

Bowel cancer sometimes referred to as colon or rectal cancer is a cancer that originates from the large bowel. The symptoms include persistent blood in the stool, persistent change in bowel habit, lower abdominal pain. The detection of bowel cancer is usually through blood test to check for iron deficiency, fecal occult blood (FOB) test in stool. The risk factors are old age, overweight or obesity, diet, alcohol, smoking, and family history.

1.1.2 SCREENING

Cancer screening is an effort made towards the detection of cancer after it has formed, but before any noticeable symptoms appear in which the individual is referred promptly for early diagnosis and treatment (NIH, 2018). The improvement of genomic and surveillance technologies for more precise imaging and for evaluating blood-based tumor markers for greater specificity, offers new opportunities in progress for cancer screening. Although, cancer screening is not available for many types of cancers, screening tests could be conducted for potentiality of breast cancer as discussed earlier using mammography. For cervical cancer screening, visual inspection with acetic acid (VIA) or cytology screening methods, including Pap smears can be adopted especially for low-income settings. Comparatively, screening program is a more complex public health intervention when compared to early diagnosis. This is because when a test is selected and implemented effectively or linked to other steps in the screening process, quality need be assured. A successful screening program is the one aimed at preventing the escalation of cancer. There are fundamental principles to consider in order to achieve a successful screening program; (1) a screening program should focus on the common types of cancer of high morbidity with a likelihood of the test correctly identifying a particular type of cancer; (2) the test procedures should be affordable, safe, and acceptable; and (3) when cancer or precancerous lesion is detected, efforts should be directed on effective treatment.

In addition to the basic principles, factors to consider when adopting screening techniques are:

- sensitivity;
- specificity;
- positive predictive value;
- negative predictive value; and
- acceptability.

1.2 BIOMARKERS IN DRUG DEVELOPMENT

The application of biomarkers[4] in basic and clinical practice or research has overtime gained wide recognition that their presence as primary endpoints is generally acceptable (Strimbu and Tavel, 2010). Biomarkers functions as medical signs. Medical signs have a long history of use in clinical practice which is as old as medical practice itself. The use of biomarkers is somewhat new, and the best approaches to this practice are still being developed and refined, although it has proven to be an objective, quantifiable medical signs of modern laboratory science (Strimbu and Tavel, 2010). Biomarkers are often cheaper and easier to measure than 'true' endpoints. For a biomarker to form an integral part of clinical usage, the following Austin Bradford Hill's guidelines should be considered (Aronson, 2005):

- **Strength:** There should be a strong relationship between marker and outcome.
- **Consistency:** The association should be consistent for different individuals, places, or circumstances and at different times.
- **Specificity:** The marker should be specific for a particular disease.
- **Temporality:** The time-courses of changes should be parallel for both the marker and outcome.
- **Biological Gradient (Dose-Responsiveness):** Increasing exposure to an intervention should produce increasing effects on the marker and the disease.

---

[4]Biomarker, a portmanteau of "biological marker" is a naturally occurring molecule, gene, or characteristic by which a particular pathological or physiological process, disease, etc. can be identified. It can also be a traceable substance that is introduced into an organism as a means to examine organ function or other aspects of health.

- **Plausibility:** There should be credible mechanisms that connect the marker, the pathogenesis of the disease, and the mode of action of the intervention.
- **Coherence:** There should be consistency between the marker and natural history of the disease.
- **Experimental Evidence:** Results from intervention should be consistent with the association
- **Analogy:** There should be a similar result to which one can adduce a relationship.

In cancer diagnosis and screening, the use of biomarkers to identify precancerous lesions at early stage has so far contributed to the successful treatments of cancer. Biomarkers are not only useful for the detection of cancer but also in the evaluation of potential drug therapies viz. drug developmental process during early drug development and safety studies or during molecular profiling and proof of studies. A biomarker could serve as surrogate endpoint when evaluating clinical benefits if treatment can alter the biomarker with a corresponding correlation for improving health status (Box 1.1). For instance, a surrogate endpoint could include a shrinking tumor or lower biomarker levels instead of stronger indicators, such as longer survival or improved quality of life. In clinical trials, the application of surrogate endpoints can be helpful in allowing early drug approvals especially for the treatment of life-threatening diseases because results can be evaluated sooner. Although surrogate endpoints are not always true indicators or signs of how well a treatment works, the FDA surrogate endpoints could be helpful for drug developers during progress evaluation (FDA, 2018).

Box 1.1: Distinction between biomarkers and endpoints	
Clinical endpoint	A clinical endpoint provides information on how a patient feels, functions or survives.
Surrogate endpoint	A surrogate endpoint is an indicator or sign used in place of clinical endpoint to tell if a treatment works.
Biomarker	A biomarker is a measurable indicator of some biological state or condition.
Prognostic biomarker	Biomarker that forecasts the likely course of disease irrespective of treatment.
Predictive biomarker	Biomarker that forecasts the likely response to a specific treatment

1.3 BIOMARKERS IN CANCER SCREENING

By identifying people at high risk of developing cancer, it had been possible to develop intervention efforts such as check for premalignant lesions for the purpose of reducing mortality from cancer (Duffy, 2015). Cancer biomarkers are found in blood, urine, cerebrospinal fluid, or other body tissues that are elevated in association with cancer (Schiffman et al., 2015). The development of many-omic tools such as genomics and proteomics, has allowed the exploration of a large number of key cellular pathways simultaneously which has enabled the identification of biomarkers and signaling molecules (Kumar et al., 2006). Some were discussed below. Although, to date, many tumor markers have demonstrated poor accuracy and efficacy, particularly among the most prevalent cancers (Schiffman et al., 2015). However, the usefulness of biomarkers in cancer screening has been highlighted by Srivastava and Gopal-Srivastava (2002):

1. Use for in monitoring or evaluation in subjects with established cancer for recurrence.
2. Helps the detection of cancer at early stage in asymptomatic patients.
3. Helps for cancer diagnosis in symptomatic patients.
4. Assists in creating surveillance measure for individuals at high risk of cancer.
5. Useful in surrogate endpoint evaluation during chemoprevention.

1.3.1 *PAPANICOLAOU TEST*

The Papanicolaou (Pap) test is an effective and a widely used test for identification of pre-cancer and cervical cancer. The test is named after doctor Aurel Babeș and doctor Georgios Papanikolaou which is often abbreviated as Pap test or Pap smear, cervical smear, cervical screening, and smear test. The procedure involves the use of speculum to widen the vagina canal and then using small brush or Aylesbury spatula to gently remove cells from the cervix so that they can be checked for cervical cancer or cell changes (called cervical intraepithelial neoplasia (CIN) or cervical dysplasia; the squamous intraepithelial lesion (SIL) system that may lead to cervical cancer. When cells are collected, Paps stains are applied and viewed under a microscope by a trained cytologist. A Pap test may also help to detect other conditions. It is sometimes done at the same time as a pelvic exam and may also be done at the same time as a test for certain types of HPV, a sexually transmitted DNA virus. If a Pap test appeared positive for a precancerous condition, the doctor can treat or remove the abnormal tissue to prevent cervical cancer.

Pap tests are generally recommended to women with high-risk factors such as:

- a history of sexual intercourse at an early age;
- multiple sex partners;
- smoking;
- a history of infection with certain HPVs.

1.3.2 PROSTATE-SPECIFIC ANTIGEN (PSA) SCREENING

The PSA blood test is an FDA approved test which is useful for the detection of prostate cancer, monitoring its treatment, or assessing its recurrence. PSA is a proteolytic glycoprotein enzyme encoded by the KLK3 gene. It is also known as gamma-seminoprotein or kallikrein-3 (KLK3). The usefulness of PSA to the human body is that it liquefies semen in the seminal coagulum and allows sperm to swim freely during ejaculation (Balk et al., 2003). An elevated level of PSA between 4 and 10 ng/mL (nanograms per milliliter) are considered to be suspicious and may indicate prostate cancer, a noncancerous condition such as prostatitis, or an enlarged prostate. A PSA value below 4.0 ng/mL is considered normal. Although it has been found that men with a PSA lower than 4.0 ng/mL had prostate cancer. However, over the time, a high level of PSA is a signal of potential abnormality in which the doctor recommends prostate biopsy or a DRE. Some imaging tests such as transrectal ultrasound, x-rays, or cystoscopy can also be recommended as well.

1.3.3 CARCINOEMBRYONIC ANTIGEN (CEA)

The CEAs are glycosylphosphatidylinositol (GPI) cell-surface-anchored glycoproteins involved in cell adhesion. The CEA blood test is a test used to diagnose and manage certain types of cancers such as ovarian, cervix, breast, colon, lung, liver, pancreatic, medullary thyroid carcinoma, and prostate cancers. Depending on the part of the body, CEA are expressed in cancerous sites. Although, CEA blood test is not a reliable indices for diagnosing cancer or as a screening test for early detection of cancer (Duffy et al., 2003), an abnormally high serum level is a warning signal. However, not all cancer produce CEA, also elevated in the serum of heavy smokers, a CEA level higher than 20 (ng/mL) could be as a result of cancer.

1.3.4 FECAL OCCULT BLOOD (FOB)

As the name implies, FOB means hidden blood in the stool. The presence of blood in the stool could be as a result of bleeding polyps, ulcers, inflammatory bowel disease, diverticulosis, hemorrhoids (piles), bleeding gums or nosebleeds, anal fissures, benign or malignant tumors. The fecal occult blood test (FOBT) is a medical test usually carried out with the aim of detecting small traces of blood in the stool, which may not be visible with the naked eye. Positive results may either indicate upper gastrointestinal (GI) bleeding or lower GI bleeding which may inform further investigation for peptic ulcers or a malignancy (such as colorectal cancer or gastric cancer). The test is a useful test for the screening of colorectal cancer. Once a positive FOBT is gotten, the next in line will be to recommend other kinds of tests such as sigmoidoscopy, colonoscopy, upper GI endoscopy, double contrast barium enema, etc. For the primary purpose of screening colorectal cancer, colon polyps (small clump of cells that forms on the lining of the colon) are usually looked up. Although, most colon polyps are harmless, some can develop into colon cancer in later stages which is harmful. A negative result (absence of blood) in the stool does not rule out the possibility of developing colon cancer in future. So it is advisable to go for the test at least every two years.

1.3.5 CA-125

Cancer antigen 125, a glycoprotein, is also known as carcinoma antigen 125 or carbohydrate antigen 125 is abbreviated as CA-125. They are also called mucin 16 or MUC16. This glycoprotein is found in the respiratory tract, the ocular surface which includes the conjunctiva and cornea, and the epithelial of the female reproductive system. Certain cancers can cause an increased level of CA-125. This protein has also been proven to play a role in tumorigenesis advancement and tumor proliferation by several mechanisms such as suppressing the response of natural killer cells, reducing the sensitivity of cancer cells to drug therapy, and cell-to-cell interactions that enable the metastasis of tumor cells. A blood test for CA-125 is useful for the identification or detection of some types of cancer including ovarian cancer (Suh et al., 2010). Although, some conditions such as such as menstruation can cause an elevated CA-125. This test is also useful in monitoring the treatment responses for patients with ovarian cancer and follow-up. A persistent high level of CA-125 during therapy is associated with high death rate (Gocze and Vahrson, 1993). The normal values for CA-125 are between 0–35 (U/mL).

1.3.6 CA 19-9

The CA 19-9 is a tumor marker primarily useful for the management of pancreatic cancer. A high CA 19-9 level indicates unresectable lesions and a poor prognosis. The CA 19-9 blood test is usually recommended after confirmed pancreatic cancer diagnosis. CA 19-9 levels are useful indices to check if a tumor is growing, staying the same or getting smaller. Although an elevated level has been linked to several other kinds of GI cancer, such as colorectal cancer, esophageal cancer and hepatocellular carcinoma (Perkins et al., 2003). Again, they noted that aside cancer, elevated levels may be as a result of diseases of the bile ducts, pancreatitis, or cirrhosis. A normal range in healthy subjects is 0–37 U/mL.

1.3.7 CA 15-3

CA 15-3 belongs to a family of glycoproteins encoded by the MUC 1 gene (Keshaviah et al., 2007). As a tumor marker for breast cancer and some forms of cancer, an increased level of CA15-3 and alkaline phosphatase (ALP) is associated with an increased tendencies of breast cancer recurrence. CA 15-3 blood test is performed on a sample of blood to measure level of CA 15-3 in blood and for monitoring response to therapy. The latter is because increasing or decreasing values show correlation with disease progression or regression (Molina et al., 1995). The test CA15-3 is not useful for early stage breast cancer screening because only small amount can be present in the blood at that stage. The normal range in females is between 0–30 U/mL.

1.3.8 CD20

CD20 is a non-glycosylated phosphoprotein expressed on the surface of all mature B-cells. Antibodies deliberately targeting CD20 are useful for the treatment of B cell non-Hodgkin lymphoma and rheumatoid arthritis.

Although not relevant for prognosis because of no characteristic differences in observation, CD20 remains a useful marker in diagnosing B-cell lymphomas and leukemia because they are present on the cells of most B-cell neoplasms, and are absent on T-cell neoplasms. Aside its usefulness as a biomarker, it has a clinical significance because is the target of most drugs for the treatment of all B cell lymphomas, leukemias, and B cell-mediated autoimmune diseases. Some monoclonal antibodies drugs targeting CD20 are rituximab, ibritumomab, tiuxetan, ocrelizumab, obinutuzumab, ublituximab, etc.

1.3.9 ALPHA-FETOPROTEIN (AFP)

Alpha-fetoprotein (AFP) is a major plasma glycoprotein produced by the yolk sac and the fetal liver during fetal development. During this stage, some quantity is released into the maternal blood. A high level may indicate birth defects. However, the function of AFP in adult humans is unknown and has been found to be high in patients with certain kinds of tumors such as hepatocellular carcinoma, germ cell tumors, yolk sac tumor, etc. AFP is the most widely used biomarker in hepatocellular carcinoma surveillance.

1.3.10 VANILLYLMANDELIC ACID (VMA) AND URINARY HOMOVANILLIC

Vanillylmandelic acid (VMA), an end-stage metabolite of the catecholamines is present in the urine, along with some catecholamine metabolites such as homovanillic acid (HVA), metanephrine, and normetanephrine. High levels of urinary VMA are associated with patients with tumors that secrete catecholamines (Magera et al., 2003). These urinalysis tests are used to diagnose an adrenal gland tumor known as pheochromocytoma, a tumor of catecholamine-secreting chromaffin cells. The test can also be used to diagnose neuroblastoma and monitor treatments.

1.3.11 CALCITONIN

Calcitonin also known as thyrocalcitonin is a hormone that is produced in the parafollicular cells (C-cells) of the thyroid gland in humans. It opposes the effects of parathyroid hormone (PTH) by lowering blood calcium (Ca^{2+}). An abnormally high level of calcitonin is associated to medullary thyroid cancer, as such a useful biomarker for its diagnosis. It is also useful in monitoring recurrence after surgery or even screening for cancer during biopsy. Normal levels are below 5 ng/L for females and 12 ng/L for males while children under 6 months of age are 40 ng/L and children between 6 months and 3 years of age is 15 ng/L.

1.3.12 LACTATE DEHYDROGENASE

Lactate dehydrogenase enzyme (LDH) is an enzyme that catalyzes the reversible conversion of lactate to pyruvic acid. LDH is found in almost all cells of the body and is present in small amount under normal circumstances. An abnormally high level in the blood could be as a result of cancer or other medical conditions such as human immune virus, lung or liver disease, anemia, heart failure, hypothyroidism or acute pancreatitis. LDH as a tumor biomarker may not be useful in identifying a specific type of cancer.

1.3.13 HUMAN CHORIONIC GONADOTROPIN (HCG)

The human chorionic gonadotropin (hCG) is a hormone secreted during implantation in the placenta. The hCG urine or blood tests are famously used to detect pregnancy and as such are referred to as the pregnancy hormone. In the absence of pregnancy, an abnormally high level could serve as a marker for germ cell tumors and gestational trophoblastic diseases.

KEYWORDS

- biomarker
- cancer
- drug development
- premalignant lesions
- surrogate endpoint
- trophoblastic diseases

REFERENCES

American Cancer Society, (2018). *Cancer Facts and Statistics*. Available at: https://www.cancer.org/research/cancer-facts-statistics.html (accessed on 20 February 2020).

Anand, P., Kunnumakkara, A. B., Kunnumakara, A. B., Sundaram, C., Harikumar, K. B., Tharakan, S. T., Lai, O. S., Sung, B., & Aggarwal, B. B., (2008). Cancer is a preventable disease that requires major lifestyle changes. *Pharmaceutical Research*, 25(9), 2097–2116.

Aronson, J. K., (2005). Biomarkers and surrogate endpoints. *Br. J. Clin. Pharmacol.*, 59(5), 491–494.

Balk, S. P., Ko, Y. J., & Bubley, G. J., (2003). Biology of prostate-specific antigen. *Journal of Clinical Oncology*, 21(2), 383–391.

Cancer Care, (2016). *Early Detection and Breast Cancer*. Available at: https://www.cancercare.org/publications/82-early_detection_and_breast_cancer. (accessed on 20 February 2020).

CDC (Centers for Disease Control and Prevention), (2017). *What Are the 12 Leading Causes of Death in the United States?* Available: https://www.healthline.com/health/leading-causes-of-death#1 (accessed on 20 February 2020).

Duffy, M. J., (2015). Use of biomarkers in screening for cancer. In: Scatena, R., (eds.), *Advances in Cancer Biomarkers. Advances in Experimental Medicine and Biology* (p. 867). Springer, Dordrecht.

Duffy, M. J., Van Dalen, A., Haglund, C., Hansson, L., Klapdor, R., Lamerz, R., Nilsson, O., Sturgeon, C., & Topolcan, O., (2003). Clinical utility of biochemical markers in colorectal cancer: European Group on Tumor Markers (EGTM) guidelines. *European Journal of Cancer*, 39(6), 718–727.

FDA, (2018). *Table of Surrogate Endpoints That Were the Basis of Drug Approval or Licensure.* Available at: https://www.fda.gov/Drugs/DevelopmentApprovalProcess/DevelopmentResources/ucm613636.htm (accessed on 20 February 2020).

GBD, (2016). *Disease and Injury Incidence and Prevalence, Collaborators*. "Global, regional, and national incidence, prevalence, and years lived with disability for 310 diseases and injuries, 1990–2015: A systematic analysis for the global burden of disease study 2015." *Lancet*, 388(10053), 1545–1602.

Gocze, P., & Vahrson, H., (1993). Ovarian carcinoma antigen (CA 125) and ovarian cancer (clinical follow-up and prognostic studies). *Orvosi. Hetilap. (in Hungarian)*, 134(17), 915–918.

Gotzsche, P. C., & Jorgensen, K. J., (2013). Screening for breast cancer with mammography. *The Cochrane Database of Systematic Reviews*, 6, CD001877.

Hanahan, D., & Weinberg, R. A., (2000). The hallmarks of cancer. *Cell*, 100(1), 57–70.

Hellquist, B. N., Czene, K., Hjalm, A., Nystrom, L., & Jonsson, H., (2015). Effectiveness of population-based service screening with mammography for women ages 40 to 49 years with a high or low risk of breast cancer: Socioeconomic status, parity, and age at birth of first child. *Cancer*, 121(2), 251–258.

Keshaviah, A., Dellapasqua, S., Rotmensz, N., Lindtner, J., Crivellari, D., Collins, J., Colleoni, M., Thurlimann, B., Mendiola, C., Aebi, S., Price, K. N., & Pagani, O., (2007). CA15-3 and alkaline phosphatase as predictors for breast cancer recurrence: A combined analysis of seven International Breast Cancer Study Group trials. *Annals of Oncology*, 18(4), 701–708.

Kumar, S., Mohan, A., & Guleria, R., (2006). Biomarkers in cancer screening, research, and detection: Present and future: A review. *Biomarkers*, 11(5), 385–405.

Magera, M. J., Thompson, A. L., Matern, D., & Rinaldo, P., (2003). Liquid chromatography-tandem mass spectrometry method for the determination of vanillylmandelic acid in urine. *Clin. Chem.*, *49*(5), 825–826.

Maurer Foundation, (2018). *The Stages of Breast Cancer and the Importance of Early Detection.* Available at: https://www.maurerfoundation.org/the-stages-of-breast-cancer-the-importance-of-early-detection/ (accessed on 20 February 2020).

Molina, R., Zanon, G., & Filella, X., (1995). Use of serial carcinoembryonic antigen and CA 15-3 assays in detecting relapses in breast cancer patients. *Breast Cancer Res Treat*, *36*, 41–48.

NIH, (2018). *Cancer Screening Overview (PDQ®)-Patient Version.* Available at: https://www.cancer.gov/about-cancer/screening/patient-screening-overview-pdq (accessed on 20 February 2020).

Nordqvist, C., (2017a). *Prostate Cancer in Detail.* Available at: https://www.medicalnewstoday.com/articles/159675.php (accessed on 20 February 2020).

Nordqvist, C., (2017b). *How do You Get Ovarian Cancer.* Available at: https://www.medicalnewstoday.com/articles/159675.php (accessed on 20 February 2020).

Onega, T., Goldman, L. E., Walker, R. L., Miglioretti, D. L., Buist, D. S., Taplin, S., Geller, B. M., Hill, D. A., & Smith-Bindman, R., (2016). Facility mammography volume in relation to breast cancer screening outcomes. *J. Med Screen*, *23*(1), 31–37.

Perkins, G., Slater, E., Sanders, G., & Prichard, J., (2003). "Serum tumor markers." *American Family Physician*, *68*(6), 1075–1082.

Plummer, M., De Martel, C., Vignat, J., Ferlay, J., Bray, F., & Franceschi, S., (2016). Global burden of cancers attributable to infections in 2012: A synthetic analysis. *Lancet Glob. Health*, *4*(9), 609–616.

Schiffman, J. D., Fischer, P. G., & Gibbs, P., (2015). *Early Detection of Cancer: Past, Present, and Future.* Available at: https://meetinglibrary.asco.org/record/104176/edbook#fulltext (accessed on 20 February 2020).

Srivastava, S., & Gopal-Srivastava, R., (2002). Biomarkers in cancer screening: A public health perspective. *J. Nutr.*, *132*(8), 2471S–2475S.

Stewart, B. W., & Wild, C. P., (2014). World cancer report (2014). *Lyon: International Agency for Research on Cancer.* Available at: http://publications.iarc.fr/Non-Series-Publications/World-Cancer-Reports/World-Cancer-Report-2014 (accessed on 20 February 2020).

Strimbu, K., & Tavel, J. A., (2010). What are biomarkers? *Current Opinion in HIV and AIDS*, *5*(6), 463–466. http://doi.org/10.1097/COH.0b013e32833ed177 (accessed on 20 February 2020).

Suh, K. S., Park, S. W., Castro, A., Patel, H., Blake, P., Liang, M., & Goy, A., (2010). Ovarian cancer biomarkers for molecular biosensors and translational medicine. *Expert Review of Molecular Diagnostics*, *10*(8), 1069–1083.

WHO (World Health Organization), (2018a). *Cancer: Key Facts.* Available at: http://www.who.int/en/news-room/fact-sheets/detail/cancer (accessed on 20 February 2020).

WHO (World Health Organization), (2018b). *Early Detection of Cancer.* Available at: http://www.who.int/cancer/detection/en/ (accessed on 20 February 2020).

CHAPTER 2

Kinase Targets in Cancer Drug Discovery

S. USHA

Department of Bioinformatics, Bharathiar University,
Coimbatore – 641046, Tamil Nadu, India, E-mail: usha@buc.edu.in

ABSTRACT

Over 20 years, researchers have been focusing on identifying cancer signaling pathways and associated protein targets to discover inhibitors of therapeutic importance with enhanced selectivity, efficacy, and reduced toxicity. Protein kinases represent such molecular targets and remarkable efforts have been carried out in developing drugs that influence the action of pathogenic kinases, and clinical studies performed so far have shown the positive effects of kinase inhibitors to treat cancer. In this chapter, the structural features of different kinases, their mechanism of action, types of inhibitors based on such binding mechanisms and the current state of the art have been discussed.

2.1 INTRODUCTION

The mechanism of drug discovery becomes successful depending on how the drug target is defined. So far, 893 human and pathogen-derived protein targets have been explored in which 667 human-genome-derived proteins on which around 1500 FDA-approved drugs act (Santos et al., 2017). Proteins of kinase families have evolved as one of the most promising drug targets. Protein kinases possess an important role in cellular metabolic activities. They act as mediators in the phosphorylation of proteins and are crucial in disease pathology (e.g., cancer). Kinases are constantly activated due to mutations and because of their transforming capability; they provide space for survival and growth of the cancer cell. These cancer cells are highly liable to specific kinase inhibitors (Paul et al., 2004; Zhang et al., 2009). The human genome contains 518 (478 typical and 40 atypicals) protein kinase genes; they constitute about 2% of all eukaryotic genes (Milanesi et al., 2005). There are three major groups of typical protein kinases which include (i) protein-serine/threonine kinases, (ii) protein-tyrosine kinases, and (iii) tyrosine-kinase like proteins based on how the –OH group of amino acid gets phosphorylated in the target proteins (Grover, 2017).

2.1.1 CURRENT STATUS OF KINASE DRUG DISCOVERY

At present, there are above 25 approved oncology drugs that target kinases and much therapeutic compounds are undergoing various clinical evaluation stages. Small-molecule kinase inhibitors (SMKIs), 28 of which are approved by the US Food and Drug Administration (FDA), are currently available as successful inhibitors.

An earlier report has shown that for more than one-fourth of FDA-approved small molecule kinase inhibitors, the molecular weight (MW) is above 500 and all possess around three and five rings (Wu et al., 2016). For example, imatinib inhibits BCR-ABL1 in chronic myelogenous leukemia (CML) and acute lymphoblastic leukemia (Kantarjian et al., 2002), crizotinib, and other ALK kinase inhibitors for cancers (Shaw et al., 2014), lapatinib for ERBB2/HER2-amplified tumors (Arteaga and Engelman, 2014), gefitinib, and erlotinib for EGFR mutated tumors (Engelman, 2014), and vemurafenib for BRAF mutant tumors (Chapman, 2011).

2.1.2 THE STRUCTURE OF PROTEIN KINASE DOMAIN

The structural details of protein kinases have provided the basis for molecular-level understanding of kinase functioning (Knighton, 1991). From that onwards, about 200 crystal structures of diverse kinase proteins were determined and about 5000 crystal structures are available in public. The protein kinases consist of a highly conserved kinase domain which contains a small N-terminal lobe and a large C-terminal lobe. A hinge region connects these two lobes and helps to open and close the kinase structure. An adenosine tri-phosphate (ATP) molecule is bound near the hinge region (Figure 2.1).

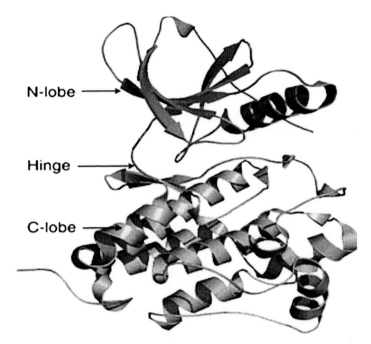

FIGURE 2.1 The three-dimensional structure of a kinase showing N-terminal lobe, hinge region, and C-terminal lobe. Source: Reprinted by permission from Zhang et al., 2009. © Springer Nature.

2.1.3 ACTIVE AND INACTIVE KINASE CONFORMATION

The kinases can interconvert between active and inactive conformational states. In active conformation, the aspartate of DFG motif projects towards the ATP-binding site and interacts with two Mg ions; thereby making the activation part as an open and extendable form. In an inactive kinase conformation, the DFG motif is formed as a flipped conformation (180°), i.e., the aspartate residue of the DFG motif points away from the ATP site.

2.1.4 KINASE INHIBITORS WITH REFERENCE TO ALLOSTERIC SITE

The ATP binding site forms a promising target for kinase inhibitors. However, due to the highly conserved ATP binding site, the inhibitor selection of kinases has been directed towards identifying allosteric inhibitors.

2.2 CLASSICAL OR ATP-COMPETITIVE INHIBITORS

2.2.1 TYPE I INHIBITORS

The type I inhibitors occupy the adenosine binding pocket and binds to the kinase active conformation and form H-bonds with the hinge region residues. The conformational nature of the key structural elements does not influence such binding.

Since the unique shape of the adenine site allows a little change in the heterocyclic system, it is very hard to design type I inhibitors with high selectivity (Dar and Shokat, 2011).

2.2.2 TYPE II INHIBITORS

The type II kinase inhibitors bind to stabilize the kinase inactive conformation thus prevent the successive events such as ATP-binding and kinase activation. These inhibitors are considered as ATP-competitive inhibitors as they bind to the ATP site (Dixit and Verkhivker, 2009).

In type II inhibitors, the DFG shift makes the amino acids surrounding the pocket more exposed and they are less conserved than those in the ATP binding pocket. This is advantageous than type I inhibitors and thus favors the selectivity of kinase (Liu, 2006; Mol et al., 2004).

2.2.3 ALLOSTERIC OR NONCOMPETITIVE INHIBITORS

The allosteric inhibitors bind to the sites other than the ATP binding site (e.g., type III inhibitors) and show activity against drug resistance based on mutations and are useful in finding more selective inhibitors in binding pockets which are not conserved. Therefore, the major focus of many drug discovery projects is to identify and develop such inhibitors.

2.2.4 TYPE III INHIBITORS

Certain inhibitors block kinase activity without displacing but binding along with ATP molecule (Gavrin and Saiah, 2013). One such target of inhibition is MEK kinase. A typical example of such inhibitor is that identified by Parke-Davis, which binds in the active site without interfering with ATP binding (Alessi et al., 1995; Ohren et al., 2004). Cobimetinib-vemurafenib combination has been approved as treatment of advanced melanoma by U.S. FDA (Rice et al., 2012; Ascierto et al., 2016).

2.2.5 TYPE IV INHIBITORS

These inhibitors bind at a site remote from the ATP-binding e.g., surface pockets and influence the binding of key regulators. These sites are present anywhere in the kinase domain other than the ATP-neighboring sites. The type IV inhibitors do not hinder the ATP site and are known as allosteric or noncompetitive inhibitors of ATP. These inhibitors induce conformational changes to make the kinase-inactive which forms their remarkable feature (Cox et al., 2011; Monad et al., 1963; Simard et al., 2009). GNF-2 was the first reported type IV inhibitor of Abl kinase which binds to the kinase C-lobe. A set of 1,3,4-thiadiazole compound series shows successful inhibition against Abl kinase (Adrian et al., 2006; Fallacara et al., 2014).

2.2.6 TYPE V INHIBITORS

These are bi substrate analog inhibitors which bind to both the ATP and protein substrate binding sites; and bivalent inhibitors that bind to the ATP binding cleft and sites other than the kinase substrate binding site.

Different protein kinases are specific in binding their own substrate, which forms the basis for the design of bisubstrate analog inhibitors (Lamba et al., 2012). For a compound analog IRS 727, a bisubstrate analog inhibitor has been designed by Parang et al. (2001). The bivalent inhibitors are designed so as to target both the kinase catalytic and regulatory sites (Profit et al., 1999).

2.2.7 TYPE VI INHIBITORS

In these inhibitors, the alkene portion of the inhibitor and the cysteine residue of the ATP-binding site covalently interact in certain kinases. For example, the alkene of afatinib interacts with the cysteine in the hinge region of the ATP binding pocket (Solca et al., 2012; Roskoskijr et al., 2016). Cohen et al. (2005) has designed and tested few selective and irreversible kinase inhibitors. Type-I, Type-II, Type-III, Type-IV, and Type-V inhibitors are reversible but Type VI inhibitors are irreversible. Afatinib and Ibrutinib inhibitors were designed by targeting a surface exposed cysteine residue.

2.3 CONCLUSION

The knowledge about the structure and physiology of protein kinases has helped to improve the understanding the molecular level functioning of the kinases. The drug discovery programs that target protein kinases have achieved substantial development since the first kinase inhibitor was discovered. The success of many kinase inhibitors has driven them and has made them the most desirable field of research (Behera, 2017).

The human kinome is classified into seven major kinase families. However, the small molecule inhibitor discovery has been so far limited to selected group of kinases. The majority of more than 30 inhibitors approved are limited to the tyrosine kinases, serine/threonine kinases, and tyrosine-like kinases families which indicate there markable possibility of drug discovery research in future.

KEYWORDS

- cancer
- drug development
- inhibitors
- kinases targets
- molecular targets
- threonine kinases

REFERENCES

Adrián, F. J., Ding, Q., Sim, T., Velentza, A., Sloan, C., Liu, Y., Zhang, G., Hur, W., Ding, S., Manley, P., & Mestan, J., (2006). Allosteric inhibitors of BCR-ABL-dependent cell proliferation. *Nat. Chem. Biol., 2*, 95–102.

Alessi, D. R., Cuenda, A., Cohen, P., Dudley, D. T., & Saltiel, A. R., (1995). PD 098059 is a specific inhibitor of the activation of mitogen-activated protein kinase kinase *in vitro* and *in vivo. J. Biol. Chem., 270*, 27489–27494.

Arteaga, C. L., & Engelman, J. A., (2014). ERBB receptors: From oncogene discovery to basic science to mechanism-based cancer therapeutics. *Cancer Cell, 25*, 282–303.

Ascierto, P. A., McArthur, G. A., Dréno, B., Atkinson, V., Liszkay, G., Di Giacomo, A. M., Mandalà, M., Demidov, L., Stroyakovskiy, D., Thomas, L., & De La Cruz-Merino, L., (2016). Cobimetinib combined with vemurafenib in advanced BRAFV600-mutant melanoma (coBRIM): Updated efficacy results from a randomized, double-blind, phase 3 trial. *Lancet Oncol., 17*, 1248–1260.

Behera, P. M., & Dixit, A., (2017). The story of kinase inhibitors development with special reference to allosteric site. In: Grover, A., (ed.), *Drug Design: Principles and Applications* (p. 57). Springer, Singapore.

Chapman, P. B., Hauschild, A., Robert, C., Haanen, J. B., Ascierto, P., Larkin, J., Dummer, R., Garbe, C., Testori, A., Maio, M., & Hogg, D., (2011). Improved survival with vemurafenib in melanoma with BRAF V600E mutation. *N. Engl. J. Med., 364*, 2507–2516.

Cohen, M. S., Zhang, C., Shokat, K. M., & Taunton, J., (2005). Structural bioinformatics-based design of selective, irreversible kinase inhibitors. *Science, 308*, 1318–1321.

Cox, K. J., Shomin, C. D., & Ghosh, I., (2011). Tinkering outside the kinase ATP box: Allosteric (type IV) and bivalent (type V) inhibitors of protein kinases. *Future Med. Chem., 3*, 29–43.

Dar, A. C., & Shokat, K. M., (2011). The evolution of protein kinase inhibitors from antagonists to agonists of cellular signaling. *Annu. Rev. Biochem., 80*, 769–795.

Dixit, A., & Verkhivker, G. M., (2009). Hierarchical modeling of activation mechanisms in the ABL and EGFR kinase domains: Thermodynamic and mechanistic catalysts of kinase activation by cancer mutations. *PLoS Comput. Biol., 5*, e1000487.

Fallacara, A. L., Tintori, C., Radi, M., Schenone, S., & Botta, M., (2014). Insight into the allosteric inhibition of Abl kinase. *J. Chem. Inf. Model., 54*, 1325–1338.

Gavrin, L. K., & Saiah, E., (2013). Approaches to discover non-ATP site kinase inhibitors. *Med. Chem. Comm., 4*, 41–51.

Grover, A., (2017). *Drug Design: Principles and Applications*. Springer, Singapore.

Kantarjian, H., Sawyers, C., Hochhaus, A., Guilhot, F., Schiffer, C., Gambacorti-Passerini, C., Niederwieser, D., Resta, D., Capdeville, R., Zoellner, U., & Talpaz, M., (2002). Hematologic and cytogenetic responses to imatinib mesylate in chronic myelogenous leukemia. *N. Engl. J. Med., 346*, 645–652.

Knighton, D. R., Zheng, J. H., Ten Eyck, L. F., Ashford, V. A., Xuong, N. H., Taylor, S. S., & Sowadski, J. M., (1991). Crystal structure of the catalytic subunit of cyclic adenosine monophosphate-dependent protein kinase. *Science, 253*, 407–414.

Lamba, V., & Ghosh, I., (2012). New directions in targeting protein kinases: Focusing upon true allosteric and bivalent inhibitors. *Curr. Pharm. Des., 18*, 2936–2945.

Liu, Y., & Gray, N. S., (2006). Rational design of inhibitors that bind to inactive kinase conformations. *Nat. Chem. Biol., 2*, 358.

Milanesi, L., Petrillo, M., Sepe, L., Boccia, A., D'Agostino, N., Passamano, M., & Paolella, G., (2005). Systematic analysis of human kinase genes: A large number of genes and alternative splicing events result in functional and structural diversity. *BMC Bioinformatics, 6*, S20.

Mol, C. D., Fabbro, D., & Hosfield, D. J., (2004). Structural insights into the conformational selectivity of STI-571 and related kinase inhibitors. *Curr. Opin. Drug Discov. Devel., 7*, 639–648.

Monad, J., Changeux, J. P., & Jacob, F., (1963). Allosteric proteins and cellular control mechanisms. *J. Mol. Biol., 6*, 306.

Ohren, J. F., Chen, H., Pavlovsky, A., Whitehead, C., Zhang, E., Kuffa, P., Yan, C., McConnell, P., Spessard, C., Banotai, C., & Mueller, W. T., (2004). Structures of human MAP kinase kinase 1 (MEK1) and MEK2 describe novel noncompetitive kinase inhibition. *Nat. Struct. Mol. Biol., 11*, 1192.

Parang, K., Till, J. H., Ablooglu, A. J., Kohanski, R. A., Hubbard, S. R., & Cole, P. A., (2001). Mechanism-based design of a protein kinase inhibitor. *Nat. Struct. Mol. Biol., 8*, 37.

Paul, M. K., & Mukhopadhyay, A. K., (2004). Tyrosine kinase-role and significance in cancer. *Int. J. Med. Sci., 1*, 101.

Profit, A. A., Lee, T. R., & Lawrence, D. S., (1999). Bivalent inhibitors of protein tyrosine kinases. *J. Am. Chem. Soc., 121*, 280–283.

Rice, K. D., Aay, N., Anand, N. K., Blazey, C. M., Bowles, O. J., Bussenius, J., Costanzo, S., Curtis, J. K., Defina, S. C., Dubenko, L., & Engst, S., (2012). Novel carboxamide-based allosteric MEK inhibitors: Discovery and optimization efforts toward XL518 (GDC-0973). *ACS Med. Chem. Lett., 3*, 416–421.

Roskoski, Jr. R., (2016). Classification of small molecule protein kinase inhibitors based upon the structures of their drug-enzyme complexes. *Pharmacol. Res., 103*, 26–48.

Santos, R., Ursu, O., Gaulton, A., Bento, A. P., Donadi, R. S., Bologa, C. G., Karlsson, A., Al-Lazikani, B., Hersey, A., Oprea, T. I., & Overington, J. P., (2017). A comprehensive map of molecular drug targets. *Nat. Rev. Drug Discov., 16*, 19.

Shaw, A. T., Kim, D. W., Mehra, R., Tan, D. S., Felip, E., Chow, L. Q., Camidge, D. R., Vansteenkiste, J., Sharma, S., De Pas, T., & Riely, G. J., (2014). Ceritinib in ALK-rearranged non-small-cell lung cancer. *N. Engl. J. Med., 370*, 1189–1197.

Simard, J. R., Klüter, S., Grütter, C., Getlik, M., Rabiller, M., Rode, H. B., & Rauh, D., (2009). A new screening assay for allosteric inhibitors of cSrc. *Nat. Chem. Biol., 5*, nchembio-162.

Solca, F., Dahl, G., Zoephel, A., Bader, G., Sanderson, M., Klein, C., Kraemer, O., Himmelsbach, F., Haaksma, E., & Adolf, G. R., (2012). Target binding properties and cellular activity of afatinib (BIBW 2992), an irreversible ErbB family blocker. *J. Pharmacol. Exp. Ther., 343*, 342–350.

Wu, P., Nielsen, T. E., & Clausen, M. H., (2016). Small-molecule kinase inhibitors: An analysis of FDA-approved drugs. *Drug Discov. Today, 21*, 5–10.

Zhang, J., Yang, P. L., & Gray, N. S., (2009). Targeting cancer with small molecule kinase inhibitors. *Nat. Rev. Cancer, 9*, 28.

CHAPTER 3

Essential Drug Properties and Stages in Anti-Cancer Drug Development

ANDREW G. MTEWA,[1,2,4] DUNCAN SESAAZI,[2,4] AMANJOTANNU,[2] and SERAWIT DEYNO[2,3,4]

[1]*Chemistry Section, Department of Applied Science, Malawi Institute of Technology, Malawi University of Science and Technology, Thyolo, Malawi, E-mails: andrewmtewa@yahoo.com, amtewa@must.ac.mw (A. G. Mtewa), ORCID: 0000-0003-2618-7451*

[2]*Department of Pharmacy and Pharmacology, Mbarara University of Science and Technology, Mbarara, Uganda*

[3]*Department of Pharmacology, School of Medicine, College of Medicine and Health Sciences, Hawassa University, Hawassa, Ethiopia*

[4]*Pharm-BioTechnology and Traditional Medicine, A World Bank-Africa Center of Excellence (ACE II), Mbarara University of Science and Technology, Mbarara, Uganda*

ABSTRACT

Cancer is a disease in which cells uncontrollably divide, destroying body tissue. Sometimes, it is regarded as a group of diseases that involves abnormal cell growth with the potential to spread to other body parts. Drug development involves the optimization of a new or existing pharmaceutical agent identified from drug discovery. This process begins from studying disease weaknesses and strengths with the aim of finding a way to reverse or stop its mechanistic strengths on body cells. Critical to an effective drug development process are drug properties which include lipophilicity, solubility, PKa, and permeability. These properties inform the developers of how they should manipulate the drug molecules through structural-activity-relationships (SARs) studies to make them as least toxic and most effective as possible. Before getting to the market, the properties guide formulation and their effects are monitored through different stages of clinical trials.

3.1 INTRODUCTION

Nitrogen Mustard was the first anti-cancer agent that was developed, and it was an accidental discovery (Hawthorn and Redmond, 2006). An observation on soldiers who had been exposed to mustard gas revealed remarkable lymphoid tissue reduction which brought up ideas leading to the development of derivatives of mustard gas as chemotherapy against lymphomas. Drug development and discovery consist of a series of inter-disciplinary activities and tools to identify source materials, bioactive ingredients and their characterization. There are several aspects of drug properties that need to be looked into when developing anti-cancer drugs. These properties include Structural properties, physicochemical properties, biochemical properties, Toxicity, and pharmacokinetics (PK). These properties interlink with each other in that one influences the other. For example, PK, and toxicity of drugs largely depend on their underlying structural properties (Kerns and Di, 2008). All of the properties mentioned in this chapter have a strong *in vivo* bearing on the efficiency and reliability of drug activities against biological targets. This shows that the properties of drug leads and drugs themselves are a critical component of drug discovery and development. The understanding of drug leads' properties in discovery and development helps to guide investors in projecting success and failure rates in coming up with viable drugs, considering the high costs that drug development has from inception to the market. As for many other drugs, cancer drug development passes through a series of stages of tests and checks before considering applications for approvals. These loosely include preclinical *in vitro* and *in vivo* tests, formulation, and clinical trials (Rangitsch, 2009). All these stages are crucial to ensure that the drug being developed is first and foremost, safe to the patient, efficient against cancer cells in humans and better than existing drugs against the same cancer cells which ensures cost-effectiveness of the drug development project.

3.2 ANTICANCER DRUG PROPERTIES

Anticancer drug properties are not necessarily different from basic drug properties in general. However, this section details specifics of anticancer drug properties which may slightly be emphasized for the disease. For a drug to function as desired, it needs to have specific properties befitting the nature and location of a tumor. Further to this, the variations to location of tumors themselves and sometimes the strength of patients from earlier studies may guide drug developers to design or modify drug properties to effective administration. For example, it is not appropriate to prescribe tablets to patients with advanced esophageal cancers. This means drug developers have to work on the properties that affect bioavailability of the drug to be as effective if they administer the same drug intravenously. This section explains four categories of properties that drug developers need to look into.

3.2.1 STRUCTURAL PROPERTIES

Structural properties are summarized by the Lipinski 'rules of 5' and the Veber rules. The BBB permeability rule and Oprea's rules have been described in fewer details as more underlying rules are covered by the Lipinski and Verber's rules. It should be noted here that these rules are not strict yardsticks but they form an effective basis of reference.

3.2.1.1 LIPINSKI 'RULES OF 5'

Work is done by Lipinski and others (Lipinski et al., 1997) showed that it is possible to predict the influence of the nature of drug structures on its permeation and absorption. They stated that the permeation and absorption of the drug will be poor if the following describes the drug structures:

- The structure having a molecular weight (MW) of greater than 500.
- The structure having more than 5 H-bond donors. This accounts for all OH and NH species that can donate Hs in the structure.
- The structure having more than 10 H-bond acceptors. This accounts for all acceptor nitrogen and oxygen available in the structure.
- The structure having lipophilicity measure, logP, of greater than 5.
- Biological transporter substrates are exceptions.

It happens that the same species counts as an H-donor and acceptor at the same time. Examples of such are primary amines (-NH$_2$) which can donate 2-Hs, at the same time, can accept H via the single N available. Carboxylic acids (-COOH) can donate 1-H and can also accept 2-Hs via the 2-Os.

Breaking one rule may not necessarily lead to poor absorption, rather, the higher the number of the virtual rules get violated, the poorer the absorption efficiency.

3.2.1.2 VEBER'S RULES

Veber and others (Veber et al., 2002) conducted their studies on structural dependent drug bioavailability in rats. Critical oral bioavailability determinants were found to include hydrogen bond count, molecular structural flexibility, where rotatable bonds (RotB) are manually or soft-ware counted and polar surface area (PSA), which is closely related to H-bonding. According to their work and findings, good oral bioavailability in rat models follows the following rules:

- The structure having 10 or less RotB;
- The structure having a PSA of 140 Å^2 or less;
- The structure having 12 or less total H-bonds, including both donors and acceptors.

3.2.1.3 BBB PERMEABILITY RULE

The blood-brain-barrier (BBB) permeability rule was proposed by Patridge (1995) and it predicts the ability of structures to permeate through the BBB.

3.2.1.4 OPREA'S 'RULE OF 3'

Oprea and others (Oprea et al., 2001; Oprea, 2002; Hann and Oprea, 2004) came up with three rules to assist drug discoverers and developers at drug-leads' optimization stage to identify appropriate leads they should proceed with.

3.2.2 LIPOPHILICITY

Lipophilicity is a nature of drug compound structures which measures its tendency to partition into a polar aqueous environment versus a non-polar lipid environment. There is a high correlation between lipophilicity and drug permeability, distribution, metabolism, absorption, plasma-protein binding and toxicity (Lombardo et al., 2002). Hansch et al. (2004) devised a way to calculate lipophilicity values as an initial compound assessment to determine its suitability as a drug lead. Depending on the ionic nature of the partition phases involved, lipophilicity values are presented either as Log P (where both aqueous and organic phases are in neutral form) and Log D at specific pH x (where part of the compound molecule is neutral and another part is ionic). The following is an expressions used to describe Log P.

$$Log\ P = \log\left(\frac{[Organic\ compound]}{[Aqueous\ compound]}\right)$$

Kern and Di (2008) suggested that Log P in the range of 0 to 3 provides a good balance of solubility and permeability for optimal gastrointestinal (GI) absorption through oral bioavailability. Drug compounds below this range have poor lipid bilayer permeability due to their being more polar. On the other hand, drugs above this range have a more non-polar nature giving them poor solubility in aqueous bilayers.

Log D takes consideration of specific pH conditions of the pharmacokinetic environments. For example, Log D at the pH of 7.4, which is the physiological pH of blood serum (CMC, 2012), is one of the conditions that draw particular interest to drug developers. The following expression is used to determine log D of drug compounds:

$$Log\ D_{pHx} = \log\left(\frac{[Organic\ compound]}{[Aqueous\ compound]}\right)$$

$Log\ D_{7.4}$ provides an important estimation of how the structure will perform on absorption, metabolism, brain penetration, solubility, and permeability. The following ranges were explained by Comer (2003):

At $Log\ D_{7.4}$ of less than 1, drug solubility is good but absorption and brain penetration are low. This is due to low passive diffusion permeability. The polarity of the drugs in this range provides high clearance in the kidney. For low MW compounds, paracellular permeation is exhibited. The range of $Log\ D_{7.4}$ between 1 and 3 is considered an ideal one with a good solubility and passive diffusion permeability balance providing excellent intestinal absorption. This range facilitates low drug molecular binding to metabolic enzymes, significantly minimizing metabolism. Low solubility in the range of 3 to 5 leads to low absorption of the drugs but good permeability. Binding to metabolic enzymes is increased leading to higher drug metabolism. At $Log\ D_{7.4}$ of above 5, solubility of the drug is very low leading to low bioavailability, absorption, and increased half-life. Due to this, drug components break down and persist in tissues (Comer, 2003) which risks unintended side effects.

3.2.3 PK_a

PK_a is defined as the negative log of the ionization constant (K_a) and it denotes the ionizability of a drug compound. PK_a is the main permeability and solubility determinant in drug discovery and development (Roda et al., 1996). A majority of drugs have ionizable components in their structures, most of which are basic with some, acidic. When a drug is noted to be less soluble or permeable, it is possible to manipulate the acidic and/or basic substructures on the drug molecular structures (MS) to obtain a modified structure with the desired PK_a. The polar nature of ionized drug molecules makes them more soluble in aqueous media than neutral drug molecules. On the other hand, neutral drug molecules tend to be much more lipophilic making them more permeable than ionized drug molecules. It should also be noted that in increased acidity (decreased PK_a) and decreased structural size, drug potency is increased as observed from decreased IC_{50} values (Miller et al., 1972).

3.2.4 PHYSICOCHEMICAL PROPERTIES

3.2.4.1 SOLUBILITY

Drug solubility can be defined as the totality of all concentration a drug reaches in a specified solvent matrix at equilibrium with solid compounds under specific conditions. It determines bioavailability of active compounds in drugs and intestinal absorption (Patel et al., 2012). Structural properties of a drug are important in predicting solubility. Factors that can affect the solubility of a drug area follow: Compound physical status, structure, solvent composition, pH, purity, solution components, and temperature, among others. It should be noted that solubility varies depending on action-site conditions in the body. Low drug solubility during discovery and development is costly if not well managed. It can compound challenges by increasing development project time and incur huge costs in formulation of a less bioavailable drug. This can also push project/manufacturer recovery costs to patients and also force prescription of high and more frequent doses to a cancer patient which is yet another burden on them. Insoluble drug compounds have poor absorption and bioavailability (Van-de-Waterbeemd et al., 2001), mainly in oral dosing and risks high residual drug presence in blood from intravenous administration. Other problems with low drug solubility include low values for bioassays and likely compromised assay results due to challenges in assay methodologies. Solubility can be increased by the reduction of Log P, reduction of drug MW, the addition of hydrogen bonding, the addition of polar substituents, and addition of ionizable substituents (Kerns and Di, 2008). This can be done by monitoring the solubility rates at every change of variables just mentioned in order to control PK of drug compounds as desired. Chirality of drug compounds has a consequence on solubility. According to the Wallach rule (Brock et al., 1991), crystals from racemates are more dense and stable than their chiral counterparts. The increased stability achieved hereafter reduces aqueous thermodynamic solubility. Bioavailability of a drug molecule can be further controlled by manipulating dissolution rates. This depends on the desired goal that drug developers are interested to achieve. A drug with all its excipients can have its dissolution rate increased by the reduction of particle sizes. If a drug is pre-dissolved in a solution, an oral administration route drug should be preferred. Salt form and surfactants made drugs are good for improved wetting of the solid material.

3.2.4.2 PERMEABILITY

Permeability is the velocity at which drug molecules move past through a biological membrane barrier (Kerns and Di, 2008). The drug is required to pass through membranes for absorption processes. This happens all the time in the blood, intestines, and cellular barriers for drugs to reach appropriate biological targets as well as in the liver and kidneys for elimination processes. *In vitro* tests provide good leads about a drug's permeability. Not all membranes are the same and behave similarly. Typically, high permeability is good for high drug bioavailability. There are different ways in which drugs permeate through membranes.

Passive diffusion: Here drug compounds move by Brownian motion from an aqueous environment to another, through a lipid bilayer membrane. This is normally influenced by concentration gradient.

Membrane transporter expression: This is also known as active transport where drug compounds get bound to trans-membrane proteins, which have a strong affinity to the drug. The trans-membrane protein carries the drug usually against concentration gradient, a process that demands energy. This method is vital for the permeation of natural ligands.

Inter-cellular junction tightness (paracellular): Very small and polar drug molecules pass through intercellular pores that are small to the ranges of around 8Å (Kerns and Di, 2008). Glomerulus and GI tract cells are sometimes called leaky because they provide for paracellular permeability between cells unlike the BBB which normally has very tight cell packing.

3.2.5 OTHER PROPERTIES

Other essential drug properties are biochemical and pharmacokinetic and toxicity properties. The structural and physicochemical properties thus far described have a direct influence on how the drug will certainly function in the body of a patient (Prentis et al., 1988). Most has already been explained but of particular interest are metabolism, protein tissue binding transportation, clearance in kidneys and liver, drug half-life, bioavailability, drug-drug interaction and toxicity of the drug. All these parameters are benchmarked by the basic properties discussed. When a drug compound is not permeable enough, it is likely not to reach its intended tumor cells, rendering its bioavailability insignificant. Solubility largely plays a role that directly affects the PK and shelf life of the drug itself. The less soluble the drug is, the higher the shelf life which if not necessary, may lead to the introduction of foreign toxins in some cells, tissue, organs, or systems of the body. Drugs residues are supposed to be eliminated at an appropriate time from body systems. Insolubility of a drug also negatively affects its clearance in the kidneys. Overburdened kidney may lead to renal failure problems and ultimately, death.

3.3 STAGES IN CANCER DRUG DEVELOPMENT

The foremost step in drug development is the determination and identification of a biological target responsible for the progenesis and/or the progression of a disease. For cancers, the mechanisms of the cancer itself have to be known and well understood. This gives drug developers ability to design drugs that will either counter or inhibit a particular cancer pathway or completely kill responsible cells. Drug development takes a few but highly involving and costly steps, starting from decision making through the science of drugs and approvals to the market, with ethical and legal implications.

3.4 PRECLINICAL ACTIVITIES

3.4.1 PRECLINICAL TESTS

Bench tests and observations are the first step that drug designers should consider to set up. Bench tests for anti-cancer drugs include bioassays on cancerous cell lines in a dish or wells. The main aim of this stage is to have an idea about the efficacy and loosely dosing and safety on normal cells of the test drug against the particular cancer of interest. It should be noted that the response that breast cancer cells will have towards a drug is different from how haptic, esophageal, or any other cancer will have on the same drug, in the same conditions. Cell-line assays help in streamlining an appropriate route that a drug developer should take as they proceed with the project. This stage saves costs.

Realizing that dish or well tests cannot be as reliable for cells in a real functioning animal, the next stage is conducted on animal models. It should be noted that any research involving animals should not be undertaken unless with a written approval by a reliable ethics board. A lot of tests have to be conducted, this time, with the main focus being safety of the drug and loosely, bioavailability of the drug. To ensure certainty on animals, the drug has to be tested on at least two animal species, usually rats and mice. Teratogenicity tests are conducted along with other tests to determine if the drug effects may not cause birth defects. Data from these studies should cover how the drug is affecting each and every vital part of the body as well as behavior. Details to be studied may include the liver, kidneys, dermatology, eyes, hair, histology, lungs, breathing rates, appetite, insomnia, hearing ability and psychology among others. This is important as it shows what side effects the drug may have if any.

3.4.2 FORMULATION

The next step in drug development is formulation. This is where the drug is given an appropriate medium based on the preferred mode of administration which depends on the properties of the drug compounds. Formulations can be solid, liquid, or aerosol.

3.4.3 DOSING ROUTE PRE-DETERMINATION

Drug compound properties assist in suggesting the dosing route vis-à-vis nature of biological target. This needs to be determined prior to advanced clinical trial stages, where the routes could be optimized further. The following are possible dosing routes that drug designers consider:

1. **Injections:** Subcutaneous (SC) method is one that administers drugs through under-skin injection. Intravenous (IV) doses are administered into the blood stream directly, either by infusion (continuously) for some minutes or rapidly, within a minute or a few seconds. Intramuscular doses are injected directly into muscle tissues. Doses directly injected into the inside parts of lower vertebrae bones are called epidural. Sometimes, drugs are injected directly into the abdominal cavity (peritoneal); this is called the intraperitoneal method (IP).
2. **Through the Skin Surface:** Topical doses are applied directly onto the surface of the skin, either as solutions or suspension. Transdermals are applied as patches on the surface of the skin and the drugs move through the skin into the affected area of the body.
3. **Through the Mouth and Nasal Cavity:** Oral doses (PO) are ingested through the mouth by swallowing. Sublingual method is where drugs are held under the tongue till they dissolve and spread on their own. This is different from the buccal method, where a drug is held between the hum and the cheek. Intranasal method applies mostly to aerosols and vaporous drugs, which are spread into the nose.
4. **Insertions:** Sometimes, doses are designed to be directly inserted into the rectum, these are called suppository doses.

3.5 CLINICAL TRIALS

It is only about 2% of all drugs that were involved in preclinical tests that successfully get to the clinical trial stage (Hawthorn and Redmond, 2006). At this stage, the preclinical tests must have confirmed the safety and efficacy of the drug in an animal. This is just a tip of an iceberg to drug development as the findings in animals do not necessarily entirely translate to the same in humans. The drug needs to be tested further systematically and ethically on humans (Rangitsch, 2009). There are three phases in clinical trials for a drug before considerations for approval applications and one last one for optimization.

3.5.1 PHASE I TRIAL

This aims at ensuring the safe (maximum tolerated doses) and effective (least cytotoxic) doses in humans with minimal side effects. It also establishes the best form of medicine and route of administration. A few (~10 to 100) volunteers, termed as cohorts, are involved and it takes between 6 and 12 months. The study is a single arm and unblinded. Sometimes, considering the high levels of allowable toxic doses in cancer drug clinical trials, it is argued to be unethical to involve healthy volunteers. In this case, cancer patients who have failed all other cancer therapies are allowed for recruitment in cohorts, otherwise phase I is completely by-passed, with justifications (Komen, 2017).

3.5.2 PHASE II TRAILS

This study aims at affirming efficacy in a more complex condition. For example, the studies may be carried out on patients with multiple cancer types to check on which ones the drug is more effective or if it can treat multiple cancers. This study has hundreds of subjects. It may be randomized and is usually unblinded. Optimum doses and short term side effects are monitored and recorded at this stage. The time frame can be between 6 to 24 months.

3.5.3 PHASE III TRAILS

At this stage, the test drug is compared against a standard available drug on the market in all aspects including safety, efficiency, efficacy, and stability. A large group of subjects (~1000) is involved in blinded randomized trials which can be placebo-controlled. Usually, only one type of cancer which might have been treated before or not is targeted. This can take between 1 and 4 years depending on the tumor type. Success at this stage leads to drug approvals.

3.5.4 PHASE IV TRIALS

This is called therapy optimization study. This phase is conducted after the drug has already been approved and is on the market. The study is randomized, usually double-blinded and can be placebo-controlled. Usually, a single type of cancer is targeted. This phase is useful to explore alternative treatment regimens, continuous safety monitoring of the drug and it may take several years.

ACKNOWLEDGMENT

The time and other resources in this work were sponsored by PHARMBIO-TRAC, Africa Center of Excellence in Pharmacology, Biotechnology, and Traditional medicine host at Mbarara University of Science and Technology in Uganda.

KEYWORDS

- **cancer drug properties**
- **clinical trials**
- **lipophilicity**
- **permeability**
- **solubility**
- **therapy optimization study**

REFERENCE

Brock, C. P., Schweizer, W. B., & Dunitz, J. D., (1991). On the validity of Wallach's rule: On the density and stability of racemic crystals compared with their chiral counterparts. *J. Am. Chem. Soc., 113*(26), 9811–9820.

CMC, (2012). Lipophilicity. In: *Drug Discovery Resources*. [Online Material]. Retrieved from: https://www.cambridgemedchemconsulting.com/resources/physiochem/logD.html (accessed on 20 February 2020).

Comer, J. E., (2003). A. High throughput measurement of logD and PKa. In: Artursson, P., Lennersnas, H., & Van-de-Waterbeemd, (eds.), *Methods and Principles in Medicinal Chemistry* (Vol. 18, pp. 21–45). Weinheim. Wiley-VCH.

Hann, M. M., & Oprea, T. I., (2004). Pursuing the lead likeness concept in pharmaceutical research. *Curr. Opinion in Chem. Biol., 8*, 255–263.

Hansch, C., Leo, A., Mekapati, S. B., & Kurup, A., (2004). QSAR and ADME. *Bio. Org. Med. Chem., 12,* 3391–3400.

Hawthorn, J., & Redmond, K., (2006). *A Guide to Cancer Drug Development and Regulation.* AstraZeneca, Cheshire, U.K.

Kerns, E. H., & Di, L., (2008). *Drug-like Properties: Concepts, Structure Design, and Methods.* Elsevier, New York, USA.

Komen, S. G., (2017). *Facts for Life, Clinical Trials.* [Online source]. Retrieved from: https://ww5.komen.org/uploadedFiles/_Komen/Content/About_Breast_Cancer/Tools_and_Resources/Fact_Sheets_and_Breast_Self_Awareness_Cards/ClinicalTrials.pdf (accessed on 20 February 2020).

Lipinski, C. A., Lombardo, F., Dominy, B. W., & Feeney, P. J., (1997). Experimental and approaches to estimate solubility and permeability in drug discovery and development settings. *Adv. Drug Delivery Rev., 23,* 3–25.

Lombardo, F., Obach, R. S., Shalaeva, M. Y., & Gao, F., (2002). Prediction of volume of distribution values in humans for neutral and basic drugs using physicochemical measurements and plasma protein binding data. *J. Med. Chem., 45,* 2867–2876.

Miller, G. H., Doukas, P. H., & Seydel, J. K., (1972). Sulfonamide structure-activity relation in a cell free system. Correlation of inhibition of folate synthesis with antibacterial activity and physicochemical parameters. *J. Med. Chem., 15,* 700–706.

Oprea, T. I., (2002). Chemical space navigation in lead discovery. *Curr. Opinion Chem. Biol., 6,* 384–389.

Oprea, T. I., Davis, A. M., Teague, S. J., & Leeson, P. D., (2001). Is there a difference between leads and drugs: A historical perspective. *J. Chem. Info. Comp. Sci., 41,* 1308–1315.

Patel, J. N., Rathod, D. M., Patel, N. A., & Modasiya, M. K., (2012). Techniques to improve the solubility of poorly soluble drugs. *Int. J. Pharm. Life Sci., 3*(2), 1459–1469.

Patridge, W. M., (1995). Transport of small molecules through the blood-brain-barrier: Biology and methodology. *Adv. Drug Delivery Rev., 15,* 5–36.

Prentis, R. A., Lis, Y., & Walker, S. R., (1988). Pharmaceutical innovation by the seven UK-owned pharmaceutical companies (1964–1985). *Brit. J. Clin. Pharm., 25,* 387–396.

Rangitsch, K., (2009). *Clinical Trials Alaska.* (pp. 1–20). Alaska Department of Health and Social Services. [Online source]. Retrieved from: http://dhss.alaska.gov/dph/Chronic/Documents/Cancer/assets/ClinicalTrials.pdf (accessed on 20 February 2020).

Roda, A., Cerre, C., Manetta, A. C., Cainelli, G., Umani-Ronchi, A., & Panunzio, M., (1996). Synthesis and physicochemical, biological, and pharmacological properties of new bile acids amidated with cyclic amino acids. *J. Med. Chem., 39,* 2270–2276.

Van-de-Waterbeemd, H., Smith, D. A., Beaumont, K., & Walker, D. K., (2001). Property-based design: Optimization of drug absorption and pharmacokinetics. *J. Med. Chem., 44,* 1313–1333.

Veber, D. F., Johnson, S. R., Cheng, H. Y., Smith, B. R., Ward, K. W., & Kopple, K. D., (2002). Molecular properties that influence the oral bioavailability of drug candidates. *J. Med. Chem., 45,* 2615–2623.

CHAPTER 4

Plant Constituent as Anti-Cancer Drugs

T. VENI,[1] T. PUSHPANATHAN,[2] and G. VIMALA[3]

[1]Department of Zoology, Ayya Nadar Janaki Ammal College (Autonomous), Sivakasi – 626 124, Tamil Nadu, India, Tel.: 099431 26685, E-mail: venientomology@gmail.com

[2]Department of Zoology, St. Xavier's College (Autonomous), Palayamkottai – 627 002, Tamil Nadu, India

[3]Aringnar Anna Government Arts College for Women, Walajapet – 632 513, Tamil Nadu, India

ABSTRACT

There are many complications and deficiencies in chemically defined drugs, especially drug resistances that appear in cancer chemotherapy. In spite of the fact that chemically defined drugs are most effective, severe adverse effects limit the use of these drugs. People have to show more attention to plant-based sources of cancer drugs to enhance the anti-cancer agents present in the drugs. Historically, plant-derived products have been the source of the active component of pharmacological agents, especially in the areas of the disease-causing organism and cancer treatment. Plant-based products play an important role in the innovation of novel drugs and new drug-lead compounds; it is a vital resource of the new structure and unique role in preventing the growth of tumors. Nowadays, more than 50% of extensively used anti-cancer drugs in the world are resulting from plant sources. Numerous active plant products and their derivatives with elevated activities and fewer side effects can be getting by structure alteration and transformation. Particularly in the field of anti-tumor and anti-AIDS drugs, the researches and improvement of new drugs from plant-based products show a booming vitality and an attractive prospect. The purpose of the study was to review the results, development, and the main anti-tumor active components of plant products in recent years.

4.1 INTRODUCTION

Cancer is a cluster of diseases where the cells grow unusually and proliferate through unrestrained cell division. They are metabolically more active than normal cells. It gradually invades and destroys nearby normal cells by forming a lump called a tumor. This is true of all cancers except leukemia (cancer of the blood) (Kaur et al., 2011; Chavan et al., 2013; Jaikumar et al., 2016). Not all tumors or lumps are cancerous; non-malignant tumors are not cancer. It is localized and of small size that tends to grow quite slowly. It does not invade other parts of the body and is rarely life-threatening. It is non-invasive and unable to metastasize. On the other hand, malignant tumors are growing at a high rate of division and have the capacity to invade and destroy nearby tissues. Both external and internal factors such as dietary fat intake, solvent, and pesticide exposure, exposure to ionizing radiation (causing acute leukemias, thyroid cancer, breast cancer, lung cancer, and others), smoking cigarette (Agarwal et al., 2012), some virus (HIV, HPV, and Hepatitis B virus) and unhealthy lifestyle (being overweight, limited physical exercise, too much alcohol, too much sugar, and red meat, not enough vegetables and fruit) can cause cancer (Patel et al., 2016; Singh et al., 2016; Thomson et al., 2014). Moreover, Cancer is a major public health problem, and it is known as the most important reason of fatality worldwide. In 2008, there were 7.6 million deaths or about 13% of all deaths worldwide, and 13.1 million deaths caused by cancer are estimated in 2030 (Globocan, 2010). In recent times, in 2012, in the United States, a total of 1,638,910 new cancer cases were detected, and 577,190 deaths caused by cancers were expected (Siegel, 2012). In Europe, there were an estimated 3.45 million new cases of cancer and 1.75 million deaths in 2012 caused by cancer; this record was estimated from 40 countries in Europe (Ferlay et al., 2013). This information underlines the significance of anticancer drug development and the need for new useful anticancer drugs for the cure of different forms of tumors.

Plant derivatives are rich sources of antitumor agents and have considerable impact on anticancer drug improvement (Cragg et al., 2009). The documentation revealed that among 155 small-molecule antitumor drugs, the number of drugs encouraged by plant-based products is 113, accounting for 72.9% (Newman et al., 2012). Plants are considered as an important source of anticancer drugs; therefore, there have been several natural compounds from higher plants currently used as anticancer drugs or being evaluated in clinical trials (Cragg et al., 2009; Shah et al., 2013). Interestingly, endophytic fungi, which live in the inner tissues of healthy host plants, are found to produce plant secondary metabolites, some of which are anticancer drugs or leads (Chandra et al., 2012; Shweta et al., 2010; Eyberger et al., 2006). Marine organisms, for example, marine sponges, tunicates, soft corals, and algae, are also wealthy sources of antitumor leads, and a few marine natural products and derivatives have been approved as anticancer drugs chemotherapy (Cragg et al., 2009; Petit et al., 2013). Moreover, numerous marine-derived components are currently on the clinical pipeline for anticancer chemotherapy (Cragg et al., 2009; Petit et al., 2013).

Even though plant-based products are the main sources of anticancer drugs, crude extracts containing cytotoxic activity could not be directly used as drugs because of detrimental side effects of the toxicity. Moreover, small amounts of antitumor agents are normally isolated from the producing organisms; as a result, the isolation of particular plant-based products for the usage as anticancer drugs is impossible. Furthermore, amounts, and numbers of bioresource, that is, higher plants and marine invertebrates, which are the producers of anticancer compounds, are also restricted for the isolation of a large quantity of anticancer drugs. Therefore, a chemical production of natural anticancer agents has take part in a crucial role in the advance of anticancer drugs, and it also helps the development of drug properties through the rational drug design.

In the earlier period, antitumor drug development heavily depends on cytotoxic efficacy of the investigated components against cancer cell lines; unfortunately, the selection of cytotoxic agents for further drug development is performed without the knowledge of mechanism of action. Even though the regular approach has obtained definite achievement in the invention of novel antitumor medicine, the rising of knowledge on cancer molecular targets and the comprehension of tumor at a molecular level have changed the paradigms in anticancer drug improvement to which the target-based approaches have been heavily employed in the drug development processes. This chapter presents the recent progress of antitumor drugs, as well as anticancer leads, which are inspired by plant-based products, and also focus on antitumor plant-based products and derivatives, which are discovered through the most recently employed anticancer drug targets.

4.2 DRUG DISCOVERY FROM NATURAL PRODUCTS

Plant-based products have been used for centuries for the management of several illnesses. There are numerous basic earliest therapeutic systems derived from dietary sources. In current civilization, financial system and expertise are more developed, conventional medicines are still used in many countries as basic healthcare. Even though many conventional pharmaceutical approaches have been replaced, however, there is an existing resurgence in the interest in plant-based products by the general public, and the make use of complementary and substitute drugs is rising rapidly in urban countries. Inadequate investigation has been completed, and more and more pharmaceutical industries are involved in examining their potential as sources of new medicinal components (Zhang et al., 2013). Several pharmacological active compounds have been invented from flora, fauna, and microbes, such as natural products and organic compounds, which have been developed into drugs to treat diseases. Traditionally, natural products in the field of anti-cancer research has made considerable attainments, over 60% of the experimental use of anti-cancer drugs derived from natural products (Seelinger et al., 2012), including plants, marine organisms, microbes, and more than 3,000 verities of plants could be used against cancer. Plants have been the major resources in conventional medicine and natural products are considered as significant sources of antitumor drugs. Possibly secondary metabolites are invented mainly by chemical, biochemical, pharmacological, and clinical research data screening. With the advance of pharmacological experiment technology, the structure of the guide, modification, and transformation of plant-based products have also support the quick improvement of synthetic drugs and with the help of modern extraction, modification, and transformation, technology is bond to find some new ideas and approaches in the healing of chronic disease.

4.3 BIOACTIVE COMPOUNDS USED AS CANCER DRUGS

Organic compounds derived from plants and marine products have served us well in fighting cancer. The components are well characterized as possessing a wide range of anti-cancer properties, for example, induction of programmed cell death, and controlled digestion of damaged organelles within a cell and inhibition of cell propagation. Secondary metabolites such as alkaloids, flavonoids, terpenoids, polysaccharide, and saponin collected from plants have effective biological properties such as anticancer, painkiller; help to reduce inflammation, Immune response is altered to a desired level and used specifically for treating viral infections, etc. Marine organisms, invertebrates, and algae present rich sources of anti-cancer agents with structurally different biologically active compounds and bioactive secondary metabolites that exhibits different anticancer activities; indole alkaloids being the most general. A number of products were first invented from microorganisms such as antimicrobial substance active against bacteria, which has anticancer activity. The anticancer efficacy of most natural anti-neoplastic drugs often do not kill cancer cells directly, but regulates the human immune function to attain the purpose or both. Separation and replication is a series of important events in cell cycle, and a deregulation of cell cycle can have effect on the development of cancer (Wang and Ren, 2010). DNA topoisomerase I (Topo I) is a crucial enzyme concerned in cell development. The inhibition of Topo I is a significant antitumor pathway. And also, a huge number of antitumor drugs fight cancers through cell cycle arrest, stimulation of cell death and differentiation as well as through inhibition of cell growth and propagation, or a combination of two or more of these mechanisms (Abdullah et al., 2012). The search for new drugs is still a main concern target for malignancy treatment due to the fact that chemotherapeutic drug resistance is becoming more and more frequent.

4.4 ANTI CANCER PROPERTIES OF BIOACTIVE SUBSTANCES ISOLATED FROM PLANTS

More than 60% of the present antitumor drugs were obtained in one way or another from plant sources (Newman et al., 2012). Nature continues to be an abundant resource of bioactive and varied chemotypes, and whereas relatively a small number of the actual extracted natural products are developed into clinically efficient drugs in their own right, these unique molecules often serve as models for the preparation of more effective analogs and prodrugs through the application of chemical methodology, such as total or combinatorial (parallel) synthesis, or the manipulation of biosynthetic pathways. As well as, developments in formulation may result in more effective administration of the drug to patients, or combination of poisonous natural molecules to monoclonal antibodies or polymeric carriers specifically targeting epitopes on tumors of interest can lead to the development of efficacious targeted therapies. The vital role played by plant products in the innovation and improvement of new anticancer agents, and the significance of multidisciplinary collaboration in the optimization of novel molecular leads from natural product sources have been extensively reviewed (Table 4.1) (Cragg and Newman, 2009; Cragg et al., 2009, 2012; Grothaus et al., 2010; Demain et al., 2011; Basmadjian et al., 2014).

Historically, plants have been primary sources of natural product drug discovery, and in the anticancer area, plant-derived agents, such as VBL and vincristine (VCR), etoposide, paclitaxel (Taxol ®), docetaxel, topotecan, and irinotecan, are among the most effective cancer chemotherapeutics currently available (Cragg et al., 2012). Oridonin is a complex ent-kaurane diterpenoid extracted from the traditional Chinese herb Isodon rubescens (Fujita et al., 1976). It has demonstrated great efficacy in the healing of various human cancers due to its unique and safe anticancer pharmacological profile (Li et al., 2011; Zhou et al., 2007; Bohanon et al., 2014), and has recently been reported to have a potent anti-mycobacterial activity (Xu et al., 2014).

Camptothecin, a pentacyclic alkaloid extracted from *Camptotheca acuminata* (Wall et al., 1966), acts as a potent Topo I inhibitor showing strong antitumor activity both in laboratory culture and living organism (Oberlies and Kroll, 2004). Largazole, a depsipeptide plant product extracted from the *cyanobacterium Symploca* sp. (Taori et al., 2008), is a histone deacetylase (HDAC) inhibitor with potent antiproliferative activity and selectivity for cancer cells. Halichondrin B, a large polyether macrolide discovered from several sponge sources, exhibits growth inhibition on a panel of various cancer cell lines at nanomolar concentrations. It suppressed the microtubule growth phase without affecting the shortening phase, and caused tubulin sequestration into non-productive aggregates (Dabydeen et al., 2006). Liphagal, a meroterpenoid natural product collected from the sponge *Aka coralliphaga*, was discovered in a screening program designed to find new isoform-selective PI3K inhibitors. It inhibited PI3Kα with an IC_{50} of 100 nM and showed an approximately 10-fold selectivity for PI3Kα com-pared with PI3Kγ in a fluorescent polarization enzyme bioassay (Marion et al., 2006)

Curcumin, a yellow principal polyphenol curcuminoid, were isolated from the *Curcuma longa* (Govindarajan, 1980). It displays a diversity of biological and cellular performance including it removes potentially damaging oxidizing agents in a living organism., reduce certain signs of inflammation, as swelling, tenderness, fever, and pain, used in the treatment of cancer and drug that lowers the amount of cholesterol in the blood (Salem, 2014). Combretastatin A-4 (CA-4, 42), a naturally microtubular-destabilizing agent extracted from the South African tree *Combretum caffrum*, used in the treatment of cancer and drug that attacks the vasculature of a tumor both in laboratory culture and living organism (Lin et al., 1988). Oleanolic acid, a triterpenoid compound isolated from many Asian herbs, such as *Fructus ligustri lucidi, Fructus forsythiae, Radix ginseng,* and *Akebia trifoliate*, exhibits to prevent damage to the liver, reduce certain signs of inflammation, used in the treatment of cancer, removes potentially damaging oxidizing agents in a living organism and antiglycative activities (Tsai and Yin, 2012).

Podophyllotoxin (podofilox) and its derivatives, etoposide, and teniposide, are all cytostatic (antimitotic) glucosides. The mechanism of action of podophyllotoxin lies in the formation of a complex with tubulin and prevention of the synthesis of microtubules. It blocks cell division in late S- and G2-phases of the cell cycle. It inhibits topoisomerase II, which results in DNA damage through strand breakage. It is useful against testicular and ovarian germ cell cancers, lymphomas, and acute myelogenous and lymphoblastic leukemia (Alam et al., 2014; Song et al., 2014).

Paclitaxel and docetaxel are very important anti-cancer drugs that have revealed various antitumor efficacies against several malignancies including metastatic breast and ovarian carcinoma, head, and neck cancer, small cell lung cancer, oesophageal, adenocarcinoma, and hormone refractory prostate cancer (Zhang et al., 2014; Barbuti and Chen, 2015). Flavopiridol is one of flavonoid phytochemical constituent that acts by blocking cell division and induction of apoptosis. Some of its antitumor efficacies include inhibition of cyclins and cyclin dependant kinases (CDK), induction of cell death and inhibition of the development of new blood vessels (Biswas et al., 2015; Li, 2016; Nahata, 2017).

Combretastatins is effective against cancers of colon, lung, and blood and also showed significant anti-angiogenic property. Betulinic acid possesses anticancer properties while silvestrol is effective against lung carcinoma, and ductal carcinoma *in situ* (DCIS). Phytochemicals of flavonoids class are known for their anticancer properties. (-)-Epigallocatechin-3-gallate (EGCG) is known to prevent growth of malignant cells by modulating various cellular signaling pathways and inducing apoptosis of cancer cells selectively without affecting normal cells (Biswas et al., 2015; Li, 2016; Nahata, 2017).

TABLE 4.1 Several Plant-Based Anticancer Bioactive Substances Isolated from Plants

Anticancer Bioactive Substances Isolated from Plants	Types of Cancer Treated	Reported Authors
Apigenin	Pancreatic cancer	Singh et al., 2016; Alam et al., 2014
Belotecan	Ovarian, small-cell lung, and refractory colorectal cancers	Biswas et al., 2015
Berberine	Tumorigenic micro-organisms and virus	Coderch et al., 2012

TABLE 4.1 *(Continued)*

Anticancer Bioactive Substances Isolated from Plants	Types of Cancer Treated	Reported Authors
Bullatacin	Human ovarian tumor cells	Qayed et al., 2015
Calcitriol	Angiogenesis of cancer cells	Singh et al., 2016
Cathechin and Epicatechin	MCF7-human breast tumor cell line	Singh et al., 2016; Sharma, 2015
Colchicine	Hepatocellular carcinoma cells, cholangiocarcinoma cells, and Gastric cancer	Coderch et al., 2012; Nahata, 2017
Combretastatin	Ovarian, colon, lung, gastric, other solid tumors, blood, and anaplastic thyroid Cancer	Biswas et al., 2015; Li, 2016; Sharma, 2015, Prakash et al., 2013
Curcumin	Melanoma cancers, Liver cancer cell lineHepG2), Leukemic monocyte lymphoma cell line (U937), Brain cancer cells and brain-derived neural stem cells, Lymphoid leukemia and skin cancer	Singh et al., 2016; Biswas et al., 2015; Schneider-Stock et al., 2012; Shakya, 2016; Fridlender et al., 2015; Kinghorn et al., 2016
Cyanidin glycosides	Breast and colon cancer cells	Singh et al., 2016
Elliptinium	Human breast tumor cell line (MCF-7)	Patel, 2016; Mohan et al., 2012; Shoeb, 2006
Emodin	Human colon cancer cells	Agarwal et al., 2012
Estramustine	Prostate cancer	Biswas et al., 2015
Exatecan	Acute myelogenous leukemia	Sharma, 2015
Fisetin	Human acute promyelocytic leukemia cell line(HL-60)	Singh et al., 2016
Genistein	Cancers of breast, uterus, cervix, lung, stomach, colon, pancreas, liver, kidney, urinary bladder, prostate, testis, oral cavity, larynx, and thyroid	Singh et al., 2016
Gingerol	Colon cancer cells	Singh et al., 2016; Schneider-Stock et al., 2012; Kinghorn et al., 2016
Halofuginone	Glioma, wilms tumer, hepatocellular carcinoma, bladder, prostate, and pancreatic cancer, Preclinical trail, Progressive advanced solid tumors	Juarez, 2014

TABLE 4.1 *(Continued)*

Anticancer Bioactive Substances Isolated from Plants	Types of Cancer Treated	Reported Authors
Hespertin	Colon, cervical, and breast cancer	Alam et al., 2014
Honokiol	Lung, colon, liver, breast, and prostate cancer	Juarez, 2014
Kaempferol	Pancreatic cancer cells	Singh et al., 2016
Licochalcone A	Gastric cancer	Song et al., 2014
Matrine	Breast, gastric, lung cancer	Song et al., 2014
Maytansine	Breast cancer	Shoeb, 2006
Montamine	HT29 (colon adenocarcinoma cell line), H460 (non-small cell lung carcinoma cell line), RXF393 (renal cell carcinoma cell line), MCF7 (human breast cancer cell line), and OVCAR3 (epithelial ovarian cell line)	Patel, 2016; Mohan et al., 2012
Naringin	TGF-β1/Smad3 (pancreatic cancer cell line)	Alam et al., 2014
9-Nitrocamptothecin A549	(lung cancer cells line)	Ramakrishnan, 2013; Song et al., 20
Piperine	Multi-drug resistance cancer, B16F10 (melanoma cancer cells line)	Singh et al., 2016; Alam et al., 2014
Plumbagin	Leukemia, myeloma, breast, prostate, ovarian, pancreatic, liver, cervical, and skin cancer	Juarez, 2014; Song et al., 2014
Protopanaxadiol	Cytotoxic against multidrug-resistant tumors	Sharma, 2017
Quercetin	Ovarian cancer cell line (OVCA433), human cervical cancer, prostate cancer, epidermal growth factor receptor-over expressing oral cancer, osteosarcoma	Li, 2016; Song et al., 2014
Resveratrol	Colon cancer, Neuroblastoma, myeloma, breast, prostate, pancreatic, and lung cancer	Juarez, 2014
Rohutikine	Ovarian and breast cancer	Patel, 2016
(R)-Roscovitine	Non-small cell lung cancer	Mariaule and Belmont, 2014

TABLE 4.1 *(Continued)*

Anticancer Bioactive Substances Isolated from Plants	Types of Cancer Treated	Reported Authors
Schischkinnin	Colon cancer cell lines	Patel, 2016; Mohan et al., 2012
Silvestrol	Lung, colon, blood, prostate, and breast cancer	Biswas et al., 2015; Sharma, 2015
Silymarin	Multi-drug resistance cancer	Biswas et al., 2015
Topotecan	Ovarian and small cell lung cancers, and colorectal cancers	Alam et al., 2014; Shoeb, 2006
Vinflunine	Metastatic transitional cell carcinoma of the urothelial tract	Ali et al., 2012; Coderch et al., 2012
Vinorelbine (Na-velbine)	Leukemia, lymphomas, advanced testicular, breast, and lung cancer	Coderch et al., 2012; Crețu et al., 2012
Paclitaxel, docetaxel	lung cancer, breast, and ovarian cancer, prostate cancer	Zhang et al., 2013
podophyllotoxin and deoxypodophyllotoxin	human colon cancer cells	Zhao et al., 2013
liquiritigenin	human cervical carcinoma	Liu et al., 2012
LicochalconeA	gastric cancer	Xiao et al., 2011
Glabridin	breast cancer	Hsu et al., 2011
Baicalin	ovarian cancer	Zhou et al., 2009; Chen et al., 2013
Phenoxodiol	Chemoresistant cancer cells	Alvero et al., 2008
Quercetin	human cervical cancer, prostate cancer, epidermal growth factor receptor-over expressing oral cancer, osteosarcoma	Gokbulut et al., 2013
Caffeic acid phenethyl ester	Pancreatic cancer, cervical cancer	Chen et al., 2013
Ginsenoside	Glioma cancer	Lee et al., 2013
Tubeimoside-1	Gastric cancer cell proliferation	Zhang et al., 2013
Platycodin	Apoptosis in prostate cancer cells	Lee et al., 2013
Fucoidan	Anti-metastatic effect	Shen et al., 2009
Andrographolide	Inhibit breast cancer cell proliferation	Kumar et al., 2012
Sesquiterpene lactones	Inhibit growth of tumor promoter-induced cell and transformation	Saikali et al., 2012

TABLE 4.1 *(Continued)*

Anticancer Bioactive Substances Isolated from Plants	Types of Cancer Treated	Reported Authors
Doxycycline	Anti-tumor effects through inhibition of FAK signaling pathway	Sun et al., 2009
Beta-lactams	Breast cancer	Frezza et al., 2008

4.5 CONCLUSION

The initiation has lead to a prosperous of looking for anticancer drugs from plant products, where persons began to show interest to the natural ingredients. In addition, to consider for high efficacy, low side effects and a broad range of anticancer medicine from natural resources, public also began to focus on cancer prevention research. Compounds extracted from natural resources can enhance immune cells or body immunity, anti-tumor effects and inhibit the proliferation of cancer cells. The development of cancer prevention and health care products from natural products has a broader prospect, greater economic and social benefits when compared with synthetic anti-cancer drugs.

KEYWORDS

- active ingredient
- cancer
- cholangiocarcinoma cells
- medicinal plants
- natural products
- synthetic anti-cancer drugs

REFERENCES

Abdullah, W. Z., Sulaiman, S. A., & Suen, A. B., (2012). A review of molecular mechanisms of the anti-leukemic effects of phenolic compounds in honey. *Int. J. Mol. Sci., 13*(11), 15054–15073.

Agarwal, N., Majee, C., & Chakraborthy, G. S., (2012). Natural herbs as anticancer drugs. *Int. J. of Pharm. Tech. Res.*, *4*, 1142–1153.

Alam, S., Satpathy, P., & Thosar, A., (2014). Plants and its parts as a source of anticancer compound: A review. *Int. Res. J. Phar.*, *5*, 244–250.

Ali, R., Mirza, Z., Ashraf, G. M., Kamal, M. A., & Ansari, S. A., (2012). New anticancer agents: Recent developments in tumor therapy. *Anticancer Res.*, *32*, 2999–3005.

Alvero, A. B., Kelly, M., Rossi, P., Leiser, A., Brown, D., Rutherford, T., & Mor, G., (2008). Anti-tumor activity of phenoxodiol: From bench to clinic. *Future Oncol.*, 475–482.

Basmadjian, C., Zhao, Q., & Djehal, A., (2014). Cancer wars: Natural products strike back. *Front Chem.*, *2*, 20.

Biswas, J., Roy, M., & Mukherjee, A., (2015). Anticancer drug development based on phytochemicals. *J. Drug Disc Develop Delivery*, *2*, 10–12.

Bohanon, F. J., Wang, X. F., Ding, C. Y., Ding, Y., Radhakrishnan, G. L., & Rastellini, C., (2014). Oridonin inhibits hepatic stellate. Cell prolifeation and fibrogenesis. *J. Surg. Res.*, *190*, 55–63.

Chandra, S., (2012). Endophytic fungi: Novel sources of anticancer lead molecules. *Appl. Microbiol. Biotechnol.*, *95*, 47–59.

Chavan, S. S., Damale, M. G., Shamkuwar, P. B., & Pawar, D. P., (2013). Traditional medicinal plants for anticancer activity. *Int. J. Curr. Pharm Res.*, *5*, 50–54.

Chen, J., Li, Z., Chen, A. Y., Ye, X., Luo, H., Rankin, G. O., & Chen, Y. C., (2013). Inhibitory effect of baicalin and baicalein on ovarian cancer cells. *Int. J. Mol. Sci.*, *14*(3), 6012–6025.

Coderch, C., Morreale, A., & Gago, F., (2012). Tubulin-based structure-affinity relationships for antimitotic Vinca alkaloids. *Anti-Cancer Agen. in Medici. Chem.*, *12*, 219–225.

Cragg, G. M., & Newman, D. J., (2009). Nature: A vital source of leads for anticancer drug development. *Phytochem. Rev.*, *8*, 313–331.

Cragg, G. M., Grothaus, P. G., & Newman, D. J., (2009). Natural products as sources of new drugs over the last 25 Years. *Chem. Rev.*, *109*, 3012–3043.

Cragg, G. M., Kingston, D. G. I., & Newman, D. J., (2012). *Anticancer Agents from Natural Products* (2nd edn.). Boca Raton, CRC/Taylor & Francis.

Crețu, E., Trifan, A., Vasincu, A., & Miron, A., (2012). Plant-derived anticancer agents-curcumin in cancer prevention and treatment. *Rev. Med. Chir. Soc. Med. Nat. Lasi.*, *116*, 1223–1229.

Dabydeen, D., Burnett, J. C., Bai, R., Verdier-Pinard, P., Hickford, S. J. H., & Pettit, G. R., (2006). Comparison of the activities of the truncated halichondrin B analog NSC 707389 (E7389) with those of the parent compound and a proposed binding site on tubulin. *Mol. Pharmacol.*, *70*, 1866–1875.

Demain, A. L., & Vaishnav, P., (2011). Natural products for cancer chemotherapy. *Microb. Biotechnol.*, *4*, 687–699.

Eyberger, A. L., Dondapati, R., & Porter, J. R., (2006). Endophyte fungal isolates from *Podophyllum peltatum* produce podophyllotoxin. *J. Nat. Prod.*, *69*, 1121–1124.

Ferlay, J., Steliarova-Foucher, E., Lortet-Tieulent, J., Rosso, S., Coebergh, D. J., Comber, D. J., Forman, D., & Bray, F., (2013). Cancer incidence and mortality patterns in Europe: Estimates for 40 countries in (2012). *Eur. J. Cancer*, *49*, 1374–1403.

Frezza, M., Garay, J., Chen, D., Cui, C., Turos, E., & Dou, Q. P., (2008). Induction of tumor cell apoptosis by a novel class of N-thiolated betalactam antibiotics with structural modifications at N1 and C3 of the lactam ring. *Int. J. Mol. Med.*, *21*(6), 689–695.

Fridlender, M., Kapulnik, Y., & Koltai, H., (2015). Plant derived substances with anti-cancer activity: From folklore to practice. *Frontiers in Plant Science*.

Fujita, E., Nagao, Y., Kaneko, K., Nakazawa, S., & Kuroda, H., (1976). The antitumor and antibacterial activity of the Isodon diterpenoids. *Chem. Pharm. Bull.*, *24*, 2118–2127.

Globocan, (2010). *IARC-2008*. http://globocan.iarc.fr/factsheets/populations/factsheet.asp?uno=?900 (accessed on 22 February 2020).

Gokbulut, A. A., Apohan, E., & Baran, Y., (2013). Resveratrol and quercetin-induced apoptosis of human 232B4 chronic lymphocytic leukemia cells by activation of caspase-3 and cell cycle arrest. *Hematology*.

Govindarajan, V. S., (1980). Turmeric—chemistry, technology, and quality. *Crit. Rev. Food Sci. Nutr.*, *12*, 199–301.

Grothaus, P. G., Cragg, G. M., & Newman, D. J., (2010). Plant natural products in anticancer drug discovery. *Curr. Org. Chem.*, *14*, 1781–1791.

Hsu, T. H., Chu, C. C., Hung, M. W., Lee, H. J., Hsu, H. J., & Chang, T. C., (2013). Caffeic acid phenethyl ester induces E2F-1-mediated growth inhibition and cell-cycle arrest in human cervical cancer cells. *FEBSJ*. doi: 10.1111/febs.12242.

Jaikumar, B., & Jasmine, R., (2016). A Review on a few medicinal plants possessing anticancer activity against human breast cancer. *International Journal of Pharm. Tech. Research*, *9*, 333–365.

Juarez, P., (2014). *Plant-Derived Anticancer Agents: A Promising Treatment for Bone Metatasis*. BoneKey Reports.

Kaur, R., Singh, J., Singh, G., & Kaur, H., (2011). Anticancer plants: A review. *J. Nat. Prod. Plant Resour.*, *1*, 131–136.

Kinghorn, A. D., De Blanco, E. J. C., Lucas, D. M., Rakotondraibe, H. L., & Orjala, J., (2016). Discovery of anticancer agents of diverse natural origin. *Anticancer Research*, *36*, 5623–5637.

Kumar, S., Patil, H. S., Sharma, P., Kumar, D., Dasari, S., Puranik, V. G., Thulasiram, H. V., & Kundu, G. C., (2012). Andrographolide inhibits osteopontin expression and breast tumor growth through down regulation of PI3 kinase/Akt signaling pathway. *Curr. Mol. Med.*, *12*(8), 952–966.

Lee, J. H., Oh, E. K., Cho, H. D., Kim, J. Y., Lee, M. K., & Seo, K. I., (2013). Crude saponins from platycodon grandiflorum induce apoptotic cell death in RC-58T/h/SA#4 prostate cancer cells through the activation of caspase cascades and apoptosis-inducing factor. *Oncol. Rep.*, *29*(4), 1421–1248.

Li, C. Y., Wang, E. Q., Cheng, Y., & Bao, J. K., (2011). Oridonin: An active diterpenoid targeting cell cycle arrest, apoptotic, and autophagic pathways for cancer therapeutics. *Int. J. Biochem. Cell Biol.*, *43*, 701–704.

Li, W. W., Johnson-Ajinwo, O. R., & Uche, F. I., (2016). Advances of plant-derived natural products in ovarian cancer therapy. *Int. J. of Cancer Res. and Pre.*, *9*, 81.

Lin, C. M., Singh, S. B., Chu, P. S., Dempcy, R. O., Schmidt, J. M., & Pettit, G. R., (1988). Interactions of tubulin with potent natural and synthetic analogs of the antimitotic agent combretastatin: A structure-activity study. *Mol. Pharmacol.*, *34*, 200–208.

Liu, Y., Xie, S., Wang, Y., Luo, K., Wang, Y., & Cai, Y., (2012). Liquiritigenin inhibits tumor growth and vascularization in a mouse model of HeLa cells. *Molecules*, *17*(6), 7206–7216.

Marion, A. F., Williams, D. E., Patrick, B. O., Hollander, I., Mallon, R., & Kim, S. C., (2006). Liphagal, a selective inhibitor of PI3 kinaser isolated from the sponge *aka coralliphaga*: Structure elucidation and biomimetic synthesis. *Org. Lett.*, *8*, 321–324.

Mohan, M., Jeyachandran, R., & Deepa, K., (2012). Alkaloids as anticancer agents. *Annals of Phytomedicine*, *1*, 46–53.

Nahata, A., (2017). Anticancer agents: A review of relevant information on important herbal drugs. *Int. J. Clin. Pharmacol. Toxicol.*, *6*, 250–255.

Newman, D. J., & Cragg, G. M., (2012). Natural products as sources of new drugs over the 30 years from 1981 to 2010. *J. Nat. Pro.*, *75*, 311–335.

Oberlies, N. H., & Kroll, D. J., (2004). Camptothecin and taxol: Historic achievements in natural products research. *J. Nat. Prod.*, *67*, 129–135.

Patel, S. G., (2016). A review on medicinal plants for cancer therapy. *International Journal of Medi. Pharm. Research*, *2*, 105–112.

Petit, K., & Biard, J. F., (2013). Marine natural products and related compounds as anticancer agents: An overview of their clinical status. *Anticancer Agents in Med. Chem.*, *13*, 603–631.

Prakash, O. M., Kumar, A., & Pawan, K. A., (2013). Anticancer potential of plants and natural products. *American Journal of Pharmacological Sciences*, *1*, 104–115.

Qayed, W. S., Aboraia, A. S., Abdel-Rahman, H. M., & Youssef, A. F., (2015). *Annonaceous acetogenins* as a new anticancer agent. *Der. Pharma. Chemica.*, *7*, 24–35.

Ramakrishnan, R., (2013). Potential clinical applications of natural products of medicinal plants as anticancer drugs: A review. *International Journal of Medicinal and Pharmaceutical Sciences*, *3*, 127–138.

Saikali, M., Ghantous, A., Halawi, R., Talhouk, S., Saliba, N., & Darwiche, N. A., (2012). Sesquiterpene lactones isolated from indigenous Middle Eastern plants inhibit tumor promoter-induced transformation of JB6 cells. *BMC Complement Altern. Med.*, *12*, 89.

Salem, M., Rohani, S., & Gillies, E. R., (2014). Curcumin, a promising anti-cancer therapeutic: A review of its chemical properties, bioactivity, and approaches to cancer cell delivery. *RSC Adv.*, *4*, 10815–10829.

Schneider-Stock, R., Ghantous, A., Bajbouj, K., Saikali, M., & Darwiche, N., (2012). Epigenetic mechanisms of plant-derived anticancer drugs. *Frontiers in Bioscience*, *17*, 129–173.

Seelinger, M., Popescu, R., Giessrigl, B., Jarukamjorn, K., Unger, C., Wallnöfer, B., et al. (2012). Methanol extract of the ethnopharmaceutical remedy *Smilax spinosa* exhibits anti-neoplastic activity. *Int. J. Oncol.*, *41*(3), 1164–1172.

Shaikh, R., Pund, M., Dawane, A., & Iliyas, S., (2014). Evaluation of anticancer, antioxidant, and possible anti-inflammatory properties of selected medicinal plants used in Indian traditional medication. *Journal of Traditional and Complementary Medicine*, *4*, 253–257.

Shakya, A. K., (2016). Medicinal plants: Future source of new drugs. *International Journal of Herbal Medicine*, *4*, 59–64.

Sharma, S., (2015). Review on phytochemicals with their biological roles?: A review. *International Journal of Advanced Research in Biological Sciences*, *4*, 69–77.

Shen, K. T., Chen, M. H., Chan, H. Y., Jeng, J. H., & Wang, Y. J., (2009). Inhibitory effects of chito-oligosaccharides on tumor growth and metastasis. *Food Chem. Toxicol.*, *47*(8), 1864–1871.

Shoeb, M., (2006). Anti-cancer agents from medicinal plants. *Bangladesh Journal of Pharmacology*, *1*, 35–41.

Shweta, S., Zuehlke, S., Ramesha, B. T., Priti, V., Mohana, K. P., Ravikanth, G., Spiteller, M., Vasudeva, R., & Uma, S. R., (2010). Endophytic fungal strains of *Fusarium solani*, from *Apodytes dimidiata* E. Mey. ex Arn (Icacinaceae) produce camptothecin, 10-hydroxycamptothecin and 9-methoxycamptothecin. *Phytochemistry*, *71*, 117–122.

Siegel, R., Naishadham, D., & Jemal, A., (2012). Cancer statistics. *Cancer J. Clin., 62*, 10–29.

Singh, S., Sharma, B., Kanwar, S. S., & Kumar, A., (2016). Lead phytochemicals for anticancer drug development: A review. *Frontiers in Plant Science, 7*, 1–13.

Song, Y., Sun, H., Zhang, A., Yan, G., & Han, Y., (2014). Plant-derived natural products as leads to anti-cancer drugs: A review. *Journal of Medicinal Plant and Herbal Therapy Research, 2*, 6–15.

Sun, T., Zhao, N., Ni, C. S., Zhao, X. L., Zhang, W. Z., Su, X., Zhang, D. F., Gu, Q., & Sun, B. C., (2009). Doxycycline inhibits the adhesion and migration of melanoma cells by inhibiting the expression and phosphorylation of focal adhesion kinase (FAK). *Cancer Lett., 285*(2), 141–150.

Taori, K., Paul, V. J., & Luesch, H., (2008). Structure and activity of largazole, a potent antiproliferative agent from the Floridian marine *cyanobacterium Symploca* sp. *J. Am. Chem. Soc., 130*, 1806–1807.

Thomson, A. K., Heyworth, J. S., Girschik, J., Slevin, T., & Saunders, C., (2014). Beliefs and perceptions about the causes of breast cancer: A case-control study. *BMC Research Notes, 7*, 558.

Tsai, S. J., & Yin, M. C., (2012). Anti-oxidative, anti-glycative and antiapoptotic effects of oleanolic acid in brain of mice treated by dgalactose. *Eur. J. Pharmacol., 68*, 981–988.

Wall, M. E., Wani, M. C., Cook, C. E., Palmer, K. H., McPhail, A. T., & Sim, G. A., (1966). Plant antitumor agents. I. The isolation and structure of camptothecin, a novel alkaloidal leukemia, and tumor inhibitor from *Camptotheca acuminate*. *J. Am. Chem. Soc., 88*, 3888–3890.

Wang, L. M., & Ren, D. M., (2010). Flavopiridol, the first cyclin-dependent kinase inhibitor: Recent advances in combination chemotherapy. *Mini Rev. Med. Chem., 10*(11), 1058–1070.

Xiao, X. Y., Hao, M., Yang, X. Y., Ba, Q., Li, M., Ni, S. J., Wang, L. S., & Du, X., (2011). Licochalcone A inhibits growth of gastric cancer cells by arresting cell cycle progression and inducing apoptosis. *Cancer Lett., 302*(1), 69–75.

Xu, S. T., Li, D. H., Pei, L. L., Yao, H., Wang, C. Q., & Can, H., (2014). Design, synthesis, and antimycobacterial activity evaluation of natural oridonin derivatives. *Bio. Org. Med. Chem. Lett., 24*, 2811–2814.

Zhang, S., Mercado, Uribe, I., & Liu, J., (2014). Tumor stroma and differentiated cancer cells can be originated directly from polyploid giant cancer cells induced by paclitaxel. *International Journal of Cancer, 134*, 508–518.

Zhang, X., Chen, L., Ouyang, Y., Cheng, B., & Liu, X., (2012). *Cell Prolif., 45*, 466–476.

Zhang, Y., Xu, X. M., Zhang, M., Qu, D., Niu, H. Y., Bai, X., Kan, L., & He, P., (2013). Effects of tubeimoside-1 on the proliferation and apoptosis of BGC823 gastric cancer cells in vitro. *Oncol. Lett., 5*(3), 801–804.

Zhoa, Y., Wu, Z., Zhang, Y., & Zhu, L., (2013). HY-1 induces G2/M cell cycle arrest in human colon cancer cells via ATR-Chk1-Cdc25C and Wee1 pathways. *Cancer Sci.* doi: 10.1111/cas.12182.

Zhou, G., Kang, H., Wang, L., Gao, L., Liu, P., & Xie, J., (2007). Oridonin, a diterpenoid extracted from medicinal herbs, targets AML1-ETO fusion protein and shows potent antitumor activity with low adverse effects on (8.21) leukemia *in vitro* and *in vivo*. *Blood, 109*, 3441–3450.

Zhou, Q. M., Wang, S., Zhang, H., Lu, Y. Y., Wang, X. F., Motoo, Y., & Su, S. B., (2009). The combination of baicalin and baicalein enhances apoptosis via the ERK/p38 MAPK pathway in human. *Sin., 30*(12), 1648–1658.

CHAPTER 5

Anticancer Properties of Some Effective Phytochemicals

MURAT ALAN[1] and BEHNAZ ASLANIPOUR[2]

[1]*Department of Obstetrics and Gynecology, Izmir Tepecik Training and Research Hospital, Tepecik – 35120, Izmir, Turkey*

[2]*Department of Bioengineering, Faculty of Engineering, Ege University, Izmir, Turkey, E-mail: Behnaz_Aslanipour@yahoo.com*

ABSTRACT

Cancer is one of the major causes of death in not only developing but also developed countries despite of achievement of advanced modern medicine. As cancer is a preventable disease, finding more effective phytochemicals in prevention of the disease can be less costly and more effective and safer at the same time. Phytochemicals, any of various biologically active compounds found in plants, are useful compounds which have been focused on tor treating cancer throughout history because of their low toxicity, safety, and widespread availability. Consumption of sufficiently high amount of fruits and vegetables is in a positive association with reducing cancer risk in humans. Helpful phytochemicals are used not only as disruption of anomalous signaling pathways heading to cancer but also as contribution with chemotherapy and radiotherapy. For this reason, some of the naturally appearing phytochemicals which are known as effective components in cancer chemoprevention having therapeutic potential have been discussed for many years. In the current book chapter, the anticancer effects of some of biologically active molecules such as polyphenols, terpenoids, organosulfur compounds, and phytosterols and their derivatives will be discussed. Finally, via this chapter, we will try to provide solid evidence to support of phytochemicals or dietary agents' application in both prevention and treatment of cancer.

5.1 INTRODUCTION

5.1.1 CANCER DEFINITION

Cancer, which is not a newly discovered disease, has attracted interest during the past. Almost 14 million new cases and up to 8 million death cases are reported as results of cancers annually in all around the world. Cancer is uncontrollable cell division, forms lumps or masses of tissue namely tumors. Tumors can interfere after a speedy growth period via nervous, digestive, and circulatory systems and release hormones that have the possibility of altering body function. There are more than 100 types of cancers reported up to now as its initially effected cell type classifies each of them. Lung, stomach, liver, colon, and breast cancer are introduced as the most common cancer disorders (Chen et al., 2014; Surh, 2003a). Tumors staying in one place with limited growth rate are considered to be known as benign while the ones considered as malignant are recognized with having one or both following criteria either a cancerous cell moves all over the body via blood or lymphatic systems and destroy healthy tissue in invasion process or cell manages to divide and grow by making new blood vessels to feed itself in angiogenesis process.

5.1.2 AN OVERVIEW OF CANCER TREATMENT BY BOTH CHEMICAL AND PHYTOCHEMICALS

The high percentage of cancer prevalence was the main reason for many campaigns to focus on the prevention of the disease along with tobacco cessation movements and vaccination efforts (Torre et al., 2015). Recently, chemotherapy, radiotherapy, and surgery are known as the most common treatments of cancer. There are some frequently used drugs in chemotherapy including antimetabolites (e.g., methotrexate), DNA-interactive agents (e.g., cisplatin, doxorubicin), anti-tubulin agents (taxanes), hormones, and molecular targeting agents are in head of chemical drugs in treatment of cancer (Nussbaumer et al., 2011). As the drugs are used clinically, they can cause numerous unwanted effects such as bone marrow suppression, hair loss, gastrointestinal (GI) lesions, neurologic dysfunction, cardiac toxicity, and drug resistance (Dropcho, 2011; Monsuez et al., 2010; Nussbaumer et al., 2011). Even though many intensive interventions have been discovered up to now, a large number of patients undergo poor prognosis. For this reason, the investigations upon discovering novel anticancer agents with much less side effects and better effectiveness has started. Researchers have focused on finding different useful phytochemicals for development of innovative remedies from several medicinal plants and herbal ingredients having potential anti-cancer properties (Sharma, 2011; Tan et al., 2011; Teiten et al., 2013). Besides, there are some other remedies such as decreasing of cell proliferation, inducing apoptosis, retarding metastasis, and inhibiting of angiogenesis reported to be the results of application of some phytochemicals isolated from medicinal plants (Sadeghnia et al., 2014). There have been some natural plant-derived compounds including taxol analogs, podophyllotoxin, and vinca alkaloids (vincristine, vinblastine) analogs have been broadly used during chemotherapy in patients with cancer (Sadeghnia et al., 2014). Effects of some phytosterols are listed in Table 5.1.

5.1.3 PHYTOCHEMICALS

The word "phyto" in phytochemicals originated from the Greek word *phyto* meaning as plant. Therefore, phytochemicals are defined as plant obtained chemicals including fruits, vegetables, grains, and other plant foods with protective properties against the risk of major chronic diseases. They are non-nutritive compounds with disease protectable properties. It has been estimated that there have been up to 10000 specific phytochemicals identified from not only fruits but also vegetables or grains 6,000 of which are contained in the class of flavonoids (Harborne, 1995). Apart from this, there is also a large amount of unidentified or need to be identified of phytochemicals as well for making the benefits of the phytochemicals in whole foods clear. More than a few reports suggest that the benefits of phytochemicals should be greater than what is recently understood (Liu, 2003). The health developing properties of phytochemicals have been notified via several epidemiological studies as there was a positive correlation between sufficient consumption of fruits and vegetables and reduction of numerous diseases' development such as variety of cancers (Key et al., 2009). Regarding to the various number of physiological roles of phytochemicals in plant tissues belonging to enzymes which are involved in metabolisms of cells as well as defense mechanisms against foreign agents such as radiations, viruses, and parasites. They also have strong association with pleiotropic effects in animal cells. Phytochemicals have gained attention of many scientists according to the involvement of phytochemicals in inflammatory and oncogenic transformation, such as cell cycle control alteration, evasion of apoptosis, angiogenesis, and metastases. In addition to this, many epidemiological studies reported

TABLE 5.1 Some *In Vitro* and *In Vivo* Anti-Cancer Effects of Plant-Derived Phytomedicals

Source	Phytochemical/ Molecule Name	Study Case	Cancer Site	Mechanism of Action	References
Tea	Polyphenols	A number of 190 gastric cancer cases	Gastric cancer	Inhibition of gastric and esophageal cancer development.	Sun et al., 2002
Tea	Polyphenols	A number of 162 colorectal cancer cases	Colorectal cancer	Inhibition of colon cancer in humans	Yuan et al., 2007
Citrus	Flavonoids	A number of 250 incident breast cancer	Breast cancer	Reduction of breast cancer risk	Dai et al., 2002
Curcuma Longa	Curcumin and Resveratrol	Lung carcinogenesis in mice	Lung cancer	Modulating p53 hyper-phosphorylation, regulating caspases as well as cellular metabolism enzymes	Malhotra et al., 2014
Grapes	Polyphenol (Resveratrol)	*In vitro* and on ACF *in vivo*.	Colon cancer	Preventing the risk of colon cancer development	Mazue et al., 2014
Grapes	Resvertrol	Human A431 SCC cells	Skin cancer	Downregulating Rictor, Suppressing UV-Induced skin cancer	Back et al., 2012
Red wine	Resveratrol	Human OVCAR-3 cells	OVCAR-3 cells	Mediating apoptosis in human OVCAR-3 cells	Lin et al., 2011
Grapeseed	GSE and resveratrol	C57BL/6 mice	Tongue tumorigenesis	Induction of apoptosis and autophagy in C57BL/6 mice	Shrotriya et al., 2015
Blueberries	Polyphenol (Quercetin)	Thyroid cancer cell lines (TPC-1, FTC-133, NPA, FRO, and ARO)	Thyroid Cancer	Intervening of therapeutic effects on thyroid cancer redifferentiation	H. J. Kang et al., 2011
Blueberry	Pterostilbene	Human Breast Cancer Cells	Breast cancer	Suppressing the development of breast cancer stem cells	Mak et al., 2013

Anticancer Properties of Some Effective Phytochemicals

TABLE 5.1 (Continued)

Source	Phytochemical/Molecule Name	Study Case	Cancer Site	Mechanism of Action	References
Blueberry	Pterostilbene	Human CSCs	CSCs	Suppression of enrichment of CD133+ in hepatoma stem cells	Lee et al., 2013
Blueberry	Pterostilbene	Human prostate cancer cells	Prostate Cancer	Inhibiting both tumor growth and metastasis progression of cancer cells	Li et al., 2013
Pholidota yunnanensis	Phoyunbene B	HepG2 liver cancer cells	Liver cancer	Induction of G2/M cell cycle arrest as well as increasing of apoptosis	Wang et al., 2012
Citrus	Monoterpenoid (Limonene)	Human colorectal cancer.	Colorectal cancer	Having chemo-preventive activity against colorectal cancer	Sun, 2007
Citrus	D-Limonene and docetaxel	Human prostate carcinoma DU-14	Prostate cancer	Treating the result of hormone-refractory prostate cancer	Bishayee and Rabi, 2009
Citrus	D-Limonene	Humans with advanced breast cancer	Breast cancer	D-Limonen, a tolerated metabolite by breast cancer, has anti-cancer property	Vigushin et al., 1998
Vegetables and fruit	Total phytosterols	A number of 463 subjects with lung cancer	Lung cancer	Reduction of cancer risk by approximately 50% after consumption of	Mendilaharsu et al., 1998
Vegetables and fruit	Total phytosterols	A case-control study with 120 patients	Stomach cancer	The positive association was Observation of a positive link between phytosterol intake and reduction of stomach cancer	De Stefani et al., 2000
Soy	Stigmasterol	Women with ovarian cancer	Ovarian cancer	Reduced risk of developing ovarian cancer at higher intakes of stigmasterol	McCann et al., 2003

TABLE 5.1 (Continued)

Source	Phytochemical/Molecule Name	Study Case	Cancer Site	Mechanism of Action	References
Plants	β-Sitosterol	Human breast cancer cells in ovariectomized mice	Breast cancer	Reducing tumor growth in 17β-estradiol-treated mice	Ju et al., 2004
Ginger	Gingerol	MDA MB-231 human breast cancer cells	Breast cancer	Inhibiting metastasis of MDA MB-231	Lee et al., 2008
Curcuma longa L	Curcumin	Breast cancer cell line MCF7	Breast cancer	Having estrogenic effects	Bachmeier et al., 2010
Parsley	Apigenin	Breast cancer cell lines MCF 7	Breast cancer	Cytotoxic activity	El-Alfy et al., 2011
Parsley	Apigenin	Colon cell line (HCT 116),	Colon cancer	Cytotoxic activity	El-Alfy et al., 2011
Apples	Fisetin	HCT-116 human colon cancer cells	Colon cancer	Anti-carcinogenesis properties	Geraets et al., 2009
Tea	Kaempferol	A case study of 529 cases of exocrine pancreatic cancer	Pancreatic cancer	Anti-carcinogenesis properties	Nothlings et al., 2007
Tea	Kaempferol	Patient-based case-control	Lung cancer	Anti-carcinogenesis properties	Grand, 1993
Wax-like coatings of fruits	Triterpenoids	In vivo and in vitro studies	Breast cancer	Inducing chemoprevention	Bishayee et al., 2011
Cruciferous vegetables	Sulforaphane	Metastatic SW620 cell line	colon cancer	Activating various concurrent proapoptic pathways	Yu et al., 1976
Tomato	Lycopene	Human blood and tissues	Colon cancer	Preventing colon carcinogenesis with low doses	Narisawa et al., 1996

that a daily consumption of phytochemicals can decrease the prevalence of various types of cancers (D'Incalci et al., 2005; Russo et al., 2005). They have the ability of putting a stop or a delay in developing of cancer cells into the malignant ones (Shen et al., 2014). Besides, they are capable of impeding the initiation or repealing the promotion step of multistep carcinogenesis (Russo et al., 2012; Surh, 2003b). Although the number of sources of useful phytochemicals is limited, it has been proved that the usage of anti-cancer drugs obtained from plants has raised through decades. Specially, some particular beneficial plant-derived phytochemicals are targeted to obtain an ever-increasing requirement on the access to modern and diverse chemical libraries. For this reason, natural products are supposed to be a developing field for drug discovery due to having numerous sources of compounds with a variety of structures (Newman et al., 2003). Considering these issues, there are some anti-cancer agents such as vincristine, irinotecan, etoposide, and paclitaxel which are found to be one of the most discussable fields of knowledge. They are all naturally derived agents which can play a crucial role dominant role in chemotherapy.

5.2 THE EFFECTS OF SOME CERTAIN PLANT SECONDARY COMPOUNDS AS IMMUNE MODULATORS

5.2.1 POLYPHENOLS

Polyphenols are a huge number of plant phytochemicals, more than 40,000 metabolites, which are effective agents to protect the plants against stress and chemical pathogens (Amararathna et al., 2016). Polyphones comprise at least one aromatic ring having one or more hydroxyl groups from which flavonoids contain two aromatic rings that are attached by three atoms of carbon confining in an oxygenated heterocycle ring (Ferrazzano et al., 2011; Tsao, 2010). Although the exact anti-cancer properties of polyphenols have not yet been completely understood, it can be indicated that polyphenols have beneficial effects against various kinds of cancers (Del Rio et al., 2013). As it is known, green tea is a nice source of polyphenols indicating that the regular using of green tea can influence the prohibition of cancer development (Rodriguez-Mateos et al., 2014). Less than a few studies have reported the effect of polyphenols in clinical trials in which both green tea and curcumin polyphenols are mostly considered. Apart from the problematic issue of polyphenol consumption which is known as the rapid metabolism and low bioavailability of them, they are mostly considered because of high accessibility, specificity of their response and their low toxicity (Brglez Mojzer et al., 2016). It is proved that polyphenols have strong power of fighting against cervical cancer and tumor initiation as well as having cytotoxic activity toward tumor cells (Lepley et al., 1996; Scalbert et al., 2005). It is also reported that polyphenols have the property of targeting viral oncogenes and also inhibiting the deregulation and gene signaling expression of the host cells (Hail et al., 2008; Sun et al., 2004). There are a variety of polyphenols that have attracted attention of many researches to have some therapeutic properties such as phenolic acids, flavonoids, stilbenes, curcuminoids, and lignans which some of them are described in the current chapter.

5.2.1.1 PHENOLIC ACIDS

Dietary polyphenols contain up to 30% of phenolic acids (Ramos, 2008). There is a widely heterogenic group in secondary metabolites of plants known as phenolic compounds having at least one aromatic ring which carries one or maybe more hydroxyl groups. Phenolic acids with hydroxybenzoic or hydroxycinnamic acids derivatives are mostly found in plants (Huang et al., 2009). It is reported that flavonoids and non-flavonoids two major subgroups of phenolic compounds. Although flavonoids comprise two aromatic rings carrying an oxygen hetrocycle, non-flavonoids have two demonstrative subunits known as benzoic and cinnamic acid (Bravo, 1998). Phenolic compounds have attracted attention of many researchers due to their beneficial effects in both health care and food industry. It has been proved that phenolic compounds have numerous treating properties over healing of some disorders such as inflammation, Parkinson, diabetes, and Alzheimer (Bravo, 1998; Mohamed, 2014). It is reported that the anticancerogenic properties of ellagic acid, one of the most common hydroxybenzoic acid derivative phenolic acids, main polyphenol in pomegranate, has been widely (Bell and Hawthorne, 2008; Tomas-Barberan and Clifford, 2000). A study suggested that over 50% potential of the antioxidant activity belongs to ellagic acid in pomegranate juice which makes it a strong anticancer property and atherosclerosis treating effect (Bell and Hawthorne, 2008). It is usually consumed as ellagitannins or it can be conjugated with a glycoside moiety such as glucose (Bell and Hawthorne, 2008) and it is generally released while digestion and metabolism (Seeram et al., 2004). The mechanism of action of ellagic acid has been investigated upon prostate cancer which has represent antiproliferative activity (Bell and Hawthorne,

2008). Apart from this, the induction of cell cycle arrest, antiproliferative effect, and apoptosis on cells in a variety of cancers (Strati et al., 2009). The effect of this widely investigated phenolic acid, ellagic acid is via inhibition of NF-κB (Bell and Hawthorne, 2008) and insulin-like growth factor 2 (IGF-II), (Narayanan and Re, 2001.) and induction of p53/p21 expression as well as modulation of pro- and anti-apoptotic proteins (Bell and Hawthorne, 2008).

5.2.1.2 FLAVONOIDS

Flavonoids, natural components that existed in not only vegetables but also fruits, have been reported to be more than 4000 which are taken part in our routine diet. They are also more than 60% of dietary polyphenols with a special chemical structure containing two benzene rings attached to three carbon atoms by which an oxygenated heterocycle is formed (Ramos, 2008). The following seven subgroups including flavones, flavonols, flavanones, isoflavones, anthocyanins, catechins, and chalcones are different classified groups derived from flavonoids. They are broadly found in plants as either glycosides or free-state. Flavonoids are normally water-soluble with a chemical structure comprising C6-C3-C6 skeleton of carbon. An important implication can be proposed that these compounds have useful properties upon treating some dysfunctions such as the ability of protection from viral infections, treating some disorders such as diabetes and cardiovascular, neurological, and inflammatory diseases. It can be added that they are preventers of oxidative cell damage and anti-cancers with the probability of protecting carcinogens, inhibiting the initiation, promotion, and tumor progression as well (Androutsopoulos et al., 2010; Murakami et al., 2008; Ramos, 2008). Flavonoids have the responsibility of enhancing the activity of many enzyme system activities as well as effecting as efficient anti-oxidants (Erlund, 2004). Regarding to many *in vitro, in vivo* and epidemiological studies reported in many studies, it is displayed that flavonoids have protective properties against different types of cancers (Ramos, 2008). (-)-epigallocatechin-3-gallate (EGCG) has been introduced as one of the most commonly investigated flavonoids with different therapeutic effects such as decreasing recurrence of colorectal polyps in humans, antitumor effects, antiangiogenic effect (Cao and Cao, 1999; Ohga et al., 2009; Shimizu et al., 2008). There are some flavonoids listed in Table 5.1 with various anti-cancer activities.

5.2.1.3 STILBENES

Stilbenes which are derived from the phenylpropanoid pathway are a small family of plant derivative secondary metabolites and they are present in numerous plant types. It is implicated that these compounds can act as plant disease protectors and human health supporters. Additional information regarding to the structure of stilbenes is suggested that they are mainly known as essential groups of nonflavonoid phytochemicals of polyphenolic structure in which they include a 1,2-diphenylethylene nucleus. Due to having high power of being consumed as beneficial anti-cancer components. It can be noted that there is evidence to suggest that these compounds have low toxicity under *in vivo* investigations (Kang et al., 2011). Malignancy is the result of multiple cancer mechanisms which starts with initial driver mutations and follow by promotion and progression. There are many helpful chemo-preventing agents who are used in prohibition of carcinogenesis (Kang et al., 2011) and stilbenes have been proved to have the capability of reducing the presence of tumorigenesis via having interfering behavior with molecular events at all steps in carcinogenesis. Although stilbenes are distributed in limited number of plants, there are some studies studied about the anticancer effects of these components (Asensi et al., 2011).

Albeit here is no sufficient data regarding to their complete effect upon cancer cells, it has been proved that Stilbenes are involved in various targets and mechanisms against cancer in cancer cell pathways. The effects of some plant-derived stilbenes and their properties in cancer cells are listed in Table 5.1.

5.2.1.4 CURCUMINOIDS

Turmeric (*Curcuma longa*) is one of the useful dietary supplements. Its powder, curcumin, is a yellow-colored polyphenol and has been commonly used by people from Asia in which the frequency of cancer is low comparing to the other continents. It can be implicated from previous investigations that this powder which is identified as 1,6-heptadiene-3,5-dione-1,7-bis(4-hydroxy-3-methoxyphenyl)-(1E,6E) has been consumed in home remedies in treatment or prevention of many disorders (Ravindran et al., 2009). Apart from curcumin, turmeric contains some other beneficial components such as curcumin II, curcumin III and the recently identified cyclocurcumin which are all considerable components in treatment of cancer (Aggarwal et al., 2007; Kiuchi et al., 1993). It is reported that the mechanism of action of curcumin as an antioxidant activity is because of attached hydroxyl groups while their anti-inflammatory and anti-proliferative activities are depending of the presence of attached methoxy groups. Apart from these, there are some additional important activities of curcumin such as modulatory agents as growth factors and belonging receptors, transcription factors, enzymes, cytokines, and apoptosis (Aggarwal et al., 2007). It is implicated from some studies that curcumin effects the cancer cells via inhibiting both the initiation process of tumors (Huang et al., 1992) and the promotion of tumors (Conney et al., 1991; Huang et al., 1988). It can be added that unrelated membrane proteins can be structurally affected throughout various signaling pathways (Bilmen et al., 2001). A study supports that induction of negative curvature in the bilayer can be caused by inserting curcumin deep into the cellular membrane in a trans-bilayer orientation way and anchoring to either hydrogen bonding or to the phosphate group of lipids by which permeabilizing activity of the apoptotic protein tBid can be increased in the result of promoting of negative curvature by curcumin report suggests that curcumin inserts (Barry et al., 2009; Epand, 2002). Additionally, curcumin is responsible for suppressing multiple signaling pathways as well as inhibiting cell proliferation, metastasis, invasion, and angiogenesis. One of the most important properties of curcumin is the ability of inducing apoptosis via several signaling pathways (Barry et al., 2009; Bilmen et al., 2001; Epand, 2002).

5.2.2 TERPENOIDS

Terpenoids are classified in a class of the biggest natural products which are known as important source of drug discovery. Recent investigations of the effective roles of terpenoids against variety cancers have led the researchers to focus on some useful drugs from terpenoids effecting the inhibition of both proliferation of cancer cells and metastasis through numerous mechanisms. It can be mentioned that their anti-cancer effects can be via modulation of the immune system such as NF-kB signaling. Terpenoids are consisting of more than 25,000 chemical structures so they are considered to be particularly used in chemical and flavor industries (Gershenzon and Dudareva, 2007). Basically, terpenoids are assembled and modified from 5 units of carbon isoprene in many ways. It is reported that not only their functional groups but also their basic carbon skeletons can be modified and differently shaped among terpenoids. They can be found in all living creatures (Jackson, 2010). The most commonly used terpenoids in traditional medicine are plant-derived terpenoids which are mostly considered because of having aromatic qualities. The red color in tomato and the yellow color in sunflower are caused by terpenoids (Ajikumar et al., 2008). It is deduced that terpenoids are sub-classified into monoterpenoids, diterpenoids, triterpenoids, sesquiterpenoids, and tetraterpenoids which all have potential properties such as anti-cancer effects (Gershenzon and Dudareva, 2007; Huang et al., 2012).

5.2.2.1 CAROTENOID TERPENOIDS

There is a number of almost 600 known carotenoids which are separated into two different classes as xanthophylls containing oxygen and carotenes which are hydrocarbons having no oxygen. Mainly, the structure of carotenoids is consisting of a polyene chain having 9–11 double bonds and possibility of having termination in rings. The structure of conjugated double bond structure makes carotenoids to become a potential electron transporter as well as a reducer (Vershinin, 1999). All of the carotenoid derivatives comprise eight molecules of isoprene and have 40 atoms of carbon. The absorption of wavelength for carotenoids ranges from 400–550 nm and are seen in yellow, orange, or red colors. Carotenoids in algae and plants have the responsibility of absorbing light energy to be used in photosynthesis and have the ability of protecting chlorophyll from photodamage (Armstrong and Hearst, 1996). Carotenoids have antioxidant activity apart from this they have vitamin A activity, convertible to retinol, which is resulted by beta-ionone rings in their structures (Bernstein et al., 2016). Carotenoids are beneficial anti-cancerogenic agents with protection effect of cells against free radicals and other substrates which are disruptive for both cell membranes and DNA. Because of the chemicals that smokers have in their blood, they are plenty of damageable free radicals which can be knockout by some phytochemicals such as carotenoids. For this reason there are many ongoing studies referring to the effects of antioxidants and their associating with lower risk of lung cancer in smokers (Ruano-Ravina et al., 2000).

5.2.3 ORGANOSULFUR COMPOUNDS

Organosulfur compounds are the compounds having sulfur unite in their structure. Although it is commonly known that they are associated with odors of foul, it can be stated that many of them are sweet compounds such as saccharin. It is approved that these compounds are strong cancer-fighting agents as many of them are consumed in diet such as allium species which are rich source of organosulfur compounds. More than a few studies reported that consumption of garlic is a supportive way to decrease cancer risk (Qian et al., 2011). There are some biologically particular properties which are resulted from the mechanism of action of organosulfur compounds such as inducting carcinogen detoxification, inhibiting tumor cell proliferation, having antimicrobial effect and free radical scavenging, inhibiting DNA adduct formation as well as inducing apoptosis (Barcelo et al., 1998; Moriarty et al., 2007). Isothiocyanate sulforaphane which is an organosulfur compounds caused induction of GSTs and inhibition of CYPs tested over Phase I and Phase II enzymes on carcinogen metabolism in primary culture of rat and human hepatocytes (Maheo et al., 1997). Allyl sulfur compounds are important example of organosulfur compounds with high potential of anti-cancer effects (Dygos and Chinn, 1975).

5.2.4 PHYTOSTEROLS

Plant-derived sterols are alcohols with 28 and 29 atom carbons (Otaegui-Arrazola et al., 2010) which are not synthesized in human body. They are reported to be found in human diet especially those with high amount of lipids (Amaral et al., 2003; Ito et al., 1973; Weihrauch and Gardner, 1978). Application of 2 to 3g of phytostrols to healthy subjects caused an induction in sitosterol and campesterol levels (30% and 70% respectively) (Hallikainen et al., 2000; Weststrate and Meijer, 1998). Studies regarding to the effect of phytosterols suggest that they have potential of fighting against colon cancer in some animal models (Baskar et al., 2012). In a case-control study, the effect of plant sterols was found to be effective to prohibit the development of stomach cancer risk (De Stefani et al., 2000). Existing data suggest that there is a positive association between the intake of a higher amount of plant sterols and a lower risk of cancer prevalence (Jemal et al., 2011). An additional important implication suggests that plant sterols are powerful agents to fight against oxidative stress caused by some cancers such as breast cancer cells (Tas et al., 2005). They can also be effective in the improvement of blood lipid and relatively be protective against cardiovascular disorder (Ramprasath and Awad, 2015).

KEYWORDS

- **anticancer**
- **carotenoid terpenoids**
- **immune modulators**
- **lycopene**
- **organosulfur compounds**
- **phytosterols**

REFERENCES

Aggarwal, B. B., Sethi, G., Baladandayuthapani, V., Krishnan, S., & Shishodia, S., (2007). Targeting cell signaling pathways for drug discovery: An old lock needs a new key. *J. Cell. Biochem., 102,* 580–592.

Ajikumar, P. K., Tyo, K., Carlsen, S., Mucha, O., Phon, T. H., & Stephanopoulos, G., (2008). Terpenoids: Opportunities for biosynthesis of natural product drugs using engineered microorganisms. *Mol. Pharm., 5,* 167–190.

Amaral, J. S., Casal, S., Pereira, J. A., Seabra, R. M., & Oliveira, B. P. P., (2003). Determination of sterol and fatty acid compositions, oxidative stability, and nutritional value of six walnut (*Juglans regia* L.) cultivars grown in Portugal. *J. Agric. Food Chem., 51,* 7698–7702.

Amararathna, M., Johnston, M., & Rupasinghe, H., (2016). Plant polyphenols as chemopreventive agents for lung cancer. *Int. J. Mol. Sci., 17,* 1352.

Androutsopoulos, V. P., Papakyriakou, A., Vourloumis, D., Tsatsakis, A. M., & Spandidos, D. A., (2010). Dietary flavonoids in cancer therapy and prevention: Substrates and inhibitors of cytochrome P450 CYP1 enzymes. *Pharmacol. Ther., 126,* 9–20.

Armstrong, G. A., & Hearst, J. E., (1996). Carotenoids 2: Genetics and molecular biology of carotenoid pigment biosynthesis. *FASEB J., 10,* 228–237.

Asensi, M., Ortega, A., Mena, S., Feddi, F., & Estrela, J. M., (2011). Natural polyphenols in cancer therapy. *Crit. Rev. Clin. Lab. Sci.,* 197–216.

Bachmeier, B. E., Mirisola, V., Romeo, F., Generoso, L., Esposito, A., Dell'eva, R., Blengio, F., Killian, P. H., Albini, A., & Pfeffer, U., (2010). Reference profile correlation reveals estrogen-like trancriptional activity of curcumin. *Cell. Physiol. Biochem., 26,* 471–482.

Back, J. H., Zhu, Y., Calabro, A., Queenan, C., Kim, A. S., Arbesman, J., & Kim, A. L., (2012). Resveratrol-mediated downregulation of rictor attenuates autophagic process and suppresses UV-Induced skin carcinogenesis. *Photochem. Photobiol., 88,* 1165–1172.

Barcelo, S., Mace, K., Pfeifer, A. M. A., & Chipman, J. K., (1998). Production of DNA strand breaks by N-nitrosodimethylamine and 2-amino-3-methylimidazo[4,5-f]quinoline in THLE cells expressing human CYP isoenzymes and inhibition by sulforaphane. *Mutat. Res. Mol. Mech. Mutagen., 402,* 111–120.

Barry, J., Fritz, M., Brender, J. R., Smith, P. E. S., Lee, D. K., & Ramamoorthy, A., (2009). Determining the effects of lipophilic drugs on membrane structure by solid-state NMR spectroscopy: The case of the antioxidant curcumin. *J. Am. Chem. Soc., 131,* 4490–4498.

Baskar, A. A., Al Numair, K. S., Gabriel, P. M., Alsaif, M. A., Muamar, M. A., & Ignacimuthu, S., (2012). β-sitosterol prevents lipid peroxidation and improves antioxidant status and histoarchitecture in rats with 1,2-dimethylhydrazine-induced colon cancer. *J. Med. Food.*, *15*, 335–343.

Bell, C., & Hawthorne, S., (2008). Ellagic acid, pomegranate, and prostate cancer: A mini review. *J. Pharm. Pharmacol.*, *60*, 139–144.

Bernstein, P. S., Li, B., Vachali, P. P., Gorusupudi, A., Shyam, R., Henriksen, B. S., & Nolan, J. M., (2016). Lutein, zeaxanthin, and meso-zeaxanthin: The basic and clinical science underlying carotenoid-based nutritional interventions against ocular disease. *Prog. Retin. Eye Res.*, *50*, 34–66.

Bilmen, J. G., Khan, S. Z., Javed, M. H., & Michelangeli, F., (2001). Inhibition of the SERCA Ca^{2+} pumps by curcumin. Curcumin putatively stabilizes the interaction between the nucleotide-binding and phosphorylation domains in the absence of ATP. *Eur. J. Biochem.*, *268*, 6318–6327.

Bishayee, A., & Rabi, T., (2009). d-Limonene sensitizes docetaxel-induced cytotoxicity in human prostate cancer cells: Generation of reactive oxygen species and induction of apoptosis. *J., Carcinog.*, *8*, 9.

Bishayee, A., Ahmed, S., Brankov, N., & Perloff, M., (2011). Triterpenoids as potential agents for the chemoprevention and therapy of breast cancer. *Front. Biosci.*, *16*, 980–996.

Bravo, L., (1998). Polyphenols: Chemistry, dietary sources, metabolism, and nutritional significance. *Nutr. Rev.*, *56*, 317–33.

Brglez, M. E., Knez, H. M., Skerget, M., Knez, Z., & Bren, U., (2016). Polyphenols: Extraction methods, antioxidative action, bioavailability, and anticarcinogenic effects. *Molecules*, *21*, 901.

Cao, Y., & Cao, R., (1999). Angiogenesis inhibited by drinking tea. *Nature*, *398*, 381.

Chen, W. J., Ho, C. C., Chang, Y. L., Chen, H. Y., Lin, C. A., Ling, T. Y., et al., (2014). Cancer-associated fibroblasts regulate the plasticity of lung cancer stemness via paracrine signaling. *Nat. Commun.*, *5*, 3472.

Conney, A. H., Lysz, T., Ferraro, T., Abidi, T. F., Manchand, P. S., Laskin, J. D., & Huang, M. T., (1991). Inhibitory effect of curcumin and some related dietary compounds on tumor promotion and arachidonic acid metabolism in mouse skin. *Adv. Enzyme Regul.*, *31*, 385–396.

D'Incalci, M., Steward, W. P., & Gescher, A. J., (2005). Use of cancer chemopreventive phytochemicals as antineoplastic agents. *Lancet. Oncol.*, *6*, 899–904.

Dai, Q., Franke, A. A., Jin, F., Shu, X. O., Hebert, J. R., Custer, L. J., Cheng, J., Gao, Y. T., & Zheng, W., (2002). Urinary excretion of phytoestrogens and risk of breast cancer among Chinese women in Shanghai. *Cancer Epidemiol. Biomarkers Prev.*, *11*, 815–821.

De Stefani, E., Boffetta, P., Ronco, A. L., Brennan, P., Deneo-Pellegrini, H., Carzoglio, J. C., & Mendilaharsu, M., (2000). Plant sterols and risk of stomach cancer: A case-control study in Uruguay. *Nutr. Cancer.*, *37*, 140–144.

Del Rio, D., Rodriguez-Mateos, A., Spencer, J. P. E., Tognolini, M., Borges, G., & Crozier, A., (2013). Dietary (poly)phenolics in human health: Structures, bioavailability, and evidence of protective effects against chronic diseases. *Antioxid. Redox Signal*, *18*, 1818–1892.

Dropcho, E. J., (2011). The neurologic side effects of chemotherapeutic agents. *Continuum (Minneap*. Minn), *17*, 95–112.

Dygos, J. H., & Chinn, L. J., (1975). Synthesis of 9,11-secoestradiol 3-methyl ether. *J. Org. Chem.*, *40*, 685–687.

El-Alfy, T. S., Ezzat, S. M., Hegazy, A. K., Amer, A. M. M., & Kamel, G. M., (2011). Isolation of biologically active constituents from *Moringa peregrina* (Forssk.) Fiori. (Family: Moringaceae) growing in Egypt. *Pharmacogn. Mag.*, 7, 109–115.

Epand, R. F., (2002). The apoptotic protein tBid promotes leakage by altering membrane curvature. *J. Biol. Chem.*, 277, 32632–32639.

Erlund, I., (2004). Review of the flavonoids quercetin, hesperetin, and naringenin. Dietary sources, bioactivities, bioavailability, and epidemiology. *Nutr. Res.*, 24, 851–874.

Ferrazzano, G. F., Amato, I., Ingenito, A., Zarrelli, A., Pinto, G., & Pollio, A., (2011). Plant polyphenols and their anti-cariogenic properties: A review. *Molecules*, 16, 1486–1507.

Geraets, L., Haegens, A., Brauers, K., Haydock, J. A., Vernooy, J. H. J., Wouters, E. F. M., Bast, A., & Hageman, G. J., (2009). Inhibition of LPS-induced pulmonary inflammation by specific flavonoids. *Biochem. Biophys. Res. Commun.*, 382, 598–603.

Gershenzon, J., & Dudareva, N., (2007). The function of terpene natural products in the natural world. *Nat. Chem. Biol.*, 3, 408–414.

Grand, A., (1993). Pregnancy and cardiac drugs. *Rev. Fr. Gynecol. Obstet.*, 88, 297–312.

Hail, N., Cortes, M., Drake, E. N., & Spallholz, J. E., (2008). Cancer chemoprevention: A radical perspective. *Free Radic. Biol. Med.*, 45, 97–110.

Hallikainen, M. A., Sarkkinen, E. S., Gylling, H., Erkkila, A. T., & Uusitupa, M. I., (2000). Comparison of the effects of plant sterol ester and plant stanol ester-enriched margarines in lowering serum cholesterol concentrations in hypercholesterolaemic subjects on a low-fat diet. *Eur. J. Clin. Nutr.*, 54, 715–725.

Harborne, J. B., (1995). The flavonoids: Advances in research since 1986. *J. Chem. Educ.*, 72, A73.

Huang, M. T., Smart, R. C., Wong, C. Q., & Conney, A. H., (1988). Inhibitory effect of curcumin, chlorogenic acid, caffeic acid, and ferulic acid on tumor promotion in mouse skin by 12-O-tetradecanoylphorbol-13-acetate. *Cancer Res.*, 48, 5941–5946.

Huang, M. T., Wang, Z. Y., Georgiadis, C. A., Laskin, J. D., & Conney, A. H., (1992). Inhibitory effects of curcumin on tumor initiation by benzo[a]pyrene and 7,12-dimethylbenz[a] anthracene. *Carcinogenesis*, 13, 2183–2186.

Huang, M., Lu, J. J., Huang, M. Q., Bao, J. L., Chen, X. P., & Wang, Y. T., (2012). Terpenoids: Natural products for cancer therapy. *Expert Opin. Investig. Drugs*, 21, 1801–1818.

Huang, W. Y., Cai, Y. Z., & Zhang, Y., (2009). Natural phenolic compounds from medicinal herbs and dietary plants: Potential use for cancer prevention. *Nutr. Cancer*, 62, 1–20.

Ito, T., Tamura, T., & Matsumoto, T., (1973). Sterol composition of 19 vegetable oils. *J. Am. Oil Chem. Soc.*, 50, 122–5.

Jackson, M. B., (2010). Nature's chemicals. The natural products that shaped our world. *Ann. Bot.*, 106, 6, 7.

Jemal, A., Bray, F., Center, M. M., Ferlay, J., Ward, E., & Forman, D., (2011). Global cancer statistics. *CA. Cancer J. Clin.*, 61, 69–90.

Ju, Y. H., Clausen, L. M., Allred, K. F., Almada, A. L., & Helferich, W. G., (2004). Beta-sitosterol, beta-sitosterol glucoside, and a mixture of beta-sitosterol and beta-sitosterol glucoside modulate the growth of estrogen-responsive breast cancer cells *in vitro* and in ovariectomized athymic mice. *J. Nutr.*, 134, 1145–1151.

Kang, H. J., Youn, Y. K., Hong, M. K., & Kim, L. S., (2011). Antiproliferation and redifferentiation in thyroid cancer cell lines by polyphenol phytochemicals. *J. Korean Med. Sci.*, 26, 893.

Kang, N. J., Shin, S. H., Lee, H. J., & Lee, K. W., (2011). Polyphenols as small molecular inhibitors of signaling cascades in carcinogenesis. *Pharmacol. Ther., 130,* 310–324.

Key, T. J., Appleby, P. N., Spencer, E. A., Travis, R. C., Allen, N. E., Thorogood, M., & Mann, J. I., (2009). Cancer incidence in British vegetarians. *Br. J. Cancer, 101,* 192–197.

Kiuchi, F., Goto, Y., Sugimoto, N., Akao, N., Kondo, K., & Tsuda, Y., (1993). Nematocidal activity of turmeric: Synergistic action of curcuminoids. *Chem. Pharm. Bull., 41,* 1640–1643.

Lee, C. M., Su, Y. H., Huynh, T. T., Lee, W. H., Chiou, J. F., Lin, Y. K., Hsiao, M., Wu, C. H., Lin, Y. F., Wu, A. T. H., & Yeh, C. T., (2013). Blueberry isolate, pterostilbene, functions as a potential anticancer stem cell agent in suppressing irradiation-mediated enrichment of hepatoma stem cells. Evidence-based complement. *Altern. Med.,* pp. 1–9.

Lee, H. S., Seo, E. Y., Kang, N. E., & Kim, W. K., (2008). [6]-Gingerol inhibits metastasis of MDA-MB-231 human breast cancer cells. *J. Nutr. Biochem., 19,* 313–319.

Lepley, D. M., Li, B., Birt, D. F., & Pelling, J. C., (1996). The chemopreventive flavonoid apigenin induces G2/M arrest in keratinocytes. *Carcinogenesis, 17,* 2367–2375.

Li, K., Dias, S. J., Rimando, A. M., Dhar, S., Mizuno, C. S., Penman, A. D., Lewin, J. R., & Levenson, A. S., (2013). Pterostilbene acts through metastasis-associated protein 1 to inhibit tumor growth, progression, and metastasis in prostate cancer. *PLoS One, 8,* e57542.

Lin, C., Crawford, D. R., Lin, S., Hwang, J., Sebuyira, A., Meng, R., Westfall, J. E., Tang, H. Y., Lin, S., Yu, P. Y., Davis, P. J., & Lin, H. Y., (2011). Inducible COX-2-dependent apoptosis in human ovarian cancer cells. *Carcinogenesis, 32,* 19–26.

Liu, R. H., (2003). Health benefits of fruit and vegetables are from additive and synergistic combinations of phytochemicals. *Am. J. Clin. Nutr., 78,* 517S–520S.

Maheo, K., Morel, F., Langouet, S., Kramer, H., Le Ferrec, E., Ketterer, B., & Guillouzo, A., (1997). Inhibition of cytochromes P-450 and induction of glutathione S-transferases by sulforaphane in primary human and rat hepatocytes. *Cancer Res., 57,* 3649–3652.

Mak, K. K., Wu, A. T., H., Lee, W. H., Chang, T. C., Chiou, J. F., Wang, L. S., et al., (2013). Pterostilbene, a bioactive component of blueberries, suppresses the generation of breast cancer stem cells within tumor microenvironment and metastasis via modulating NF-κB/microRNA 448 circuit. *Mol. Nutr. Food Res., 57,* 1123–1134.

Malhotra, A., Nair, P., & Dhawan, D. K., (2014). Study to evaluate molecular mechanics behind synergistic chemo-preventive effects of curcumin and resveratrol during lung carcinogenesis. *PLoS One, 9,* e93820.

Mazue, F., Delmas, D., Murillo, G., Saleiro, D., Limagne, E., & Latruffe, N., (2014). Differential protective effects of red wine polyphenol extracts (RWEs) on colon carcinogenesis. *Food Funct., 5,* 663–70.

McCann, S. E., Freudenheim, J. L., Marshall, J. R., & Graham, S., (2003). Risk of human ovarian cancer is related to dietary intake of selected nutrients. Phytochemicals and food groups. *J. Nutr., 133,* 1937–1942.

Mendilaharsu, M., De Stefani, E., Deneo-Pellegrini, H., Carzoglio, J., & Ronco, A., (1998). Phytosterols and risk of lung cancer: A case-control study in Uruguay. *Lung Cancer, 21,* 37–45.

Mohamed, S., (2014). Functional foods against metabolic syndrome (obesity, diabetes, hypertension, and dyslipidemia) and cardiovasular disease. *Trends Food Sci. Technol., 35,* 114–128.

Monsuez, J. J., Charniot, J. C., Vignat, N., & Artigou, J. Y., (2010). Cardiac side-effects of cancer chemotherapy. *Int. J. Cardiol., 144,* 3–15.

Moriarty, R. M., Naithani, R., & Surve, B., (2007). Organosulfur compounds in cancer chemoprevention. *Mini Rev. Med. Chem., 7,* 827–838.

Murakami, A., Ashida, H., & Terao, J., (2008). Multitargeted cancer prevention by quercetin. *Cancer Lett., 269,* 315–325.

Narayanan, B. A., & Re, G. G., (2001). IGF-II down regulation associated cell cycle arrest in colon cancer cells exposed to phenolic antioxidant ellagic acid. *Anticancer Res., 21*(1A), 359–364.

Narisawa, T., Fukaura, Y., Hasebe, M., Ito, M., Aizawa, R., Murakoshi, M., Uemura, S., Khachik, F., & Nishino, H., (1996). Inhibitory effects of natural carotenoids, alpha-carotene, beta-carotene, lycopene and lutein, on colonic aberrant crypt foci formation in rats. *Cancer Lett., 107,* 137–142.

Newman, D. J., Cragg, G. M., & Snader, K. M., (2003). Natural products as sources of new drugs over the period 1981–2002. *J. Nat. Prod., 66,* 1022–1037.

Nothlings, U., Murphy, S. P., Wilkens, L. R., Henderson, B. E., & Kolonel, L. N., (2007). Flavonols and pancreatic cancer risk: The multiethnic cohort study. *Am. J. Epidemiol., 166,* 924–931.

Nussbaumer, S., Bonnabry, P., Veuthey, J. L., & Fleury-Souverain, S., (2011). Analysis of anticancer drugs: A review. *Talanta., 85,* 2265–2289.

Ohga, N., Hida, K., Hida, Y., Muraki, C., Tsuchiya, K., Matsuda, K., Ohiro, Y., Totsuka, Y., & Shindoh, M., (2009). Inhibitory effects of epigallocatechin-3 gallate, a polyphenol in green tea, on tumor-associated endothelial cells and endothelial progenitor cells. *Cancer Sci., 100,* 1963–1970.

Otaegui-Arrazola, A., Menendez-Carreno, M., Ansorena, D., & Astiasarán, I., (2010). Oxysterols: A world to explore. *Food Chem. Toxicol., 48,* 3289–3303.

Qian, M. C., Fan, X., & Mahattanatawee, K., (2011). *Volatile Sulfur Compounds in Food, ACS Symposium Series.* American Chemical Society, Washington, DC.

Ramos, S., (2008). Cancer chemoprevention and chemotherapy: Dietary polyphenols and signaling pathways. *Mol. Nutr. Food Res., 52,* 507–526.

Ramprasath, V. R., & Awad, A. B., (2015). Role of phytosterols in cancer prevention and treatment. *J. AOAC Int., 98,* 735–738.

Ravindran, J., Prasad, S., & Aggarwal, B. B., (2009). Curcumin and cancer cells: How many ways can curry kill tumor cells selectively? *AAPS J., 11,* 495–510.

Rodriguez-Mateos, A., Vauzour, D., Krueger, C. G., Shanmuganayagam, D., Reed, J., Calani, L., Mena, P., Del Rio, D., & Crozier, A., (2014). Bioavailability, bioactivity, and impact on health of dietary flavonoids and related compounds: An update. *Arch. Toxicol., 88,* 1803–1853.

Ruano-Ravina, A., Figueiras, A., & Barros-Dios, J. M., (2000). Diet and lung cancer: A new approach. *Eur. J. Cancer Prev., 9,* 395–400.

Russo, M., Spagnuolo, C., Tedesco, I., Bilotto, S., & Russo, G. L., (2012). The flavonoid quercetin in disease prevention and therapy: Facts and fancies. *Biochem. Pharmacol., 83,* 6–15.

Russo, M., Tedesco, I., Iacomino, G., Palumbo, R., Galano, G., & Russo, G., (2005). Dietary phytochemicals in chemoprevention of cancer. *Curr. Med. Chem. Endocr. Metab. Agents,* 61–72.

Sadeghnia, H. R., Ghorbani, H. T., Mortazavian, S. M., Mousavi, S. H., Tayarani-Najaran, Z., & Ghorbani, A., (2014). Viola tricolor induces apoptosis in cancer cells and exhibits antiangiogenic activity on chicken chorioallantoic membrane. *Biomed Res. Int.*

Scalbert, A., Manach, C., Morand, C., Remesy, C., & Jimenez, L., (2005). Dietary polyphenols and the prevention of diseases. *Crit. Rev. Food Sci. Nutr., 45,* 287–306.

Seeram, N. P., Lee, R., & Heber, D., (2004). Bioavailability of ellagic acid in human plasma after consumption of ellagitannins from pomegranate (*Punica granatum* L.) juice. *Clin. Chim. Acta, 348,* 63–8.

Sharma, S., (2011). Non-B DNA secondary structures and their resolution by RecQ helicases. *J. Nucleic Acids, 72,* 42–15.

Shen, T., Khor, S. C., Zhou, F., Duan, T., Xu, Y. Y., Zheng, Y. F., Hsu, S., De Stefano, J., Yang, J., Xu, L. H., & Zhu, X. Q., (2014). Chemoprevention by lipid-soluble tea polyphenols in diethylnitrosamine/phenobarbital-induced hepatic pre-cancerous lesions. *Anticancer Res., 34,* 683–693.

Shimizu, M., Fukutomi, Y., Ninomiya, M., Nagura, K., Kato, T., Araki, H., Suganuma, M., Fujiki, H., & Moriwaki, H., (2008). Green tea extracts for the prevention of metachronous colorectal adenomas: A pilot study. Cancer epidemiol. *Biomarkers Prev., 17,* 3020–3025.

Shrotriya, S., Tyagi, A., Deep, G., Orlicky, D. J., Wisell, J., Wang, X. J., Sclafani, R. A., Agarwal, R., & Agarwal, C., (2015). Grape seed extract and resveratrol prevent 4-nitroquinoline 1-oxide induced oral tumorigenesis in mice by modulating AMPK activation and associated biological responses. *Mol. Carcinog., 54,* 291–300.

Strati, A., Papoutsi, Z., Lianidou, E., & Moutsatsou, P., (2009). Effect of ellagic acid on the expression of human telomerase reverse transcriptase (hTERT) α+β+ transcript in estrogen receptor-positive MCF-7 breast cancer cells. *Clin. Biochem., 42,* 1358–1362.

Sun, C. L., Yuan, J. M., Lee, M. J., Yang, C. S., Gao, Y. T., Ross, R. K., & Yu, M. C., (2002). Urinary tea polyphenols in relation to gastric and esophageal cancers: A prospective study of men in Shanghai, China. *Carcinogenesis, 23,* 1497–1503.

Sun, J., (2007). D-Limonene: Safety and clinical applications. *Altern. Med. Rev., 12,* 259–264.

Sun, S. Y., Hail, N., & Lotan, R., (2004). Apoptosis as a novel target for cancer chemoprevention. *J. Natl. Cancer Inst., 96,* 662–672.

Surh, Y. J., (2003a). Cancer chemoprevention with dietary phytochemicals. *Nat. Rev. Cancer, 3,* 768–780.

Surh, Y. J., (2003b). Cancer chemoprevention with dietary phytochemicals. *Nat. Rev. Cancer, 3,* 768–780.

Tan, W., Lu, J., Huang, M., Li, Y., Chen, M., Wu, G., Gong, J., Zhong, Z., Xu, Z., Dang, Y., Guo, J., Chen, X., & Wang, Y., (2011). Anti-cancer natural products isolated from Chinese medicinal herbs. *Chin. Med., 6,* 27.

Tas, F., Hansel, H., Belce, A., Ilvan, S., Argon, A., Camlica, H., & Topuz, E., (2005). Oxidative stress in breast cancer. *Med. Oncol., 22,* 11–15.

Teiten, M. H., Gaascht, F., Dicato, M., & Diederich, M., (2013). Anticancer bioactivity of compounds from medicinal plants used in European medieval traditions. *Biochem. Pharmacol., 86,* 1239–1247.

Tomas-Barberan, F. A., & Clifford, M. N., (2000). Dietary hydroxybenzoic acid derivatives: Nature, occurrence and dietary burden. *J. Sci. Food Agric., 80,* 1024–1032.

Torre, L. A., Bray, F., Siegel, R. L., Ferlay, J., Lortet-Tieulent, J., & Jemal, A., (2015). Global cancer statistics, (2012). *CA. Cancer J. Clin., 65,* 87–108.

Tsao, R., (2010). Chemistry and biochemistry of dietary polyphenols. *Nutrients, 2,* 1231–1246.

Vershinin, A., (1999). Biological functions of carotenoids: Diversity and evolution. *Bio. Factors, 10,* 99–104.

Vigushin, D. M., Poon, G. K., Boddy, A., English, J., Halbert, G. W., Pagonis, C., Jarman, M., & Coombes, R. C., (1998). Phase I and pharmacokinetic study of D-limonene in patients with advanced cancer. Cancer research campaign phase I/II clinical trials committee. *Cancer Chemother. Pharmacol., 42,* 111–117.

Wang, G., Guo, X., Chen, H., Lin, T., Xu, Y., Chen, Q., Liu, J., Zeng, J., Zhang, X., & Yao, X., (2012). A resveratrol analog, phoyunbene B, induces G2/M cell cycle arrest and apoptosis in HepG2 liver cancer cells. *Bioorg. Med. Chem. Lett., 22,* 2114–2118.

Weihrauch, J. L., & Gardner, J. M., (1978). Sterol content of foods of plant origin. *J. Am. Diet. Assoc., 73,* 39–47.

Weststrate, J. A., & Meijer, G. W., (1998). Plant sterol-enriched margarines and reduction of plasma total- and LDL-cholesterol concentrations in normocholesterolaemic and mildly hypercholesterolaemic subjects. *Eur. J. Clin. Nutr., 52,* 334–343.

Yu, H. L., Giammarco, R., Goldstein, M. B., Stinebaugh, D. J., & Halperin, M. L., (1976). Stimulation of ammonia production and excretion in the rabbit by inorganic phosphate. Study of control mechanisms. *J. Clin. Invest., 58,* 557–564.

Yuan, J. M., Gao, Y. T., Yang, C. S., & Yu, M. C., (2007). Urinary biomarkers of tea polyphenols and risk of colorectal cancer in the Shanghai cohort study. *Int. J. Cancer, 120,* 1344–1350.

CHAPTER 6

Therapeutic Aspects of Some Extracts and Purified Cardiac Glycosides Obtained From *Nerium oleander* L.

BEHNAZ ASLANIPOUR[1] and MURAT ALAN[2]

[1]*Department of Bioengineering, Faculty of Engineering, Ege University, Izmir, Turkey, E-mail: Behnaz_aslanipour@yahoo.com*

[2]*Department of Obstetrics and Gynecology, Izmir Tepecik Training and Research Hospital, Tepecik – 35120, Izmir, Turkey*

ABSTRACT

Plant-derived phytomedicals are being gradually considered as beneficial complementary treatments for various diseases such as cancer. Substances obtained from plants which are ranged in a diverse group of compounds can be found in fruits, vegetables, spices, as well as medicinal plants naturally. The species *Nerium oleander* L. (Apocynaceae), which is originally distributed in Asia and Europe, is broadly cultivated worldwide. Anvirzel, which is an obtained extract from *Nerium oleander* has so far been undergoing for clinical trials to test whether there are treating properties over some, disorders specially cancer. *N. oleander* includes some important cardiac glycosides with therapeutical potentials. It has also become important for anticancer activity due to one of its cardio-tonic constituents, namely oleandrin which was found to be an active stimulator of heart function with diuretic effect. Oleandrin also possesses anti-inflammatory, tumor cell growth-inhibitory effects as well. The target of this chapter was to discuss the current state of knowledge concerning the different therapeutical aspects of few extracts of *N. Oleander*. Apart from this, further information was added regarding to the useful effects of oleandrin upon some special cancers in the present chapter.

6.1 INTRODUCTION

The leaves of *N. oleander* having pharmacological effect contain two groups of glycosides. They are introduced as glycosides (cardiac ones) and flavonoids. Cardiac glycosides are known as secondary metabolites, which have been considered as essential agents for congestive heart disease treatment and arrhythmias, and are existed in some plants and rarely in animals (for instance the milkweed butterflies). The natural cardiac glycosides are compounds that have the ability of inhibiting Na$^+$/K$^+$ ATPase and they have been clinically used for a long period due to their positive inotropic properties in Heart disorders and atrial arrhythmias. According to this mechanism of action, Research over the therapeutic properties of many different diseases has recently increased (Aizman and Aperia, 2003; Aperia, 2007; Kometiani et al., 2005; Xie and Askari, 2002). Apoptotic and anti-proliferative effects of these compounds were evaluated in chest (Bielawski et al., 2006; Kometiani et al., 2005; Lopez-Lazaro et al., 2005) prostate (McConkey et al., 2000), pancreas (Newman et al., 2007), lung (Frese et al., 2006; Mijatovic et al., 2006) and leukemia (Masuda et al., 1996; Pollyea et al., 2014; Raghavendra et al., 2007; Watanabe et al., 1997). Digoxin and ouabain have been utilized as cardiac glycosides from ancient time to date. Digoxin, obtained from foxglove plant, has been clinically used while ouabain has only been used experimentally. The major cardiac glycoside named oleandrin has been recorded to have many activities (Kumar et al., 2013). Oleandrin, a toxic cardiac glycoside of *N. oleander* L., has the power of inhibition of the action of nuclear factor kappa-light-chain-enhancer from activated B chain (NF-κB) in different cell lines (U937, CaOV3, human epithelial cells, and T cells). Oleander extract has been studied to understand if it has the potential of acting as chemotherapeutical agent (Newman et al., 2007).

6.2 CHEMISTRY OF OLEANDRIN

Oleandrin with the chemical formula, $C_{32}H_{48}O_9$, has a molecular weight (MW) of 576.727 g/mol with the structure given in Figure 6.1. Its IUPAC name is acetic acid [(3S,5R,10S,13R,14S,16S,17'R)-14-hydroxy-3-[[(2R,4S,5S,6S)-5-hydroxy-4-methoxy-6-methyl-2-tetrahydropyranyl]oxy]-10,13-dimethyl-17-(5-oxo-2H-furan-3-yl)-1,2,3,4,5,6,7,8,9,11,12,15,16,17-tetradecahydrocyclopenta[a]phenanthren-16-yl] ester. Oleandrin comprises a steroid nucleus in center along with an unsaturated lactone ring structure as well as a dideoxy arabinose unite on C17 and C3 respectively. Furthermore, there is an acetyloxy group substitute on C16 of the steroid' ring (Kumar et al., 2013).

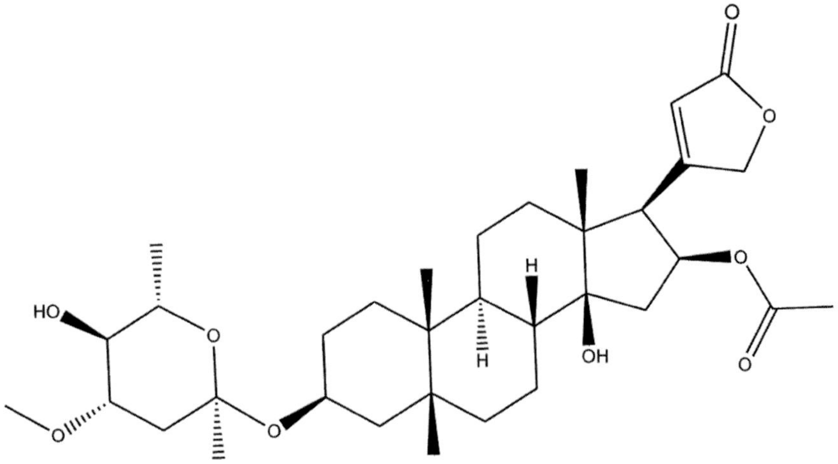

FIGURE 6.1 Structure of oleandrin.

6.2.1 BOTANICAL DESCRIPTION OF NERIUM OLEANDER L. AS SOURCE OF OLEANDRIN

N. oleander L., with a common name oleander, is known as an evergreen highly toxic ornamental shrub with greygreen leaves belonging to the family of Apocynaceae. It is endemic to northern Africa, the eastern Mediterranean basin and Southeast Asia. It has been classified in the genus. *N. Oleander* has a high dosage of poison and is commonly grown in garden (Li et al., 1964). Subtropical regions as well as warm temperate regions are very suitable areas for oleander cultivation due to its long-lasting flowering as well as having tolerance to heat, salinity, and drought. Oleander grows up to the height of 6.6–20 feet. Comparing the 1st year and the mature time flowering, it could be stated that stems of the 1st year have a glaucous bloom whereas stems which are mature have a grayish bark (Yang et al., 2009). The leaves, grown up to the length of 5–21 cm (2.0–8.3 inches) and to the width of 1–3.5 cm (0.39–1.4 inches), can be dark green, thick with an entire margin. The flowers, grown as 0.98–2.0 inches) diameter is seen as white, pink to red. It has a long narrow capsule as fruit grown up to 5–23 cm (2.0–9.1 inches) long and in mature time, they become open to release many downy seeds.

It can also grow in cold climate with high toleration. There are more than 400 cultivators of *N. oleander* classified by color of their flowers (simple or doubled) and cultivars (completely green and variegated). It is also introduced as poisonous shrub. Oleander is a well-known Chinese medicine with cardioprotective and cytotoxic effects.

6.2.2 N. OLEANDER EXTRACTS AND THEIR THERAPEUTIC APPLICATIONS

At the end of 1970's, Dr. Ziya Ozel announced that his medical preparation, no extract from *Nerium oleander*, showed extraordinary anti-carcinogenic effects on the patients with solid tumors. However, low therapeutic index of the cardiac glycosides and a number of intoxication cases resulted in deaths made scientific community dealing with natural products to oppose to use of the NO extract. Because of this opposition, Dr. Ozel decided to develop the project abroad, and then he assigned his rights to a USA based company, in which the hot water extract of *N. oleander* known as Anvirzel™ was developed. There are some compounds with high therapeutic effects in Anvirzel from which two of the most active components are oleandrin and oleandrigenin (both are cardiac glycosides). It is being used in clinic for treating cancer, HIV, Lupus, Hepatitis C, and some other viral diseases in South American countries such as the Republic of Honduras and Brazil. Moreover, the oleander extract reduces side effects of chemotherapy and boosts the immune system. Since Anvirzel's clinical trials have not yet been completed inferring no FDA approval, it cannot be marketed in the USA as a drug. Nerium oleander extracts were patented on US 5,135,745 for being used in treating of uncontrolled proliferative diseases. Later, the company, which granted its rights to conduct clinical trials, completed phase I trials with *N. oleander* hot water extract Anvirzel (TM) and developed Phase II trials as well. The effect of another oleandrin extract named Breastin (in cold water extract) has been checked in more than 400 patients since 1988. Up to now, many studies on cardiac glycosides were parallel with these results and revealed that these molecules are selective for tumor cells at low concentrations (Mijatovic et al., 2007; Slingerland et al., 2013). Phase I experiments in solid tumors was completed with addition of another effective *N. oleander* extract named PBI-05204, a supercritical CO_2 extraction of *N. oleander*, and phase II clinical trials with patients in metastatic pancreatic adenocarcinoma at stage 4 will be started (Hong et al., 2014). Almost 23 µg/ml PBI-05204 (approximately containing 1 µM oleandrin) was utilized in fluorescent protein-tagged coronal brain slices and it caused protection from glucose and oxygen deprivation, moreover, addition of 230 µg/ml of PBI-05204, increased levels of subunits of α1 and α2 in Na+/K+-ATPase of the rat brain slices. On the other hand, *in vitro* anti-tumor effect of the cold water extract originated from *N. oleander* extract, Breastin, was studied against 63 cell lines and the results of the study showed selective toxicity towards colon, prostate, pancreas as well as cell lung tumors (non-small ones). PBI-05204 has passed phase I trials on solid tumors, and it is about to be taken into phase II trials against metastatic pancreatic cancer. It is botanical anti-cancer drug candidate in both *in vivo* models of ischemic stroke and brain slice. Rashan et al. (2011) investigated anti-tumor properties of *Streptocaulon tomentosum* and *N. oleander* plant extracts, accordingly, the results displayed that the four carderolytes obtained from *Streprocaulon tomentosum* significantly reduced the rate of proliferation of MCF7 human breast cancer cells, whereas, it reduced only a small effect in L929 mouse fibroblast cells. It was also reported that known chemotherapeutic drugs for instance camptothecin and doxorubicin have not been the reason for a significant selectivity in comparative experiments with mentioned extracts. In the same study, breastin, in which oleandrin' amount is very high, was tested in 36 different normal human and tumor cells and it was revealed that the tested extract showed selective anti-tumor activity. In the case of LC70 value examination, it was discovered that the extract had a marked selectivity on the SF268 brain tumor cell, LXF1121L lung carcinoma, PANC 1 pancreatic carcinoma and DU145 prostate carcinoma cells (Rashan et al., 2011). Two cardiac glycosides including UNBS1450 and oleandrin were tested on autophagy induction in human glioblastoma and pancreatic cells after exposure 24–72 h respectively. The results exhibited that both tested components induced autophagy in both cells occurred within 24 h (Lefranc et al., 2008; Newman et al., 2007). Some studies reported the comparison between oleandrin and other cardiac glycosides as well as extracts including oleandrin so far. The effect of oleandrin and anvirzel on Pro4 cells was compared and the results represented that oleandrin could have the ability to arrest concentration-dependent growth 50 times more than Anvirzel which was measured by mitochondrial reduction of MTT. Oleandrin treatment caused the processing of procaspase-8 and procaspase-3 in the Pro4 cells. In the same study, both Anvirzel and oleandrin induced concentration-dependent DNA fragmentation in the Pro4 and LN4 cells, besides oleandrin also caused marked accumulation of cells in the cell cycle (G2-M phase) (McConkey et al., 2000). Both Anvirzel and oleandrin, caused an inhibition in- *in vitro*

FGF-2 export from DU145 and PC3 prostate cancer cells in a dose- and time-dependent manner, therefore, it was reported that there should be the anti-tumor effect of this newly tested treatment in cancer therapy (Smith et al., 2001). In a study, the effect of some cardiac glycosides including oleandrin as well as both ouabain and digoxin molecules were examined and the results confirmed that the tested molecules could prove relevant to the treatment of metastatic prostate cancer. Accordingly, it was obtained from the same study that oleandrin could be able to cause an induction in apoptosis of prostate cancer cells 50 times more than Anvirzel (McConkey et al., 2000).

6.3 CYTOTOXIC EFFECTS AND ANTI-CANCER PROPERTIES OF OLEANDRIN

6.3.1 OLEANDRIN AND CANCER

Oleandrin, soluble in lipid, has potential anti-cancer activity. It is obtained from previous studies that oleandrin attaches to its receptor and has the ability of inhibiting the alpha 3 subunit belonging to the Na/K-ATPase pump in cancer cells of humans by which it may develop inhibiting of Akt phosphorylation, upregulation of MAPK, inhibition of NF-kb activation and inhibition of FGF-2 export and might also stimulate down-regulation of Mtor. As an outcome of all these mechanisms, both p70S6K and S6 protein expression are inhibited. Therefore, all of these can have the probability to empower the apoptosis' induction. Cancer cells are remarkably delicate to oleandrin compared to normal cells due to having higher levels of apha3 subunit and lower levels of apha1 subunit with respect to normal cells. Oleandrin, which is related to expression of tumors in pump Na/K-ATPase protein subunit, can be a suitable way for treating of cancer cells (Aizman and Aperia, 2003; Aperia, 2007; Kometiani et al., 2005; Xie and Askari, 2002). Apart from this, tumor proliferation can be caused by alpha3 subunit over expression in tumor cells. All in common, lipid-soluble cardiac glycosides are known as potential components for treating of a variety of cancers (Aizman and Aperia, 2003; Aperia, 2007; Kometiani et al., 2005; Pan et al., 2017; Xie and Askari, 2002).

Utilization of the higher doses of these molecules in cancer treatment is near to the toxic doses of the therapeutic drugs, therefore they cause problems with their clinical usage. For instance, they can cause cardiac arrhythmia because of the increase in Ca^{2+} concentration. Many articles and compilations dealing with anti-cancer effects of oleandrin and other related cardiac glycosides underline the need to develop new methods to reduce the cytotoxicity of these molecules, which will increase the bioavailability of these molecules. Consequently, many investigators have focused on two solutions: the progress of less toxic semi-synthetic cardiac glycoside derivatives or different formulations (Bonelli et al., 1988). Meanwhile, there are some *in vivo* and *in vitro* studies according to investigations upon cardiac glycosides and their therapeutic activities. Epidemiologic studies have revealed that the low mortality of cancer patients using cardiotonic glycosides has increased the quantity of investigations regarding to the use of these drugs in cancer therapy (Mekhail et al., 2006; Lopez-Lazaro, 2007; Mijatovic et al., 2007; Winnicka et al., 2006; Nobili et al., 2009a, 2009b). In a study, referring to Stenkvist et al. women who received digitalis after diagnosis of breast cancer seemed to develop tumors with a lower growth potential than those patients who did not receive the same medication (Stenkvist et al., 1981). An internal dose-response analysis revealed a relationship between high plasma concentration of digitoxin and a lower risk for leukemia/lymphoma and for cancer of the kidney/urinary tract on 9271 patients (Haux, 1999; Haux et al., 2001). The first *in vitro* study about the properties of cardiac glycosides were conducted on the cancer therapy by Shiratori (1967) in which oleandrin was tested on human pancreatic tumor cells PANC-1. The results disclosed that oleandrin paused the cell proliferation of PANc-1 and arrested cells at G (2)/M stage of the cell cycle (Shiratori, 1967). It was obtained from another study that the properties of oleandrin on PANC-1 cells were via apoptotic pathway rather than autophagy processes (Newman et al., 2006). Oleandrin (Raghavendra et al., 2007) and UNBS1450 (a hemi-synthetic derivative of 2″-oxovuscharin) (Juncker et al., 2011) could induce apoptosis in the different cancer cell tested models. Oleandrin also could activate the serine/threonine phosphatase calcineurin with the ability of promoting the upregulation of Fas in hematopoietic cancer cell lines (Raghavendra et al., 2007). TNF-related apoptosis inducing ligand (TRAIL) was used to provide synergy with the extrinsic apoptotic pathway. Oleandrin induced the initiation of TRAIL-induced apoptosis in different resistant lung cancer cell models (Frese et al., 2006). It can be obtained that Anvirzel can have the ability of inhibiting the export of FGF-2 from prostate cancer cells through sodium-pump inhibition by oleandrin (Smith et al., 2001). In another study, the effect of an extract obtained from *N. oleander'* leaves (NOE-4) was tested on the susceptibility of Raji cells (human Burkitt's lymphoma), natural killer cells and mediated cytotoxicity. After 24 h incubation, in a dose-dependent manner, the NOE-4 treated cells were strongly affected by human mononuclear cell (MNCs) mediated cytotoxicity while the control Raji cells were unaffected (Ghoneum et al., 2006).

Formation of ROS in melanoma cells were stimulated by oleandrin leading to a secretion of cytochrome c from mitochondria with the power of activating caspases leading to apoptosis (Newman et al., 2006). Oleandrin represented cytostatic activity in leukemia tumor cells while it did not affect the growth of normal cells (Mijatovic et al., 2006). Yang et al. examined the effect of oleandrin on the proliferation of human tumor cells and the position of alpha subunits in NA/K-ATPase in many types of colon, oral, lung, and pancreatic cancer cells and the sensitivity to oleandrin changes with the α subunit of Na/K-ATPase. The results exhibited that the expression of the α-3 subunit was higher in pancreatic cancer cells of human (Panc-1, MiaPaca) than in pancreatic cancer cells of mouse (Panc 2), thus oleandrin was taken into cells in human pancreatic cancer cells while a resistance for uptake oleandrin was detected in mice (Yang et al., 2009). Relative expression of the α-3 subunit in tumor cells correlates with proliferation. Oleandrin was represented as a principle cytotoxic component and it was evaluated in a clinical trial (phase I) in the United States against refractory human cancers (Li et al., 2010; Newman et al., 2006).

6.3.2 ANTI-TUMOR PROMOTING EFFECTS OF OLEANDRIN

Some anti-tumor properties of oleandrin have been reported so far. Oleandrin and some other cardiac glycosides have the power of inhibiting the Na+, K+-ATPase pump which ends to an increase in calcium influx of the heart muscle (Yang et al., 2009). Oleandrin could be able to suppress the activation of NF-κB (nuclear factor-κB), besides it contributed to the induction of cell death. Although, cell type specificity plays a vital role in the cell death-promoting activity of cardiac glycosides, it was represented in a few studies that cardiac glycosides could ban apoptosis' multiple pathways in cells of vascular smooth muscle (Menger et al., 2012; Zhao et al., 2011).

In 2000, Manna et al. displayed that oleandrin prohibited NF-κB and AP-1 activation and their associated kinases. In the first human base study, the researchers investigated upon determination of the maximum tolerated doses/recommended phase II dose and definition of the pharmacodynamics (PD) and pharmacokinetics (PK) of PBI-05204 in advanced cancer patients. Cardiac glycosides were able to regulate fibroblast growth factor-2 (FGF-2) which was identified as one of the most effective angiogenesis promoting substances, and could possibly obstruct the transcription factor NF-κB activation. FGF-2 and NF-κB were supposed to be related targets to anticancer drugs. Oleandrin inhibited FGF-2 export, NF-κB activation, and phosphorylation of Akt, p70S6K, and decreases mTOR activity. This sensitivity triggered a wide range of IC_{50} values in contradiction of different pancreatic cells ranging from 5.6 nM to 500 nM (Yang et al., 2009). It was implicated that despite the selective anti-tumor activity demonstrated by cardiac glycosides *in vitro, in vivo* bioavailability of cardiac glycosides included accumulating in normal tissues of the heart, lung, and digestive tract in vascular or oral recipients was low (Smith, 1985).

Farrukh et al. studied on the topical application of oleandrin to CD-1 mice to fight 12-O-tetradecanoylphorbol-13-acetate (TPA)-induced markers of skin tumor promotion. The results of the above-mentioned study represented that application of oleandrin to the mice skin exactly before TPA application to CD-1 mice concluded in a remarkable decrease in skin edema, hyperplasia, epidermal ODC activity and protein expression of ornithine decarboxylase (ODC) as well as cyclooxygenase-2 (COX-2), classical markers of inflammation and tumor promotion (Afaq et al., 2004). Furthermore, cardenolide-oleandrin could be used for mutagenesis prevention (Manna et al., 2000).

6.3.3 TUMOR CELL GROWTH-INHIBITORY EFFECT OF OLEANDRIN

6.3.3.1 OLEANDRIN AND PROLIFERATION

There are some compounds with high therapeutic effects in Anvirzel from which two of the most active components are oleandrin and oleandrigenin (the cardiac glycosides). The oleandrin effects on cells were examined in another study and it was reported that oleandrin could inhibit the Na/K-ATPase system leading to an increase in intracellular Ca^{2+} concentration and could prohibit cell proliferation and induce apoptosis. Therefore, oleandrin treatment changed the fluidity of the cellular lipid membranes and it could cause modification of cell signaling and consequent selective death of malignant cells (Raghavendra et al., 2007). All these mechanisms led to a modification of cell signaling and consequent selective death of malignant. In addition, fibroblast banned the Na/K-ATPase pump by inhibiting growth factor-2' effect, therefore, the Na/K-ATPase pump has been the target of cancer drugs. Furthermore, inhibition of the Na/K-ATPase pump could lead to increase intracellular acidity and consequent apoptosis by enhancing the intracellular Na^+

concentration (Cerella et al., 2013; Lin et al., 2010; McConkey et al., 2000; Yang et al., 2009). Oleandrin was able to inhibit the secretion of IL-8-, formylpeptide (FMLP)-, EGF-, or nerve growth factor (NGF) but it was not able to affect the activation of IL-1- or TNF induced NF-κB in macrophages. Oleandrin also could affect inhibit IL-8 binding to the IL-8 receptors. They concluded that oleandrin inhibited the biological responses of IL-8-mediated in diverse cell types via modulating IL-8Rs through alteration of membrane fluidity and microviscosity. According to this, oleandrin might have the power of regulating IL-8-mediated biological responses in neovascularization, angiogenesis, tumorigenesis, inflammation, and metastasis (Hashimoto et al., 1997). Afaq et al. (2004) investigated on the effect of oleandrin over the tumor growth-inhibitory after 12-O-tetra decanoylphorbol-13-acetate (TPA) induction of skin carcinogenesis and they found that application of 2 mg oleandrin per mouse in half an hour time period before TPA induction (3.2 nmol per mouse) remarkably inhibited skin carcinogenesis in a time-dependent manner (Afaq et al., 2004). Oleandrin was responsible for inhibiting FGF-2 export from PC-3 and prostate cancer cells (DU145) in a dose and time-dependent manner (Smith et al., 2001). It was approved that oleandrin enhanced the sensitivity of PC-3 human prostate cells to radiation and by which it led the PC-3 cells to apoptosis that was dependent on activation of caspase 3 (Nasu et al., 2002). The evaluation of the effects of oleandrin and underlying its mechanisms on osteosarcoma (OS) cells was investigated on an *in vitro* study and the results stressed that oleandrin meaningfully inhibited both the invasion and the proliferation of OS cells via suppressing the Wnt/β-catenin signaling pathway and a final induction of apoptosis. Besides, lower invasiveness of the cells resulted by down regulating matrix' expression and activities (metalloproteinase 2) MMP-2 and (metalloproteinase 9) MMP-9 (Ma et al., 2015).

6.3.3.2 NEUROPROTECTIVE ACTIVITY OLEANDRIN

Oleandrin has the potential to be as blood-brain-barrier (BBB) penetrant. It was indicated that the neuroprotective effect of the oleandrin component of PBI-05204 was mediated by induction of neural expression of CNS neurotrophic factor brain-derived neurotrophic factor BDNF (Dunn et al., 2011; Van Kanegan et al., 2014).

6.4 OLEANDRIN POISONING

N. oleander includes oleandrin, folineriin, adynerin, and digitoxigenin as the most common cardiac glycosides in stems, leaves, young shoots, flowers, nectar, sap, and products induced by combustion (Langford and Boor, 1996). The highest percentage of cardiac glycosides was reported in seeds and roots of *N. oleander* followed by fruits and leaves. The total cardiac glycoside content was even higher in the oleander plants with red flowers in comparison to those with white flowers in any stage of flowering period (Karawya et al., 1973). Despite of existing many pharmaceutical effects in oleandrin, it can be represented that there are some studies referring to the toxic side effects of this plant. The frequency of intoxication cases of oleandrin has not been fully understood because of its narrow therapeutic index. Wasfi et al (2008) reported the detection of oleandrin in the blood sample of patients treated with oleandrin at a dose of 10 ng/ml by LC-MS/MS, a useful method for pharmacokinetic and experimental toxicological studies. Another cardiac glycoside with pseudo-molecular ion of m/z 577, a likely structural isomer of oleandrin, was also distinguished in the blood with a very low concentration. Self-Medication attempt of oleandrin in one patient with Hashimoto's thyroiditis was reported and oleandrin was identified in the urine at a dose of 3.2 ng/mL and in the serum at a dose of 8.4 ng/mL by chromatographic analysis at the time of admission. By same research, the authors could possibly let public be aware of the toxicity of oleander and help them reduce the incidence of poisoning due to Nerium species (Bavunoglu et al., 2016). In one case report, oleandrin was ingested to a 77-year-old woman with permanent atrial fibrillation for over 5 years, and a possible antiarrhythmic effect of oleandrin in the final test was discovered (Al et al., 2010).

6.4.1 TOXIC DOSE

Osterloh et al. (1982) reported a fatal poisoning via Ingestion of 5–15 *N. oleanders* leaves (Osterloh, 1982). There were also some other fatalities reported after the ingestion of unknown amounts of *N. oleander* (Blum and Rieders, 1987; Wasfi et al., 2008). In another report referring to the toxicity of oleander over children, a seven-year-old child who ingested three *N. oleander* leaves only had mild poisoning and was recovered without complications. The authors of the same study suggested that even one leaf of *N. oleander* could be toxic to children (Shaw and Pearn, 1979).

Apart from this, there are some studies reporting that ingestion of varying amounts of *N. oleandrin* leaves have reported nonfatal comparing to its leaves, flowers, and root extracts (Durakovic et al., 1996; Le Couteur and Fisher, 2002; Pietsch et al., 2005; Shumaik et al., 1988). An adult female was tested for toxicity after ingestion of five handfuls of leaves of *N. oleander* in a suicide attempt and there was moderate toxicity reported (Tracqui et al., 1998). Therefore, no certain report is available according to the exact concentration of fatality. There can be several factors causing fatality such as plant part, toxin concentration in the plant part ingested, the amount as well as age and health of the patient at the time of ingestion.

KEYWORDS

- **blood-brain-barrier**
- **cancer**
- ***Nerium oleander* L.**
- **neuroprotective activity**
- **oleandrin**
- **toxic dose**

REFERENCES

Afaq, F., Saleem, M., Aziz, M. H., & Mukhtar, H., (2004). Inhibition of 12-O-tetradecanoylphorbol-13-acetate-induced tumor promotion markers in CD-1 mouse skin by oleandrin. *Toxicol. Appl. Pharmacol., 195,* 361–369.

Aizman, O., & Aperia, A., (2003). Na, K-ATPase as a signal transducer. *Ann. N.Y. Acad. Sci., 986,* 489–496.

Al, B., Yarbil, P., Dogan, M., Kabul, S., & Yildirim, C., (2010). *A Case of Non-Fatal Oleander Poisoning.* Case reports, bcr0220091573–bcr0220091573.

Aperia, A., (2007). New roles for an old enzyme: Na, K-ATPase emerges as an interesting drug target. *J. Intern. Med., 261,* 44–52.

Bavunoglu, I., Balta, M., & Turkmen, Z., (2016). Oleander poisoning as an example of self-medication attempt. *Balkan Med. J., 33,* 559–562.

Bielawski, K., Bielawska, A., Sosnowska, K., Miltyk, W., Winnicka, K., & Pałka, J., (2006). Novel amidine analogue of melphalan as a specific multifunctional inhibitor of growth and metabolism of human breast cancer cells. *Biochem. Pharmacol., 72,* 320–331.

Blum, L. M., & Rieders, F., (1987). Oleandrin distribution in a fatality from rectal and oral Nerium oleander extract administration. *J. Anal. Toxicol., 11,* 219–221.

Bonelli, J., Waginger, W., & Gazo, F., (1988). The bioavailability of methylepoxyproscillaridin (P35): A new semisynthetic cardiac glycoside. *Int. J. Clin. Pharmacol. Ther. Toxicol., 26,* 297–299.

Cerella, C., Radogna, F., Dicato, M., & Diederich, M., (2013). Natural compounds as regulators of the cancer cell metabolism. *Int. J. Cell Biol.,* 1–16.

Dunn, D. E., He, D. N., Yang, P., Johansen, M., Newman, R. A., & Lo, D. C., (2011). In vitro and in vivo neuroprotective activity of the cardiac glycoside oleandrin from Nerium oleander in brain slice-based stroke models. *J. Neurochem., 119,* 805–814.

Durakovic, Z., Durakovic, A., & Durakovic, S., (1996). Oleander poisoning treated by resin hemoperfusion. *J. Indian Med. Assoc., 94,* 149–150.

Farrukh, A., Mohammad, S., Moammir, H. A., & Hasan, M., (2004). Inhibition of 12-O-tetradecanoylphorbol-13-acetate-induced tumor promotion markers in CD-1 mouse skin by oleandrin. *Toxicology and Applied Pharmacology, 195,* 361–369.

Frese, S., Frese-Schaper, M., Andres, A. C., Miescher, D., Zumkehr, B., & Schmid, R. A., (2006). Cardiac glycosides initiate Apo2L/TRAIL-induced apoptosis in non-small cell lung cancer cells by up-regulation of death receptors 4 and 5. *Cancer Res., 66,* 5867–5874.

Ghoneum, M., Ozel, H., & Gollapudi, S., (2006). Su.82. nerium oleander leaf extract (Noe-4) sensitizes human burkett cell lymphoma (Raji) to human cytoxcity mediated by natural killer cells. *Clin. Immunol., 119,* S188.

Hashimoto, S., Jing, Y., Kawazoe, N., Masuda, Y., Nakajo, S., Yoshida, T., Kuroiwa, Y., & Nakaya, K., (1997). Bufalin reduces the level of topoisomerase II in human leukemia cells and affects the cytotoxicity of anticancer drugs. *Leuk. Res., 21,* 875–883.

Haux, J., (1999). Digitoxin is a potential anticancer agent for several types of cancer. *Med. Hypotheses, 53,* 543–548.

Haux, J., Klepp, O., Spigset, O., & Tretli, S., (2001). Digitoxin medication and cancer, case control and internal dose-response studies. *BMC Cancer, 1,* 11.

Hong, D. S., Henary, H., Falchook, G. S., Naing, A., Fu, S., Moulder, S., et al., (2014). First-in-human study of pbi-05204, an oleander-derived inhibitor of AKT, FGF-2, NF-κB and p70s6k, in patients with advanced solid tumors. *Invest. New Drugs, 32,* 1204–1212.

Juncker, T., Cerella, C., Teiten, M. H., Morceau, F., Schumacher, M., Ghelfi, J., Gaascht, F., Schnekenburger, M., Henry, E., Dicato, M., & Diederich, M., (2011). UNBS1450, a steroid cardiac glycoside inducing apoptotic cell death in human leukemia cells. *Biochem. Pharmacol., 81,* 13–23.

Karawya, M., Balbaa, S., & Khayyal, S., (1973). Estimation of cardenolides in nerium oleander. *Planta Med., 23,* 70–73.

Kometiani, P., Liu, L., & Askari, A., (2005). Digitalis-induced signaling by Na+/K+-ATPase in human breast cancer cells. *Mol. Pharmacol., 67,* 929–936.

Kumar, A., De, T., Mishra, A., & Mishra, A., (2013). Oleandrin: A cardiac glycosides with potent cytotoxicity. *Pharmacogn. Rev., 7,* 131. https://doi.org/10.4103/0973-7847.120512 (accessed on 22 February 2020).

Langford, S. D., & Boor, P. J., (1996). Oleander toxicity: An examination of human and animal toxic exposures. *Toxicology, 109,* 1–13.

Le Couteur, D. G., & Fisher, A. A., (2002). Chronic and criminal administration of nerium oleander. *J. Toxicol. Clin. Toxicol., 40,* 523–524.

Lefranc, F., Mijatovic, T., Kondo, Y., Sauvage, S., Roland, I., Debeir, O., Krstic, D., Vasic, V., Gailly, P., Kondo, S., Blanco, G., & Kiss, R., (2008). Targeting the alpha 1 subunit of the sodium pump to combat glioblastoma cells. *Neurosurgery, 62,* 211–212.

Li, C. T., Deng, S. H., & Ho, G. B., (1964). Comparison of cardiotonic actions between oleandrin and digitoxin. *Yao Xue Xue Bao., 11,* 540–544.

Li, S., Yuan, W., Yang, P., Antoun, M. D., Balick, M. J., & Cragg, G. M., (2010). Pharmaceutical crops: An overview. *Pharm. Crop, 1,* 1–17.

Lin, Y., Ho, D. H., & Newman, R. A., (2010). Human tumor cell sensitivity to oleandrin is dependent on relative expression of Na+, K+ -ATPase subunits. *J. Exp. Ther. Oncol., 8,* 271–286.

Lopez-Lazaro, M., (2007). Dual role of hydrogen peroxide in cancer: Possible relevance to cancer chemoprevention and therapy. *Cancer Lett., 252,* 1–8.

Lopez-Lazaro, M., Pastor, N., Azrak, S. S., Ayuso, M. J., Austin, C. A., & Cortes, F., (2005). Digitoxin inhibits the growth of cancer cell lines at concentrations commonly found in cardiac patients. *J. Nat. Prod., 68,* 1642–1645.

Ma, Y., Zhu, B., Liu, X., Yu, H., Yong, L., Liu, X., Shao, J., & Liu, Z., (2015). Inhibition of oleandrin on the proliferation and invasion of osteosarcoma cells *in vitro* by suppressing WNT/β-catenin signaling pathway. *J. Exp. Clin. Cancer Res., 34,* 115.

Manna, S. K., Sah, N. K., Newman, R. A., Cisneros, A., & Aggarwal, B. B., (2000). Oleandrin suppresses activation of nuclear transcription factor-kappaB, activator protein-1, and c-Jun NH2-terminal kinase. *Cancer Res., 60,* 3838–3847.

Masuda, M., Masuda, M., Hanson, C. A., Hoffman, P. M., & Ruscetti, S. K., (1996). Analysis of the unique hamster cell tropism of ecotropic murine leukemia virus PVC-211. *J. Virol., 70,* 8534–8539.

McConkey, D. J., Lin, Y., Nutt, L. K., Ozel, H. Z., & Newman, R. A., (2000). Cardiac glycosides stimulate Ca^{2+} increases and apoptosis in androgen-independent, metastatic human prostate adenocarcinoma cells. *Cancer Res., 60,* 3807–3812.

Mekhail, T., Kaur, H., Ganapathi, R., Budd, G. T., Elson, P., & Bukowski, R. M., (2006). Phase 1 trial of Anvirzel in patients with refractory solid tumors. *Invest. New Drugs, 24,* 423–427.

Menger, L., Vacchelli, E., Adjemian, S., Martins, I., Ma, Y., Shen, S., et al., (2012). Cardiac glycosides exert anticancer effects by inducing immunogenic cell death. *Sci. Transl. Med., 4,* 143ra99.

Mijatovic, T., Op De Beeck, A., Van Quaquebeke, E., Dewelle, J., Darro, F., De Launoit, Y., & Kiss, R., (2006). The cardenolide UNBS1450 is able to deactivate nuclear factor kappaB-mediated cytoprotective effects in human non-small cell lung cancer cells. *Mol. Cancer Ther., 5,* 391–399.

Mijatovic, T., Roland, I., Van Quaquebeke, E., Nilsson, B., Mathieu, A., Van Vynckt, F., Darro, F., Blanco, G., Facchini, V., & Kiss, R., (2007). The α1 subunit of the sodium pump could represent a novel target to combat non-small cell lung cancers. *J. Pathol., 212,* 170–179.

Nasu, S., Milas, L., Kawabe, S., Raju, U., & Newman, R., (2002). Enhancement of radiotherapy by oleandrin is a caspase-3 dependent process. *Cancer Lett., 185,* 145–151.

Newman, R. A., Kondo, Y., Yokoyama, T., Dixon, S., Cartwright, C., Chan, D., Johansen, M., & Yang, P., (2007). Autophagic cell death of human pancreatic tumor cells mediated by oleandrin, a lipid-soluble cardiac glycoside. *Integr. Cancer Ther., 6,* 354–364.

Newman, R. A., Yang, P., Hittelman, W. N., Lu, T., Ho, D. H., Ni, D., Chan, D., Vijjeswarapu, M., Cartwright, C., Dixon, S., Felix, E., & Addington, C., (2006). Oleandrin-mediated oxidative stress in human melanoma cells. *J. Exp. Ther. Oncol., 5,* 167–181.

Nobili, V., Parkes, J., Bottazzo, G., Marcellini, M., Cross, R., Newman, D., Vizzutti, F., Pinzani, M., & Rosenberg, W. M., (2009a). Performance of ELF serum markers in

predicting fibrosis stage in pediatric non-alcoholic fatty liver disease. *Gastroenterology, 136,* 160–167.
Nobili, V., Parkes, J., Bottazzo, G., Marcellini, M., Cross, R., Newman, D., Vizzutti, F., Pinzani, M., & Rosenberg, W. M., (2009b). Performance of ELF serum markers in predicting fibrosis stage in pediatric non-alcoholic fatty liver disease. *Gastroenterology, 136,* 160–167.
Osterloh, J., (1982). Oleander interference in the digoxin radioimmunoassay in a fatal ingestion. *JAMA J. Am. Med. Assoc., 247,* 1596.
Pan, L., Zhang, Y., Zhao, W., Zhou, X., Wang, C., & Deng, F., (2017). The cardiac glycoside oleandrin induces apoptosis in human colon cancer cells via the mitochondrial pathway. *Cancer Chemother. Pharmacol., 80,* 91–100.
Pietsch, J., Oertel, R., Trautmann, S., Schulz, K., Kopp, B., & Dressler, J., (2005). A non-fatal oleander poisoning. *Int. J. Legal Med., 119,* 236–240.
Pollyea, D. A., Gutman, J. A., Gore, L., Smith, C. A., & Jordan, C. T., (2014). Targeting acute myeloid leukemia stem cells: A review and principles for the development of clinical trials. *Haematologic., 99,* 1277–1284.
Raghavendra, P. B., Sreenivasan, Y., Ramesh, G. T., & Manna, S. K., (2007). Retracted article: Cardiac glycoside induces cell death via FasL by activating calcineurin and NF-AT, but apoptosis initially proceeds through activation of caspases. *Apoptosis, 12,* 307–318.
Rashan, L. J., Franke, K., Khine, M. M., Kelter, G., Fiebig, H. H., Neumann, J., & Wessjohann, L. A., (2011). Characterization of the anticancer properties of monoglycosidic cardenolides isolated from Nerium oleander and Streptocaulon tomentosum. *J. Ethnopharmacol., 134,* 781–788.
Shaw, D., & Pearn, J., (1979). Oleander poisoning. *Med. J. Aust., 2,* 267–269.
Shiratori, O., (1967). Growth inhibitory effect of cardiac glycosides and aglycones on neoplastic cells: *In vitro* and *in vivo* studies. *Gan., 58,* 521–528.
Shumaik, G. M., Wu, A. W., & Ping, A. C., (1988). Oleander poisoning: Treatment with digoxin-specific Fab antibody fragments. *Ann. Emerg. Med., 17,* 732–735.
Slingerland, M., Cerella, C., Guchelaar, H. J., Diederich, M., & Gelderblom, H., (2013). Cardiac glycosides in cancer therapy: From preclinical investigations towards clinical trials. *Invest. New Drugs, 31,* 1087–1094.
Smith, J. A., Madden, T., Vijjeswarapu, M., & Newman, R. A., (2001). Inhibition of export of fibroblast growth factor-2 (FGF-2) from the prostate cancer cell lines PC3 and DU145 by Anvirzel and its cardiac glycoside component, oleandrin. *Biochem. Pharmacol., 62,* 469–472.
Smith, T. W., (1985). Pharmacokinetics, bioavailability, and serum levels of cardiac glycosides. *J. Am. Coll. Cardiol., 5,* 43A–50A.
Stenkvist, B., Bengtsson, E., Eklund, G., Eriksson, O., Holmquist, J., Nordin, B., & Westman-Naeser, S., (1980). Evidence of a modifying influence of heart glucosides on the development of breast cancer. *Anal. Quant. Cytol., 2,* 49–54.
Tracqui, A., Kintz, P., Branche, F., & Ludes, B., (1998). Confirmation of oleander poisoning by HPLC/MS. *Int. J. Legal Med., 111,* 32–34.
Van Kanegan, M. J., He, D. N., Dunn, D. E., Yang, P., Newman, R. A., West, A. E., & Lo, D. C., (2014). BDNF mediates neuroprotection against oxygen-glucose deprivation by the cardiac glycoside oleandrin. *J. Neurosci., 34,* 963–968.
Wasfi, I. A., Zorob, O., Al Katheeri, N. A., & Al Awadhi, A. M., (2008). A fatal case of oleandrin poisoning. *Forensic Sci. Int., 179,* 31–36.

Watanabe, R., Murata, M., Takayama, N., Tokuhira, M., Kizaki, M., Okamoto, S., Kawai, Y., Watanabe, K., Murakami, H., Kikuchi, M., Nakamura, S., & Ikeda, Y., (1997). Long-term follow-up of hemostatic molecular markers during remission induction therapy with all-trans retinoic acid for acute promyelocytic leukemia. Keio hematology-oncology cooperative study group (KHOCS). *Thromb. Haemost., 77,* 641–645.

Winnicka, K., Bielawski, K., & Bielawska, A., (2006). Cardiac glycosides in cancer research and cancer therapy. *Acta Pol. Pharm., 63,* 109–115.

Xie, Z., & Askari, A., (2002). Na+/K+-ATPase as a signal transducer. *Eur. J. Biochem., 269,* 2434–2439.

Yang, W. K., Hseu, J. R., Tang, C. H., Chung, M. J., Wu, S. M., & Lee, T. H., (2009). Na+/K+-ATPase expression in gills of the Euryhaline sailfin molly, *Poecilia latipinna*, is altered in response to salinity challenge. *J. Exp. Mar. Bio. Ecol., 375,* 41–50.

Zhao, Q., Guo, Y., Feng, B., Li, L., Huang, C., & Jiao, B., (2011). Neriifolin from seeds of *Cerbera manghas* L. induces cell cycle arrest and apoptosis in human hepatocellular carcinoma HepG2 cells. *Fitoterapia, 82,* 735–741.

CHAPTER 7

Challenging Compounds of Some Medicinal Plants Against Cancer

N. SHAISTA JABEEN,[1] L. JAGAPRIYA,[1] B. SENTHILKUMAR,[2] and K. DEVI[2]

[1]Department of Zoology, Dhanabagyam Krishnaswamy Mudaliar College for Women (Autonomous) Vellore – 632 001, Tamil Nadu, India

[2]Department of Zoology, Thiruvalluvar University, Vellore, Tamil Nadu, India, E-mail: sj2khan16@gmail.com

ABSTRACT

Cancer is an outrageous disease and has a great concern in the global population. Scientists, Researchers, and Physicians discover various drugs which give hopes to the victims. On this note, herbal medicine leads ahead in treating chronic diseases. Herbal medicines are traditionally used in worldwide. According to WHO, most of the developed countries adopted the traditional system of herbal medicines. According to the survey, the data revealed that herbal medicines proved to be an elixir even for those in chemotherapy with minimal and nontoxic antitumor effect in contrast to the standard tumor therapy. The present review is to highlight on those antitumor bioactive compounds which are cytotoxic and apoptosis effect on tumor cells.

7.1 INTRODUCTION

Phytonutrients of culinary spices extolled in the field of traditional medicine systems like Ayurveda, (Indian system of medicine) Unani (Persian system of medicine) and in Traditional (Chinese medicine). The spices use in Indian cuisine possess several benefits it not only boost the relish and flavor of the food but also protect the human body from pathogens. A collection of spices, their active ingredient, molecular structure (MS), and therapeutics are described in Table 7.1.

TABLE 7.1 Medicinal Plant and Their Major Constituents

S. No.	Medicinal Plants/ Common Names	Bioactive Compounds	Molecular Structure	Therapeutic Properties
1.	*Nigella Sativa*/Black cumin seeds	Thymoquinone		Anti-tumor and anti-cancer, cytotoxic, anti-inflammatory, Anti-diabetic, hypoglycemic, etc.
2.	*Ciminumcyminum*/ Cumin seeds	Cuminaldehyde		Anti-tumor activity, anti-diabetic, an inhibitor of squamous Carcoma and hepatomas
3.	*Moringa peregrine*/ Parsley	Apigenin		Cytotoxic activity, Breast, and Colon Cancer Cell line
4.	*Carumcarvi*/ Caraway seeds	Carvone		Anti-tumor/ anti-mutagenic, anticolitis, hepatoprotective, inhibition and suppression of colon and rectal tumor cells
5.	*Curcuma longa L*/ Turmeric	Curcumin		Anti-tumor activities colon, breast, lung, and brain tumors induce apoptosis in tumor cells
6.	*Saffron crocus*/ Saffron	Crocetin		Pancreatic, skin carcinoma, colorectal breast and lung carcinoma (Bathaie and Mousavi, 2007)

TABLE 7.1 *(Continued)*

S. No.	Medicinal Plants/ Common Names	Bioactive Compounds	Molecular Structure	Therapeutic Properties
7.	*Zingi*	Gingerol		Antiemetic, Anti-tumor activity (Skin carcinoma)
8.	Broccoli, tea, W-itch-hazel	Kaempferol		Chemopreventive agent, antioxidant prevent arteriosclerosis (Jan Kowalski et al., 2005)
9.	Various berries, grapes	Cyanidins		Antioxidant, inhibit breast metastasis
10.	Vegetables	Quercetin		Antioxidant, anti-inflammatory activity,
11.	*Acacia greggii, Acacia berlandieri*	Fisetin		Antihyperalgesic, Nephrotoxicity, Anti-oxidant, Anti-inflammatory, Anti-cancer activity, hypolipidemic
12.	Flemingiavestita, Soybean, Kudzu, Psoralea	Genistein		Anti-cancer activity, Antihelminthic, antiangiogenic
13.	Tomatoes, red-carrots. Watermelons	Lycopene		Anti-cancer activity (prostate cancer)
14.	Red grapes skin, Peanuts	Resveratrol		Cancer, Chemopreventive activity, anti-initiation activity

7.2 MEDICINAL PLANTS AND THEIR ACTIVE COMPOUNDS ACT AS ANTITUMOR ACTIVITY

7.2.1 NIGELLA SATIVA SEEDS (BLACK CUMIN)

Nigella sativa L. commonly called Kalonji in India and Pakistan, originated from Turkey, Pakistan, and India. The seeds contain numerous compounds moreover; each and every compound possesses therapeutic properties. The seeds and its oil uses for treating various ailments.

Anticancer activity of *N. sativa* Seeds: The major constituent of *Nigella sativa* seeds and its oil possess many chemical compounds but most importantly the seeds and oil exhibited antitumor activity because of its major active constituent thymoquinone. Thymoqinone exert anti-prolifertive, anti-angiogenic, anti-metastatic, and antioxidant effects on various tumor cells. The compound thymoquinone seems to mediate its anti-cancer effects by targeting a number of cellular pathways involving p53, NF-kB, PPARg, STAT3, MAPK, and PI3K/AKT. Moreover, thymol, thymohydroquinone, dithymoquinone, nigellimine-N-oxide, nigellicine, nigellidine, and carvacrol are phytoconstituents of *N. sativa* that have been demonstrated that anti-cancer and cytotoxic functions (Randhawa and Alghamdi, 2011).

7.2.2 CUMINUM CYMINUM SEEDS (ZEERA)

Cumin seeds (Apiaceae) are a popular culinary spice. Besides cumin seeds are uses in folklore medicines because of the presence of therapeutic compounds. Cumin has been used as anti-inflammatory, diuretic, carminative, and antispasmodic, treatment of toothaches and epilepsy and also as an aid for treating dyspepsia, jaundice, diarrhea, flatulence, and indigestion (*Sahih, 2014*).

Antitumor activity of *Cuminum Cyminum*: Cumin seeds are prevented colorectal carcenoma. It decreases the β-glucoronidase and mucinase enzyme activity. The major constituent of cumin seeds the cuminaldehyde attributes the cytotoxic activity. The aroma of cumin is due to the presence of cuminaldehyde. Cumin seeds inhibit 79% HeLa cells which is proved to be chemopreventive potentials (*Sahih, 2014*).

7.2.3 CARUM CARVI SEEDS

The culinary spice *Carum carvi* L. seeds are possessed anticancer activity. The *Carum carvi* seeds histopathological studies of colon tissues showed a tremendous inhibition (70%-90%). The apoptotic activities were observed in *Carum carvi* seeds. Caraway oil contains 50%-60% of carvone specific Tunisian *Carum carvi* seeds that exhibit chemopreventive effects. Carvone is a highly potent element against tumor cells (Mehta et al., 2010).

7.2.4 CURCUMIN FROM TURMERIC

Curcumin is a major chemical constituent of *Curcuma longa L (Turmeric)* of Indian spices. Turmeric possesses wide medicinal properties and it has great apoptotic activities it not inhibits the tumor cells but suppressed the target cells without damaging the normal cells. Curcumin modulates the growth of cancer cells by regulating multiple cell signaling pathways. Curcumin subjected to non-toxic to humans apart from the antitumor activity it has many other activities antioxidant anti-inflammatory and cytotoxic activities. It has tremendous genomic suppression of liver and intestinal tumor cells (Gullett et al., 2010).

7.2.5 FESITIN FROM BERRIES

Fesitin is family of flavanol and found in various fruits and vegetables especially berry categories like strawberries, blueberries, raspberries, cranberries onion, apples. It bears anti-carcinogenic activities and anti-inflammatory effects. It is highly potent against human colon cancer cells and lungs (Limdo, 2009).

7.2.6 GENISTEIN FROM SOYBEAN

Genistein is an isoflavone compound found in many plants and beans categories. Genistein has antioxidant, antihelminthic, and anti-carcinogenic activities. According to Akiyama et al (1987) reported that the isoflavone content of soy as an anticancer element food caught the attentions of many researchers.

7.2.7 GINGEROL FROM GINGERS

Gingerol is an active ingredient of ginger. Ginger has many pharmacological activities. Its main constitutes gingerol has been studied its effect on tumor cells. A recent review by Oyagbemi et al. (2010) stated the therapeutic properties of gingerol. In short, gingerol has demonstrated antioxidant, anti-inflammation, and antitumor promoting properties. Gingerol exhibited immense impact on human hepatic carcinoma cells.

7.2.8 KAEMPFEROL FROM BEVERAGES TEA, BROCCOLI, GRAPEFRUIT

Kaempferol is natural compound in beverages, vegetable, and in grapes. Kaempferol has been studies for it anticarcinogenic activity and exerting free radical scavenging activity. Luo et al. (2011) found that kaempferol induces apoptosis in ovarian cancer cells through the activation of p53 in the intrinsic pathway. Yang et al. (2013) reported that kaempferol inhibited quinone reductase 2 with an IC (50) value of 33.6 μM for NF-κB activity.

7.2.9 LYCOPENE FROM TOMATO

Lycopene is the chief constituent of tomatoes and carrots, watermelons, etc. It has been showed antioxidant and chemo-preventive effect reported by many studies. Lycopene also possesses inhibitory activity of breast and endometrial cancer cells (Nahum et al., 2001), prostate cancer (Giovannucci et al., 1995) and colon cancer. Treatment with whole dried tomato powder to the xenograft prostate tumors into rats exhibited that tumors grew more slowly. It might be due to the major constituent, lycopene of tomato which suppresses the prostate cancer cells (Feldman et al., 1996).

7.3 CONCLUSION

The present study is to bring out the bioactive compounds present in the Indian spice plants used for the prevention of cancer. Some of the compounds like thymoquinone and gingerolcuminaldhyde proved to be more potent against tumor cells. Furthermore, studies required to investigate their activities against cancer.

KEYWORDS

- anti-tumor activity
- bioactive compounds
- colon cancer
- medicinal plants
- prostate cancer cells
- thymoquinone

REFERENCES

Akiyama, T., Ishida, J., Nakagawa, S., Ogawara, H., Watanabe, S., Itoh, N. M., Shibuya, M., & Fukami, Y., (1987). Genistein, a specific inhibitor of tyrosine-specific protein kinases. *J. Biol. Chem., 262*, 5592–5595.

Banerjee, S., Azmi, A. S., Padhye, S., Singh, M. W., Baruah, J. B., & Philip, P. A., (2010). Structure-activity studies on therapeutic potential of thymoquinone analogs in pancreatic cancer. *Pharm. Res., 27*(6), 1146e58.

Chen, J., & Xu, X., (2010). Diet, epigenetic, and cancer prevention. *Adv Genet., 71*, 237–255.

Clere, N., Faure, S., Martinez, M. C., & Andriantsitohaina, R., (2011). Anticancer properties of flavonoids: Roles in various stages of carcinogenesis. *Cardiovasc. Hematol. Agents Med. Chem., 9*(2), 62–77.

Feldman, E. J., Seiter, K. P., Ahmed, T., Baskind, P., & Arlin, Z. A., (1996). Homoharringtonine patients with myelodysplastic syndrome (MDS) and MDS evolving to acute myeloid leukemia. *Leukemia. 10*, 40–42.

Giovannucci, E., Liu, Y., Rimm, E. B., Hollis, B. W., Fuchs, C. S., Stampfer, M. J., & Willett, W. C., (2006). Prospective study of predictors of vitamin D status and cancer incidence and mortality in men. *J. Natl. Cancer Inst., 98*, 451–459.

Gullett, N. P., Ruhul, A. A. R., Bayraktar, S., Pezzuto, J. M., Shin, D. M., Khuri, F. R., Aggarwal, B. B., Surh, Y. J., & Kucuk, O., (2010). Cancer prevention with natural compounds. *Semin. Oncol., 37*(3), 258–281.

Huang, J., Plass, C., & Gerhauser, C., (2011). Cancer chemoprevention by targeting the epigenome. *Curr. Drug Targets, 12*(13), 1925–1956.

Khare, C. P., (2007). *Indian Medicinal Plants: An Illustrated Dictionary* (pp. 124–125). Springer Science and Business Media, LLC.

Kowalski, M., (2001). *Helicobacter pylori* (*H. pylori*) infection in coronary artery disease: Influence of *H. pylori* eradication on coronary artery lumen after percutaneous transluminal coronary angioplasty. The detection of *H. pylori* specific DNA in human coronary atherosclerotic plaque. *J. Physiol. Pharmacol., 52*, 3–31.

Limdo, Y., & Park, J. H., (2009). Induction of p53 contributes to apoptosis of HCT-116 human colon cancer cells induced by the dietary compound fisetin. *Am. J. Physiol. Gastrointest. Liver Physiol., 296*, 1060–1068.

Luo, H., Rankin, G. O., Li, Z., Depriest, L., & Chen, Y. C., (2011). Kaempferol induces apoptosis in ovarian cancer cells through activating p53 in the intrinsic pathway. *Food Chem., 128*(2), 513–519.

Mehta, R. G., Murillo, G., Naithani, R., & Peng, X., (2010). Cancer chemoprevention by natural products: How far have we come? *Pharm. Res., 27*(6), 950–961.

Nahum, A., Hirsch, K., Danilenko, M., Watts, C. K., Prall, O. W., Levy, J., & Sharoni, Y., (2001). Lycopene inhibition of cell cycle progression in breast and endometrial cancer cells is associated with reduction in cyclin D levels and retention of p27(kipI) in the cyclin E-cdk2 complexes. *Oncogene, 20*, 3428–3436.

Oyagbemi, A. A., Saba, A. B., & Azeez, O. I., (2010). Molecular targets of [6]-gingerol: Its potential roles in cancer chemoprevention. *Biofactors., 36*(3), 169–178.

Pan, M. H., Lai, C. S., Wu, J. C., & Ho, C. T., (2011). Molecular mechanisms for chemoprevention of colorectal cancer by natural dietary compounds. *Mol. Nutr. Food Res., 55*(1), 32–45.

Randhawa, M., & Alghamdi, M., (2011). Anticancer activity of *Nigella sativa* (black seed): A review. *Am. J. Chin. Med., 39*(6), 1075e91.

Sahih, B., (2014). There is Healing in Black Cumin for all Diseases Except Death, 7(71), 592.

Scorei, R. I., & Popa, R. J., (2010). Boron-containing compounds as preventive and chemotherapeutic agents for cancer. *Anticancer Agents Med. Chem., 10*(4), 346–351.

Yang, J. H., Kondratyuk, T. P., Jermihov, K. C., Marler, L. E., Qiu, X., Choi, Y., et al., (2011). Bioactive compounds from the fern Lepisorus contortus. *J. Nat. Prod., 74*(2), 129–136.

CHAPTER 8

Medicinal Plants as Therapeutic Agents in the Treatment of Cancer

JAYA VIKASKURHEKAR

Department of Microbiology, Dr. Patangrao Kadam Mahavidyalaya, Sangli – 416416, Maharashtra, India

ABSTRACT

Cancer is a dreaded disease that is a cluster of many similar disease symptoms; characterized by unprecedented, limitless, reasonless division of body cells, which later on spread into its vicinity. Cancers can originate anywhere and disturb the systematic process of cell division. Cancers are treated in numerous ways, aggressive as well as non-aggressive, as per the situation. These include radiation, chemotherapy, surgery, and others. These are known to have serious side effects, which overshadow the treatment. In this background, various medicinal plants have been studied, which show promising results as therapeutic agents in combating cancer. This article elaborates on eleven such plants, which may help us in overcoming the dread and wrath of cancer, besides having little or no side effects, being easily accessible and cost-effective.

8.1 INTRODUCTION

The word "Cancer" frightens the life out of a human being. It is a dreaded disease that tops the list of most dangerous diseases. Though, a layman knows very little about this disease. Cancer is unprecedented, limitless, reasonless division of body cells, which later on spread into its vicinity. The human body is build with millions, billions, trillions of cells, which are periodically rejuvenated, as, and when required. Old and injured cells are replaced by new cells. Cancer can originate anywhere and disturb this systematic process. Cells become abnormal and deformed, cells which should die, survive, and unnecessary new cells are generated. Cells which are in excess, continuously go on dividing, forming cell tissue masses called tumors. These tumors may turn out to be non-cancerous (benign) or cancerous (malignant). If malignant, they keep dividing, spread into the tissues in the vicinity, attack nearby tissues and may break off from origin and generate tumors elsewhere in the body. Cancers may arise because of errors occurring during division of cells or because of damage to DNA, due to specific certain environmental factors or situations, which may be tobacco chemicals, impermissible food preservatives or colors, exposure to certain radiations emitted from explosions or sun. Cancer changes genetically taking up various permutations and combinations, adding more and more changes, as it grows. DNA may mutate, normal cells become abnormal, having varied genetic changes. In cancerous cases, some mutations are commonly observed.

Cancer has been treated in many ways depending on cancer types and its stage. Combination of treatments is preferred over single line of treatment. Choice of treatment is confusing, more so because of its side effects. Treatment is done with reference to various factors like type, location, and stage of the disease. Common treatment methods employed are chemotherapy, surgery, radiation, hormonal therapy, targeted therapy, immunotherapy, monoclonal antibody therapy, and synthetic lethality. All cancer treatment research is aimed at removing the cancerous part totally, without having any effect on the body, with no adverse effects. This has been difficult because of the spread of cancer or its metastasis which is the spread of cancer to distant tissues. Cancer treatment is known to be notorious, with reference to its side effects, which can result from the normal tissues being affected. Side effects may be numerous; loss of appetite, anemia, thrombocytopenia, diarrhea, edema, fatigue, constipation, delirium, peripheral neuropathy, sexual, and fertility issues, alopecia, nausea, throat infections, urinary problems, etc. On this background, attempts are being made to use naturopathy for cancer treatment. Since ancient times, a number of plants are used to cure many ailments. Various options regarding using medicinal plants as anti-cancer drugs are being explored. Some of them have been found beneficial while some are under trial. Rig-Veda, one of the world's oldest medical documentaries in the world, contains intelligent observations on the use of plants and their as cancer drugs (Greenspan and Aruoma, 1994; Cragg et al., 1996; Rao et al., 1997; Hallock et al., 1998; Das et al., 2001; Raul and Heble, 2001). Medicinal value of indigenous plants has been established worldwide as excellent anti-cancerous agents (Kurhekar, 2014). Secondary metabolites of medicinal plants are observed to possess cancer alleviating properties (Mondal and Mukherjee, 1992). Flavonoids are reported to exert anti-cancerous activity (Havsteen, 1983), a major class of phenolic compounds present in many fruits and vegetables. They play a significant role in preventing various classes of cancers because of hydrogen donating free radicals with free hydroxyl groups and catechol moiety in the ring B of flavonoid nucleus (Shetgiri and D'Mello, 2003).

The National Cancer Institute (NCI) has screened ~35,000 plant species for their potential cancer alleviating activities and only 3000 odd plant species have identified as potential anticancer plants. However, a small number of these have received scientific evaluation for their efficacy. The World Health Organization (WHO) has recommended that the medicinal plants be intensively investigated for their medicinal properties. Thus, this review is an alarm to identify such plant species which possess anticancer activities.

8.2 MEDICINAL PLANTS SHOWING GREAT POTENTIAL AS ANTI-CANCEROUS AGENTS

The medicinal plants used for the treatment of cancer and its related beneficial effects, and scientific validation are given below.

8.2.1 AEGLE MARMELOS (SANSKRIT: ADHARARUHA, SIVADRUMAH, TRIPATRA)

Aegle marmelos is commonly called as Golden Apple or Stone Apple or Wood Apple. It is a huge tree found in India, Nepal, Andaman, and Nicobar Islands, Myanmar, Sri Lanka, Thailand, and Malaysia (*Orwa, 2009*). Fruits of this plant are employed traditionally as medicine for the treatment of cancer and are known to possess xanthotoxol, furocoumarins, flavonoids, methyl ester of alloimperatorin, alkaloids (O-Methylhafordinol, Allocryptopine, and O-isopentenylhalfordinol), essential oils and aegle marmelosine (*Laphookhieo, 2011*). Further, he stated that the property of anticancer activity is due to the presence of skimmianine in the hydroalcoholic leaf extract. Proliferation of tumor cell lines *viz.,* leukemic K562, T-lymphoid Jurkat, B-lymphoid Raji, erythroleukemic HEL, melanoma Colo38, and breast cancer MCF7 and MDA-MB-231 were highly inhibited by Butylp-tolylsulfide, 6-methyl-4-chromanone and 5-methoxypsoralen of *A. marmelos* (Chockalingam et al., 2012). A great cytotoxic effect of compound 1-hydroxy-5,7-dimethoxy-2-naphthalene-carboxaldehyde (HDNC, marmelin) isolated from *A. marmelos* was observed against BT 220, H116, Int 407, HL 251 and HLK 210 human cancer cells. It activates apoptosis of epithelial cancer cells by activating tumor necrosis factor-α (TNF-α), TNF receptor (TNFR)-associated death domain (TRADD) and caspases during G1 phase of the cell cycle (Rahman and Parvin, 2014).

8.2.2 ALLIUM CEPA (SANSKRIT: PALANDU)

Allium cepa is commonly known as onion, bulb onion or common onion. The plant has a fan-like appearance, with hollow, bluish-green leaves, and bulbs with short, compressed, underground stems enclosed in fleshy modified leaves, enveloping a central bud at the stem tip. Food reserves start accumulating in the leaf bases and onion bulbs, as they mature (Brickell, 1992). The plants are abundant in Asia, Iran, China, Egypt, and Persia (Cumo, 2015; Ansari, 2007). Consuming onions is shown to decrease cancers and a reduction in risk of prostate, colorectal, renal, oesophageal, mouth, breast, and ovarian cancers, as they are an excellent source of antioxidants (Wang et al., 2012). Onions contain a high amount of polyphenols and flavonoids, quercetin oligosaccharides, arginine, selenium, and sulfur-containing compounds (Block, 1985). Ravanbakhshian and Behbahani (2018) reported that fermented onion might be good anticancer drug candidate against breast cancer.

8.2.3 ALLIUM SATIVUM (SANSKRIT: LASHUNAM)

Allium sativum is commonly called Garlic. It is found in Central Asia and northeastern Iran and Egypt. It is a bulbous plant, growing up to 4 ft. in height, with hermaphrodite flowers. *A. sativum* is a natural medicinal agent preventing cancer (Kurian, 1999). Regular intake of garlic reduces cancer of the upper digestive tract (*Guercio et al., 2016*), stomach (*Zhou et al., 2011*) and prostate (*Zhou et al., 2013*). The garlic derivative S-allyl-mercapto cysteine (SAMC) inhibits growth, arrests cell cycle at G2-M, and induces apoptosis in human colon cancer cells (Choi and Park, 2012). Whole cloves of garlic possess steroidal saponins as well as selenium compounds of organic nature, possessing significant anticancer efficacy (Tariq et al., 1988). Ajoene produced from the heating of crushed garlic is a major compound exhibiting cytotoxicity towards cancer cells (Siyo et al., 2017). Methanolic extract of *A. sativum* shows anticancer activity in DU145, A549, and MCF7 cell lines (Zhang et al., 2005). Fresh or ground garlic shows the presence of sulfur-containing compounds alliin, ajoene, diallyl polysulfides, vinyldithiins, S-allylcysteine, and non-sulfurous compounds like enzymes, saponins, and flavonoids. It is generally reported that sulfur-containing compounds are responsible for most of the beneficial effects of the garlic (Tariq et al., 1988; Bat-Chen et al., 2010).

8.2.4 ALOE VERA (SANSKRIT: GHRIT KUMARI)

Aloe Vera is an evergreen perennial and succulent plant, growing in Arabian Peninsula and Asia. It is commonly called Indian Aloe. Leaves are thick, fleshy, and greyish green. Aloe leaf gel contains phytochemicals like acetylated mannans, polymannans, anthraquinone C-glycosides, anthrones, anthraquinones (King et al., 1995; Boudreau and Beland, 2006). Since ancient times, the health benefits of *A. vera* are well known and it especially has the promising capacity to fight cancer as well. *A. Vera* exhibits anti-cancer activity both *in vitro* and *in vivo* conditions. Anthraquinone present in *Aloe Vera* kills human bladder cancer cells (Thomson, 1971). Aloin from *A. vera* showed an antiproliferative effect in HeLaS3 human cervix carcinoma cells (Niciforovic et al., 2007). Aloe-emodin of *A. vera* selectively inhibited the growth of human neuroectodermal tumors in mice (Pecere et al., 2000). *A. Vera* shows excellent applications in skin cancers and great anti-oxidant effect protecting skin damage from radiations (Dweck, 1994).

8.2.5 ANDROGRAPHIS PANICULATA (SANSKRIT: BHUNIMBA AND KALMEGHA)

Andrographis paniculata is commonly called as king of bitters. It is an annual herbaceous plant belonging to family Acanthaceae. It is native to India and Sri Lanka. It is grown in South and Southeast Asia. The leaves, roots, and whole plant are used for medicinal purposes. *A. paniculata* is used in Siddha and Ayurvedic medicine and promoted as a dietary supplement for cancer prevention and cure. Bitter principle of this plant is due to the presence of andrographolide, abicyclic, and diterpenoid lactone (Gorter, 1911). Known constituents of this plant are; "14-Deoxy-11-dehydroandrographolide, 14-Deoxy-11-oxoandrographolide, 5-Hydroxy-7,8,2,'3'-Tetramethoxy flavone, 5-Hydroxy-7,8,2'-Trimethoxyflavone, Andrographolide, Neoandrographolide, Panicoline, Paniculide-A, Paniculide-B, Paniculide-C, (Hossain et al., 2014). *A. paniculata* stimulates immune system by activating non-specific and antigen-specific responses (Puri et al., 1993). Thus, it acts as an efficient chemoprotective agent protecting against cancer (Puri et al., 1993). Andrographolide exhibits cytotoxic activity against different types of cancer cells (Kumar et al., 2004). It exhibits inhibition of growth in HT-29 cell lines of colon cancer, enhancement of division and growth of peripheral human blood lymphocytes, and exertion of supportive differentiating impact on M1 cell line of myeloid leukemia (Kumar et al., 2004; Matsuda et al., 1994).

8.2.6 CATHARANTHUS ROSEUS (SANSKRIT: NITHYAKALYANI)

Catharanthus sroseus is commonly known as the Rose Periwinkle. It is a species of flowering plant in the family Apocynaceae. It is native and endemic to Madagascar, grown as a medicinal and ornamental plant. It is a known source of the vincristine and vinblastine used as cancer cure. Vinblastine and vincristine are chemotherapy medications used to combat various types of cancers. They are biosynthesized from the coupling reactions of alkaloids, catharanthine, and vindoline (Cooper et al., 2016). The newer semi-synthetic chemotherapeutic agent, vinorelbine is useful to treat non-small-cell lung cancer (Keglevich et al., 2012). Rosinidin is an anthocyanidin pigment found in the flowers of *C. Roseus* is proved to be a chemotherapeutic agent against cancer (Redkar and Jolly, 2003). Alkaloids, vinblastine, and vincristine interfere present in leaf of this plant inhibit the cancer cells division while other chemicals have shown prospective prevention of growth of new blood vessels which supporting tumor growth (Mandal et al., 2002).

8.2.7 CEDRUS DEODARA (SANSKRIT: DEVDAR)

Cedrus deodara is commonly called as deodar or "wood of the gods." It is native to the Northern Pakistan, Eastern part of Afghanistan, India, Southwestern Tibet, and Western Nepal. *C. deodara* is used in Ayurvedic medicine (McGowan and Chris, 2008) The bark of *C. deodara* contains large amounts of taxifolin and the wood contains cedeodarin, ampelopsin, cedrin, cedrinoside, and *deodarin (3,'4′,6-tetrahydroxy-8-methyl dihydroflavonol)* (Adinarayana and Seshadri, 1965; Agrawal et al., 1980). The main components of its essential oil include α-terpineol, linalool, limonene, anethole, caryophyllene, and eugenol (Zeng et al., 2012). The *C. deodara* also contains lignans (Agrawal and Rastogi, 1982) and the phenolic sesquiterpene himasecolone (Agarwal and Rastogi, 1981) A lignan composition extracted from the stem wood of *C. deodara* exhibited cytotoxicity against human cancer cell lines (Shashi et al., 2006; Singh et al., 2007). It also induced suppression of tumors in models of murine (Singh et al., 2007).

8.2.8 *CURCUMA LONGA (SANSKRIT: HARIDRA)*

Curcuma longa is popularly known as Turmeric is a rhizomatous herbaceous perennial plant of the ginger family, Zingiberaceae (Priyadarsini, 2014). It is native to the Indian subcontinent and Southeast Asia. The phytochemicals present in this plant are diarylheptanoids, curcumin, curcuminods bisdemethoxycurcumin, and demethoxycurcumin (Nelson et al., 2017). Besides that, 34 essential oils exist in turmeric, majority of them are turmerone, germacrone, atlantone, and zingiberene (Hong et al., 2014; Hu et al., 2014). Turmeric has been an essential component of Grandmother's purse and traditional Ayurvedic medicine since ancient times, owing to its awesome medicinal values. Turmeric is used widely as a spice and its major benefit is its anti-cancer effect. Curcumin is known as an anti-cancer herb as it stops cancer formation, replication, and spread, reduces inflammation and stress related to oxidation (Kumar et al., 2015). Further, they reported that curcumin stimulates some anti-cancer drug activity, simultaneously protecting healthy organs and cells. Curcumin is the active organic compound found in turmeric, which is responsible for the death of cancerous cells, as observed by several researchers (Ravindran et al., 2009; Yodkeeree et al., 2009, 2010; Lin et al., 2015). Advantage of curcumin is that it attacks only the cancerous cells, and protecting healthy cells. Curcumin has been studied for suppressing initiation of tumors, promotion of tumors, metastasis, and has proved its potential to treat as well as prevent cancer. It controls gene signaling, affecting a wide range of cells of tumor, down-regulating factors related to transcription process, COX-2 like enzymes and few inflammatory agents like cytokines, chemokines, cell-surface adhesion molecules, and growth factors (Wilken et al., 2011).

8.2.9 VITIS VINIFERA (SANSKRIT: DRAKSHA)

Vitis vinifera is commonly called as grape vine, found in Mediterranean region, Central Europe, Southwestern Asia, Morocco, Portugal north to southern Germany and east to northern Iran. Grapes contain many phenolic compounds *like a*nthocyanins and stilbenoids, hydroxycinnamic acids and tannins (*Aizpurua-Olaizola et al., 2015*). Resveratrol, a phytochemical found in grapes skin is responsible for the anticancer property (Burns et al., 2002). Grape seeds, skin, and whole grape are a potential source of anticancer and cancer chemopreventive phytochemicals and demonstrated their potential anticancer efficacy in various preclinical and clinical studies (Scarlatti et al., 2008; Zahid et al., 2008). Grapes and grape-based products are excellent sources of various anticancer agents and consumption of grapes or grape-related products has both practical and translation potential in the fight against cancer (Kaur, 2009).

8.2.10 WITHANIA SOMNIFERA (SANSKRIT: ASHWAGANDHA)

Withania somnifera is commonly called Ashwagandha or Indian ginseng or Poison Gooseberry. It is a plant belonging to Solanaceae family. *W. somnifera* is cultivated in drier regions of India (*Mirjalili et al., 2009*), Nepal, China (*Pandit et al., 2013*) and Yemen (Scott and Mason, 1946). Its principle constituents are steroidal lactones (tropine, cuscohygrine, and alkaloids). Leaves contain steroidal lactones and Withaferin A. and roots contain withanolide A (Malik et al., 2007). Leaves *of W. Somnifera* contain Withaferin A, which brings about enhanced and fast apoptosis of cancerous cells (Malik et al., 2007). The cell signaling pathways induced by *W. Somnifera* formulation is credited to its high content of withaferin A (Malik et al., 2007). Root extract of *W. Somnifera* exhibited as a potential source of new drug leads that can curtail cancer growth (Dredge et al., 2003). Leaves have also been shown to inhibit the growth of human cancer cell lines (Yadav et al., 2010).

8.2.11 ZINGIBER OFFICINALE (SANSKRIT: SRUNGAVERA)

Zingiber officinale is commonly termed as Ginger and is a flowering plant whose rhizome; root is used as a spice and as a folk medicine. Ginger originated in the tropical rain forests from the Indian subcontinent to Southern

Asia (*Everett, 1982*). The characteristic fragrance and flavor of ginger result from volatile oils. Fresh ginger contains zingerone, shogaols, and gingerols (*An et al., 2016*). Gingerol is responsible for effective in killing tumors, ovarian cancer cell death, improving immune system and reducing inflammation. It is rich in anti-oxidants, preventing cancerous cell development. Gingerol showed strong anticancer activity against human retinoblastoma cancer cells (RB355) and this effect was mediated by apoptosis induction, cell cycle arrest, and modulation of the PI3K/Akt signaling pathway (Meng et al., 2018).

8.3 CONCLUSION

Although the present deliberation includes eleven plants belonging to various families, genera, and species in nature, many more plants, and their parts exist, which may be playing an important role as anti-cancerous agents. Potentially, aqueous extracts of medicinal plants, could be considered as prospective alternatives as therapeutic agents (Kurhekar et al., 2013). Besides, the plant as a whole has many parts, each of which is significant, in its own way. Every one of the above plant, has at least one part, which is effective in one way or the other. In addition, the method of preparation of extracts is different for different plants. That is another issue, which can be tackled separately. Each of the above plant is unique, in its own way, helping us to combat cancer. What is necessary is a more detailed study on how and in what amount; these plants will be useful, for use as anti-cancerous agents? In the future, with the threats of the severe side effects caused by chemotherapy and radiotherapy, on the human body, a return to nature cure or naturotherapy, may prove to be a better alternative!

KEYWORDS

- **cancer**
- **medicinal plants**
- **naturotherapy**
- **reasonless division**
- **side effects**
- **systematic process**

REFERENCES

Adinarayana, D., & Seshadri, T. R., (1965). Chemical investigation of the stem-bark of Cedrusdeodara. *Tetrahedron, 21*(12), 3727–3730.

Agarwal, P. K., & Rastogi, R. P., (1981). Terpenoids from Cedrusdeodara. *Phytochemistry, 20*(6), 1319–1321.

Agrawal, P. K., & Rastogi, R. P., (1982). Two lignans from Cedrus deodar. *Phytochemistry, 21*(6), 1459–1461.

Agrawal, P. K., Agarwal, S. K., & Rastogi, R. P., (1980). Dihydroflavonols from Cedrusdeodara. *Phytochemistry, 19*(5), 893–896.

Aizpurua-Olaizola, O., Ormazabal, M., Vallejo, A., Olivares, M., Navarro, P., Etxebarria, N., & Usobiaga, A., (2015). Optimization of supercritical fluid consecutive extractions of fatty acids and polyphenols from Vitis vinifera grape wastes. *Journal of Food Science, 80*(1), E101–E107.

An, K., Zhao, D., Wang, Z., Wu, J., Xu, Y., & Xiao, G., (2016). Comparison of different drying methods on Chinese ginger (Zingiber officinale Roscoe): Changes in volatiles, chemical profile, antioxidant properties, and microstructure. *Food Chem., 197*(Part B), 1292–1300. PMID 26675871.

Ansari, N. A., (2007). Onion cultivation and production in Iran. *Middle Eastern and Russian Journal of Plant Science and Biotechnology, 1*(2), 26–38.

Bat-Chen, W., Golan, T., Peri, I., Ludmer, Z., & Schwartz, B., (2010). Allicin purified from fresh garlic cloves induces apoptosis in colon cancer cells via Nrf2. *Nutr. Cancer, 62*(7), 947–957.

Block, E., (1985). The chemistry of garlic and onions. *Sci. Am., 252*(3), 114–119.

Boudreau, M. D., & Beland, F. A., (2006). An evaluation of the biological and toxicological properties of Aloe barbadensis (Miller), Aloe Vera. *J. Environ Sci. Health C Environ. Carcinog. Ecotoxicol. Rev., 24*(1), 103–154.

Brickell, C., (1992). *The Royal Horticultural Society Encyclopedia of Gardening* (p. 345). Dorling Kindersley.

Burns, J., Yokota, T., Ashihara, H., Lean, M. E., & Crozier, A., (2015). Plant foods and herbal sources of resveratrol. *J. Agric. Food Chem., 50*, 3337–3340. [PubMed], Catharanthusroseus. *Orpheus Island Research Station-James Cook University.*

Chockalingam, V., Suryakiran, K. S. D. V., & Pratheesh, G., (2002). Antiproloferative and antioxidant activity of *Aegle marmelos* (Linn) leaves in Dalton's lymphoma ascites transplanted mice. *Indian Journal of Pharmacology, 44*(2), 225–229.

Choi, Y. H., & Park, H. S., (2012). Apoptosis induction of U937 human leukemia cells by diallyl trisulfide induces through generation of reactive oxygen species. *J. Biomed. Sci.*, 19.

Cooper, R., & Deakin, J. J., (2016). Africa's gift to the world. *Botanical Miracles: Chemistry of Plants That Changed the World* (pp. 46–51). CRC Press.

Cragg, G. M., Simon, J. E., Jato, J. G., & Snader, K. M., (1996). Drug discovery and development at the national cancer institute: Potential for new pharmaceutical crops. In: Janick, J., (ed.), *Progress in New Crops* (pp. 554–560) ASHS Press, Arlington, V. A.

Cumo, C. E., (2015). Onion. In: Foods That Changed History: How Foods Shaped Civilization from the Ancient World to the Present (pp. 248–250). ABC-CLIO LLC (American Bibliographic Center, CLIO Press).

Das, B., Venkakaiah, B., & Das, R., (2001). Lignans: Promising anti-cancer agents. In: Role of Biotechnology in Medicinal and Aromatic Plants, Special Volume on Diseases (pp. 42–49). Ukaaz Publication, Hyderabad, IV.

Dredge, K., Dalgleish, A. G., & Marriott, J. B., (2003). Angiogenesis inhibitors in cancer therapy. *Curr. Opin. Investig. Drugs, 4,* 667–774.

Dweck, A. C., (1994). The Green Pharmacy Herbal Handbook. www.mothernature.com (accessed on 20 February 2020).

Everett, & Thomas, H., (1982). The New York Botanical Garden Illustrated Encyclopedia of Horticulture (Vol. 10, p. 3591). Taylor & Francis.

Gorter, (1911). *Rec. Trav., Chim, Pays-Bas, 30,* 151.

Greenspan, H. C., & Aruoma, O. I., (1994). Oxidative stress and apoptosis in HIV infection, role of plant-derived metabolites with synthetic antioxidant activity. *Immunology Today, 15,* 209–213.

Guercio, V., Turati, F., La Vecchia, C., Galeone, C., & Tavani, A., (2016). Allium vegetables and upper aerodigestive tract cancers: A meta-analysis of observational studies. *Molecular Nutrition and Food Research, 60*(1), 212–222.

Hallock, Y. F., Cardillina, J. H., Schaffer, M., Bringmann, G., Francois, G., Boyd, M. R., & Korundamine, A., (1998). A novel HIV-inhibitory and anti-malarial hybrid naphthyl isoquinoline alkaloid heterodimer from anistrocladuskorupensis. *Bio Org. Med. Chem. Lett., 8,* 1729–1734.

Havsteen, B., (1983). Flavonoids, a class of natural products of high pharmacological potency. *Biochem. Pharmacol., 32,* 1141–1148.

Hong, S. L., Lee, G. S., Syed, A. R. S. N., Ahmed, H. O. A., Awang, K., Aznam, N. N., & AbdMalek, S. N., (2014). Essential oil content of the rhizome of Curcumapurpurascens Bl. (Temu Tis) and its antiproliferative effect on selected human carcinoma cell lines. *The Scientific World Journal, 397–430.*

Hossain, M. S., Urbi, Z., Sule, A., & Hafizur, R. K. M., (2014). Andrographispaniculata (Burm. f.) Wall. Ex Nees: A review of ethnobotany, phytochemistry, and pharmacology. *Scientific World Journal.*

Hu, Y., Kong, W., Yang, X., Xie, L., Wen, J., & Yang, M., (2014). GC-MS combined with chemometric techniques for the quality control and original discrimination of curcumaelongae rhizome: Analysis of essential oils. *Journal of Separation Science, 37*(4), 404–411.

Kaur, M., Chapla, A., & Rajesh, A., (2009). Anticancer and cancer chemopreventive potential of grape seed extract and other grape-based products. *J. Nutr., 139*(9), 1806S–1812S.

Keglevich, P., Hazai, L., Kalaus, G., & Szantay, C., (2012). Modifications on the basic skeletons of vinblastine and vincristine. *Molecules, 17, 5893–5914.*

King, G. K., Yates, K. M., Greenlee, P. G., Pierce, K. R., Ford, C. R., McAnalley, B. H., & Tizard, I. R., (1995). The effect of acemannan immunostimulant in combination with surgery and radiation therapy on spontaneous canine and feline fibro sarcomas. *J. Am. Anim. Hosp. Assoc., 31*(5), 439–447.

Kumar, P., Kadakol, A., Shasthrula, P. K., Mundhe, N. A., Jamdade, V. S., Barua, C. C., & Gaikwad, A. B., (2015). Curcumin as an adjuvant to breast cancer treatment. *Anticancer Agents Med Chem., 15*(5), 647–656.

Kumar, R. A., Sridevi, K., Kumar, N. V., Nanduri, S., & Rajagopal, S. J., (2004). *Ethnopharmacol., 92*(2/3), 291–295.

Kurhekar, J. V., (2013). Comparative evaluation of response of burn wound pathogens to medicinal plant extracts and standard antibiotics. *Indian Streams Research Journal, 3*(7), 1–3.

Kurhekar, J. V., (2014). Conservation of biodiversity with reference to indigenous herbal therapeutic agents. *Journal of Applied and Environmental Microbiology, 2*, 42–45,

Kurian, J. C., (1999). *Plants That Heal*. Oriental Watchman Publishing House, Pune: 14, 15, 38, 63, 112, 117, 138, 111.

Laphookhieo, S., (2011). Chemical constituents from Aegle marmelos. *J. Braz. Chem. Soc., 22, 176–178.*

Lin, H., Lin, J., & Ma, J., (2015). Demethoxycurcumin induces autophagic and apoptotic responses on breast cancer cells in photodynamic therapy. *J. Funct. Foods, 12,* 439–449.

Malik, F. A., Kumar, A., Bhushan, S., Khan, S., Bhatia, A., Suri, K., Qazi, G. N., & Singh, J., (2007). *Apoptosis, 12*(11), 2115–2133.

Malik, F. B., Singh, J., Khajuria, A., Suri, K. A., Satti, N. K., Singh, S., Kaul, M. K., Kumar, A., Bhatia, A., & Qazi, G. N., (2007). *Life Sci., 80*(16), 1525–1538.

Mandal, S. S., Akhtar, S. A., Pandey, A., Verma, D. K., & Sinha, N. K., (2002). Plant biotechnology in modeling of biopharmaceuticals for cancer-role of biotechnology in medicinal and aromatic plants. *Special Volume on Diseases* (p. 235). Ukaaz Publication, Hyderabad, VI.

Matsuda, T., Kuroyanagi, M., Sugiyama, S., Umehara, K., Ueno, A., & Nishi, K., (1994). *Chem. Pharm. Bull., (Tokyo), 42*(6), 1216–1225.

McGowan, C., (2008). *The Deodar Tree: The Himalayan "Tree of God."* Cited in Wikipedia (https://en.wikipedia.org/wiki/Cedrus_deodara#cite_ref-gowan_4-0).

Meng, B., Ii, H., Qu, W., & Yuan, H., (2018). Anticancer effects of gingerol in retinoblastoma cancer cells (RB355 Cell Line) are mediated via apoptosis induction, cell cycle arrest and up regulation of PI3K/Akt signaling pathway. *Med. Modi., 24,* 1980–1987.

Mirjalili, M. H., Moyano, E., Bonfill, M., Cusido, R. M., & Palazón, J., (2009). Steroidal lactones from Withaniasomnifera, an ancient plant for novel medicine. *Molecules, 14(7), 2373–2393.*

Mondal, P., & Mukherjee, P. K., (1992). Notes of ethnobotany of Keonjhar dist., Orissa. *J. Econ. Tax. Bot. Addl. Ser., 10,* 7–18.

Nelson, K. M., Dahlin, J. L., & Bisson, J., (2017). The essential medicinal chemistry of curcumin: Mini perspective. Journal of Medicinal Chemistry, 60(5), 1620–1637.

Niciforovic, A., Adzic, M., Zabric, B., & Radojcic, M. B., (2007). Adjuvant antiproliferative and cytotoxic effect of aloin in irradiated HeLaS3 cells. *Biophys. Chem., 81,* 1463–1466.

Orwa, C., (2009). Aeglemarmelos. Agrofores Tree Database: A Tree Reference and Selection Guide Version 4.0.

Pandit, S., Chang, K. W., & Jeon, J. G., (2013). Effects of Withania somnifera on the growth and virulence properties of Streptococcus mutans and Streptococcus sobrinus at sub-MIC levels. *Anaerobe, 19,* 1–8.

Pecere, T., Gazzola, M. V., Mucignat, C., Parolin, C., Dalla, V. F., Cavaggioni, A., Basso, G., Diaspro, A., Salvato, B., Carli, M., & Palu, G., (2000). Aloe-emodin is a new type of anticancer agent with selective activity against neuroectodermal tumors. *Cancer Res., 60,* 2800–2804.

Priyadarsini, K. I., (2014). The chemistry of curcumin: From extraction to therapeutic agent. *Molecules, 19*(12), 20091–20112.

Puri, A., Saxena, R., Saxena, R. P., Saxena, K. C., Srivastava, V., & Tandon, J. S., (1993). *J. Nat. Prod., 56*(7), 995–999.

Rahman, S., & Parvin, R., (2014). Therapeutic potential of *Aegle marmelos* (L.): An overview. *Asian Pac J. Trop. Dis., 4*(1), 71–77.

Rao, K. V., Lakshmi, N. M., & Kavi, K. P. B., (1997). Anti malarials of plant origin. In: *Role of Biotechnology in Medicinal and Aromatic Plants, Special Volume on Diseases* (pp. 34–49). Ukaaz Publication, Hyderabad, VI.

Raul, U. B., & Heble, M. R., (2001). Plant constituents in AIDS therapy, In: Role of Biotechnology in Medicinal and Aromatic Plants, Special Volume on Diseases, Ukaaz Publication (pp. 1–14). Hyderabad, VI.

Ravanbakhshian, R., & Mandana, B., (2018). Evaluation of anticancer activity of lacto- and natural fermented onion cultivars. *Iranian Journal of Science and Technology, Transactions A: Science, 42*(4), 1735–1742.

Ravindran, J., Sahdeo, P., & Bharat, B., A., (2009). Curcumin and cancer cells: How many ways can curry kill tumor cells selectively? *A.A.P.S.J., 11*(3), 495–510. Published online 2009 Jul 10. PMCID: PMC2758121.

Redkar, R. G., & Jolly, C. I., (2003). Natural products as anticancer agents. *Ind. Drugs, 40*, 619–626.

Scarlatti, F., Maffei, R., Beau, I., Codogno, P., & Ghidoni, R., (2008). Role of non-canonical Beclin 1-independent autophagy in cell death induced by resveratrol in human breast cancer cells. *Cell Death Differ., 15*, 1318–1329.

Scott, H., & Kenneth, M., (1946). *Western Arabia and the Red Sea, Naval Intelligence Division* (p. 597). London. ISBN 0-7103-1034-X.

Shashi, S., Jaswant, S., Madhusudana, R. J., Kumar, S. A., & Nabi, Q. G., (2006). *Nitric Oxide, 14*, 72–88.

Shetgiri, P. P., & D'Mello, P. M., (2003). Antioxidant properties of flavonoids, a comparative study. *Indian Drugs, 40*, 567–569.

Singh, J., (2007). In indo-US symposium on Botanicals organized by CSIR, IIIM, Jammu & NCNPR, University of Mississippi. New Delhi: IGH, NASC complex, 2007. A Novel Standardized Herbal Formulation of *Withania somnifera* Useful for Anti-Cancer Land Th-1 Immune Upregulation. *Indian Patent: 0202NF2006*, Del 01321 dated 19.06.2007.

Siyo, V., Schafer, G., Hunter, R., Grafov, A., Grafova, I., Nieger, M., Katz, A. A., Parker, M. I., & Kaschula, C. H., (2017). The cytotoxicity of the ajoene analog BisPMB in WHCO1 oesophageal cancer cells is mediated by CHOP/GADD153. *Molecule, 22*(6), 892–911.

Tariq, H. A., Kandil, O., Elkadi, A., & Carter, J., (1988). Garlic revisited: Therapeutic for the major diseases of our times. *J. Natl. Med. Assoc., 80*, 439–445.

Thomson, R. H., (1971). *Naturally Occurring Quinines* (2nd edn.). Academy Press, London.

Wang, H., Tin, OoKhor., Limin, S., Zhengyuen, S., Francisco, F., Jong-Hun, L., Ah-Ng, & Tony, K., (2012). Plants against cancer: A review on natural phytochemicals in preventing and treating cancers and their druggability. *Anticancer Agents Med. Chem., 12*(10), 1281–1305.

Wilken, R., Veena, M. S., Wang, M. B., & Srivatsan, E. S., (2011). Curcumin: A review of anti-cancer properties and therapeutic activity in head and neck squamous cell carcinoma. *Mol. Cancer*, 10–12.

Yadav, B., Bajaj, A., Saxena, M., & Saxena, A. K., (2010). *In vitro* anticancer activity of the root, stem and leaves of *Withania somnifera* against various human cancer cell lines. *Indian J. Pharm. Sci., 72*(5), 659–663.

Yodkeeree, S., Chaiwangyen, W., Garbisa, S., & Limtrakul, P., (2009). Curcumin, demethoxycurcumin, and bisdemethoxycurcumin differentially inhibit cancer cell invasion through the down-regulation of MMPs and UPA. *J. Nutr. Biochem.*, *20*, 87–95.

Yodkeree, S., Ampasavate, C., Sung, B., Aggarwal, B. B., & Limtrakul, P., (2010). Demethoxycurcumin suppresses migration and invasion of MDA-MB-231 human breast cancer cell line. *Eur. J. Pharmacol.*, *627*, 8–15.

Zahid, M., Gaikwad, N. W., Ali, M. F., Lu, F., Saeed, M., Yang, L., Rogan, E. G., & Cavalieri, E. L., (2008). Prevention of estrogen-DNA adducts formation in MCF-10F cells by resveratrol. *Free Radic. Biol. Med.*, *45*, 136–145.

Zeng, Wei-Cai., Zhang, Z., Gao, H., Jia, Li-Rong., & He, Q., (2012). Chemical Composition, antioxidant, and antimicrobial activities of essential oil from pine needle (Cedrusdeodara). *Journal of Food Science,* *77*(7), C824–829.

Zhang, S., Won, Y. K., Ong, C. N., & Shen, H. M., (2005). Anti-cancer potential of sesquiterpene lactones: Bioactivity and molecular mechanisms. *Curr. Med. Chem. Anticancer Agents*, *5*(3), 239–249.

Zhou, X. F., Ding, Z. S., & Liu, N. B., (2013). Allium vegetables and risk of prostate cancer: Evidence from 132,192 subjects. *Asian Pacific Journal of Cancer Prevention: APJCP,* *14*(7), 4131–4134.

Zhou, Y., Zhuang, W., Hu, W., Liu, Guan-Jian., Wu, Tai-Xiang., & Wu, Xiao-Ting., (2011). Consumption of large amounts of allium vegetables reduces risk for gastric cancer in a meta-analysis. *Gastroenterology, 141*(1), 80–89.

CHAPTER 9

Cytotoxic, Apoptosis Inducing Effects and Anti-Cancerous Drug Candidature of Jasmonates

PARTH THAKOR,[1] RAMALINGAM B. SUBRAMANIAN,[1]
SAMPARK S. THAKKAR,[2] and VASUDEV R. THAKKAR[1]

[1]P.G. Department of Biosciences, Sardar Patel Maidan, Bakrol-Vadtal Road, Satellite Campus, Bakrol, Sardar Patel University, Vallabh Vidyanagar, Gujarat, India, E-mails: parth7218@gmail.com (T. Parth), vasuthakkar@gmail.com (T. Vasudev)

[2]Department of Advanced Organic Chemistry, P.D. Patel Institute of Applied Sciences, Charusat, Changa – 388421, Gujarat, India

ABSTRACT

Natural products have tremendous potential as active pharmaceutically lead molecules. Jasmonates are such molecules widely distributed in plant kingdom. They have reported for myriad activities, as signaling molecules to antiproliferative compounds. Cytotoxic and apoptosis inducing effects of Jasmonates on (*Schizosaccharomyces pombe*) fission yeast cells was carried out in the present study. Moreover, its cytotoxic potential was further checked on three renal cell carcinoma cell lines (ACHN, Caki-1, SN12K1) and normal cell line (HK-2). Jasmonic acid (JA) showed characteristic apoptosis inducing effect at IC_{50} concentration in *S. pombe* cells. The ability of three jasmonates (JA, cis-jasmone (CJ), methyl jasmonate (MJ)) to become potential drug candidate has revealed from the pharmacokinetic study.

9.1 INTRODUCTION

Natural products have regarded as sources of potential chemotherapeutic agents (Tan et al., 2006). Over 50% of the drugs in clinical trials for anticancer properties were isolated from natural sources (Cragg and Newman, 2000). Vinblastine, Vincristine, Navelbine, Etoposide, Teniposide, Taxol, Taxotere, Topotecan, and Irinotecan are clinically approved anticancer drugs derived from medicinal plants. Secondary metabolites derived from fungi have also been contributed significantly to the field of medicine. Array of fungal compounds including Brefeldin A, Cytochalasin E, Gliotoxin, Irofulven, Leptomycins, Tricyclic acid A and Wortmannin have been investigated for their anticancer activity (Cragg and Newman, 2005; Sulkowska-Ziaja K et al., 2005).

Jasmonates have reported for *in vitro* and *in vivo* anticancer activities (Flescher, 2007). Jasmonate family consists of cis-jasmone (CJ), jasmonic acid (JA), methyl jasmonate (MJ) and fatty acid-derived cyclo-pentanones (Sembdner and Parthier, 1993). Jasmonates and their synthetic derivatives reported for inhibition of cell proliferation alongside with the induction of cell death in various breast, prostate, melanoma, lymphoblastic leukemia and lymphoma cells (Fingrut and Flescher, 2002). Moreover, Jasmonates reported for selective cytotoxicity towards cancer cells even when they mixed with leukemic and normal cells (Flescher, 2005). Jasmonates are plant stress hormones and occur ubiquitously in the plant kingdom.

In this work, we have studied cytotoxic and apoptosis-inducing effects of jasmonates on *Schizosaccharomyces pombe* cells and further confirmed their anticancer effects on RCC cell lines, normal cell line and evaluated pharmacokinetic properties for its drug candidature. RCC is the most common type of kidney cancer. It is standing in third position after prostate and bladder cancer in the category of genitourinary cancer. RCCs are resistant to many types of the cancer treatments. RCC is enormously metastatic and assorted disease with 16 different histologic subtypes (Lopez-Beltrana et al., 2006). There is an immense need for the investigation of the alternative effective compound in RCC treatment along with its drug candidature.

9.2 MATERIALS AND METHODS

9.2.1 CHEMICALS

JA, MJ, and CJ (≥97% purity) compounds were purchased from Sigma-Aldrich (St. Louis, MO, USA). Compounds were diluted further in cell culture medium to the desired concentration.

9.2.2 EFFECT OF JA ON VIABILITY OF THE S. POMBE CELLS

For the determination of the IC_{50} value of JA, treatment of JA in varying concentration (0.5–10 mM) was given to the *S. pombe* cells. IC_{50} value calculated from the dose-response curve generated in the GraphPad prism 6.

9.2.3 STUDY OF THE APOPTOTIC TRAITS OF JA ON S. POMBE CELLS

9.2.3.1 EFFECT OF JA ON THE INTEGRITY OF THE DNA OF S. POMBE CELLS

The alkali lysis method along with the mechanical shearing with glass beads employed for the genomic DNA isolation of treated and untreated *S. pombe* cells (Patel and Thakkar, 2013) and electrophoresed on 1% agarose gel. Images of DNA captured in BioRad gel doc system. Treatment time for the experiment was 17–18 hours.

9.2.3.2 AO/ETBR STAINING AND DAPI STAINING

To check whether JA induces apoptosis or not, untreated, and JA-treated cells in the exponential phase of growth were harvested and washed with 1x PBS. Apoptotic traits were analyzed with the help of AO/EtBr staining as described by Thakor et al. (2016) In addition to this DAPI staining was also performed to determine the effect of the JA on nuclei fragmentation by the method described by Thakor et al. (2016).

9.2.4 CELL LINES AND CELL CULTURE CONDITIONS

Human RCC cell lines (ACHN, Caki-1, SN12K1) and normal kidney epithelial cell line (HK-2) were maintained as a monolayer in DMEM/F12 (Gibco, Invitrogen, CA, USA) containing 10% fetal bovine serum (Gibco, Invitrogen, CA, USA) supplemented with 50 U/ml Penicillin, 50 µg/ml (streptomycin) at 37°C in a humidified atmosphere of 5% CO_2/95% air (Thakor et al., 2017a).

9.2.4.1 MEASUREMENT OF ANTIPROLIFERATIVE ACTIVITY OF JASMONATES BY MTT ASSAY

In brief, per well 5000 cells seeded in 96 well culture plates. After 24 hours, cells were treated with the various concentrations of jasmonates (JA, MJ, CJ) to find out IC_{50} value. After 24 hours of the treatment period, 5 µl of MTT was added to each well and incubated further for ninety minutes at 37°C in CO_2 incubator. The volume of culture was 100 µl/well. After the incubation period, the culture medium was discarded and the purple crystals formed were dissolved in 100 µl of dimethyl sulfoxide (DMSO). The absorbance was recorded with the help of microplate reader at 570 nm with a reference wavelength of 690 nm (Thakor et al., 2017a).

9.2.5 IN SILICO PHARMACOKINETIC EVALUATION

The pharmacokinetic parameters of JA, MJ, and CJ were obtained using programs Vlife Molecular Design Suit (MDS) 4.6 (Vlife Sciences, Pune, India) and ADMET software. These programs computed pharmacokinetic parameters such as logP, molecular weight (MW), PSA, HBA, HBD, BBB, %HIA, and %PPB (Thakor et al., 2017b; Thakkar et al., 2017a, b).

9.3 RESULTS

9.3.1 EFFECT OF JA ON VIABILITY OF S. POMBE CELLS

The cytotoxic effect of various concentrations of JA was studied on *S. pombe* cells by trypan blue viability assay. The percentage viability was found to be 67% at 5 mM concentration and it decreased with the increase in the concentration of JA. IC_{50} value of the JA was calculated from the dose-response curve, which was found to be 9.65 mM (Figure 9.1).

9.3.2 STUDY OF APOPTOTIC TRAITS OF JA

To investigate the mechanism of cytotoxicity of JA on *S. pombe* cells, whether JA is inducing apoptosis or not was studied. As it is a well-known fact that apoptosis involves fragmentation of DNA before cell death, DNA isolated from JA-induced and control cells of *S. pombe* were electrophoresed on agarose gel. It was found that the integrity of DNA was affected by the treatment of JA. Untreated cells showed the genomic DNA as an intact band, while DNA from JA-treated *S. pombe* cells showed a smear on an agarose gel (Figure 9.2).

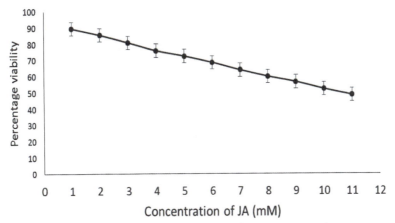

FIGURE 9.1 Percentage viability of *S. pombe* cells in response to various concentrations of JA.

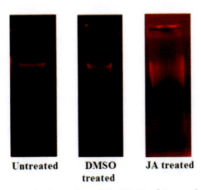

FIGURE 9.2 Effect of JA on the integrity of the DNA of *S. pombe* cells.

Effect of JA on the integrity of DNA in *S. pombe* cells was also confirmed by AO/EtBr staining. AO stained cells showed green fluorescence under the fluorescence microscope at 40x. EtBr binds with the DNA and EtBr stained *S. pombe* cells showed red or orange color; while AO-EtBr combined staining revealed the difference between the live and apoptotic cells. Live cells showed intact chromatin while apoptotic cells showed condensed or fragmented chromatin (Figure 9.3).

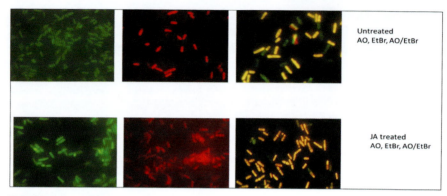

FIGURE 9.3 AO/EtBr staining of *S. pombe* cells.

DAPI (4,6-diamidino-2-phenylindole) staining was also performed to determine the effect of JA on nuclei fragmentation. The DAPI bound DNA region showed blue fluorescence. Untreated *S. pombe* cells showed densely stained nuclei with an intact DNA, while JA-treated *S. pombe* cells showed fragmented nuclei with degraded DNA (Figure 9.4). All these combined results on the integrity of DNA confirmed that JA induces fragmentation of DNA and its mechanism of cytotoxicity is by inducing apoptosis. As apoptosis-inducing compounds are useful for the treatment of cancer, we thought to study the ability of JA as an antineoplastic agent. Therefore, we studied the effect of JA and its analogs on RCC cell line along with the normal cell line.

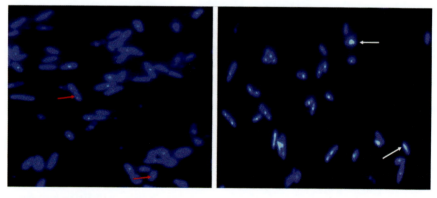

FIGURE 9.4 DAPI staining of *S. pombe* cells. Red arrows indicate the intact nuclei while white arrows indicate the damaged nuclei.

9.3.3 CYTOTOXIC EFFECT OF JASMONATES (JA, MJ, CJ) ON RCC CELL LINES

As JA showed good apoptosis-inducing ability on *S. pombe* cells, we were interested in studying cytotoxicity of JA and its analogs (MJ and CJ) on normal as well as on RCC cell lines. Jasmonates did not show any effect on HK-2, the most sensitive normal cells. Various concentrations of MJ, CJ, and JA (0.27 mM to 0.0042 mM of MJ, 0.30 mM to 0.0047 mM of CJ and 0.31 mM to 0.0024 mM of JA) were applied in MTT assay. At these concentrations, none of the jasmonates showed any cytotoxicity. As a result, higher concentration of jasmonates ranging between 0.5 mM to 5 mM was applied further (Figure 9.5). All Jasmonates except JA showed cytotoxic effect in all three RCC cell lines. Their respective IC_{50} values are summarized in Table 9.1. As JA did not show cytotoxicity even at higher concentration, its further study was skipped.

TABLE 9.1 IC_{50} Values of Jasmonates on RCC and HK-2 Cells

Compound Name	ACHN, Caki-1, SN12K1
Jasmonic Acid	At 2.5 mM concentration no effect
Methyl Jasmonate	2.7 mM
Cis-Jasmone	2.4 mM

All three Jasmonates have no effect on HK-2 cells.

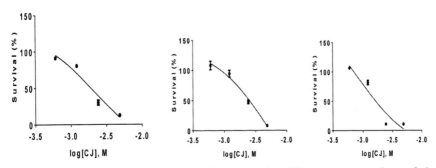

FIGURE 9.5 Percentage survival of cells against the different concentrations of the jasmonates in three different cell lines.

9.3.4 PHARMACOKINETIC EVALUATION OF JASMONATES

To avoid late-stage failure, the preliminary pharmacokinetic study is important for molecules which are expected to be a drug. The pharmacokinetic parameters were obtained by Vlife MDS 4.6 and ADMET software. The data found are presented in Tables 9.2 and 9.3.

TABLE 9.2 Evaluation of Pharmacokinetic Parameters

Compound ID(s)	Mol. Wt	HBA	HBD	RotB	logP	PSA (°A²)
Jasmonic acid	210.13	3	1	5	2.39	42.43
Methyl Jasmonate	224.14	3	0	6	2.74	34.62
Cis Jasmone	164.12	1	0	3	2.92	14.04

TABLE 9.3 Evaluation of ADMET Parameters

Compound ID(s)	BBB	Caco$_2$ (nm/sec)	% HIA	%PPB	MDCK (nm/sec)	hERG Inhibition
Jasmonic acid	1.5512	15.0291	93.9440	73.6983	85.0512	Low Risk
Methyl Jasmonate	0.2760	22.1524	97.6172	78.0208	69.5846	Low Risk
Cis Jasmone	1.6597	53.5917	100.000	95.7471	68.0596	Low Risk

BBB (blood-brain-barrier); High absorption CNS >2.0; Middle absorption CNS 2.0–0.1; Low absorption to CNS <0.1; CaCo$_2$; High permeability >70; Middle permeability 4–70; Low permeability <4; %HIA (human intestinal absorbance); Well absorbed compounds 70–100%; Moderately absorbed compounds 20–70%; Poorly absorbed compounds 0–20%; %PPB (plasma protein binding); Strongly Bound >90%; Weakly Bound <90%, MDCK: Higher permeability >500; Medium Permeability 25–500; lower permeability <25.

9.4 DISCUSSION

This is the first report of the effect of JA on *S. pombe*. In the past, many research groups studied the effect of all jasmonates (JA, CJ, MJ) on several cancer cell lines. All of them reported the IC$_{50}$ for JA and other jasmonates (i.e., MJ, and CJ) in the range of 2–4 mM. No matter to the higher IC$_{50}$; we extended the study of JA as an apoptosis-inducing agent on model organism *S. pombe*. JA-treated *S. pombe* cells (with IC$_{50}$) showed the characteristic genomic DNA degradation pattern. Moreover, results of AO/EtBr and DAPI staining of JA-treated *S. pombe* cells showed the nuclei fragmentation. Such apoptotic traits were not observed in untreated *S. pombe* cells. Although, the IC$_{50}$ value and effect of the compound varied from cell type to cell type and organism to organism. Thus, we contemplated the effect of JA and its structural analogs (jasmonates) on RCC cell lines. MTT assay was performed after 24 h of the treatment to check out the dosage concentration of jasmonates on respective cell lines. Different IC$_{50}$ values represented the response of cell types to the drug. However, the effect of jasmonates on RCC has not been addressed yet. Jasmonates studied in the present study did not show any cytotoxic effect on normal cells. HK-2 cells are the normal cells. Jasmonates showed the cytotoxic effect on RCC cell lines which supported the potential of jasmonates as anticancer agents. Jasmonates exhibited selective cytotoxicity towards cancer cells even when they were mixed with leukemic and normal cells made from the origin of chronic lymphocytic leukemia (CLL) patients (Fingrut et al., 2002; Flescher, 2005). Thus, jasmonates could be utilized in the treatment of cancer without affecting the normal cells.

Lipinski's rule of 5 is a combination of molecular descriptors created by Lipinski (Thakor et al., 2017b; Thakkar et al., 2017a; Karad et al., 2017). The rule indicates that an orally active drug has no more than one violation of following criteria: logP $\leq$ 5, MW (Mol. Wt.) $\leq$ 500, number of hydrogen bond acceptors (HBA) $\leq$ 10, number of hydrogen bond donors (HBD) $\leq$ 5, rotatable bonds (RotB) $\leq$ 10, polar surface area (PSA) $\leq$ 140°A^2. None of the compounds selected for this study showed any violation (Vide Table 9.2).

The parameters obtained from ADMET software was compared with those reported (Thakkar et al., 2017b; Malani et al., 2017; Thakor, 2017). The low value for BBB in molecule MJ suggested low permeability of this molecule to CNS. The molecules JA and CJ have a moderate value of BBB indicating more ability of these molecules to cross BBB and consequently greater transport of them to CNS (Vide Table 9.3).

The high %HIA values suggest higher absorption of the molecules in the intestine. CJ had good CaCo$_2$ permeability while JA and MJ had low permeability. The higher values of %PPB suggest that the compound. CJ can bind strongly to plasma protein. Values of MDCK represent the excretion of the drugs by kidney (Table Vide 9.3). All the molecules had reasonably good MDCK permeability. hERG is best known for its contribution to the electrical activity of the heart that coordinates the heart's beating. In the present investigation, all the molecules had low hERG inhibition risk.

9.5 CONCLUSION

The present study on effect of jasmonates on *S. pombe* cells reports that jasmonates have anticancerous effects on RCC cell lines by inducing apoptosis. Although, anticancerous effect of jasmonates has been reported in other cell lines, the present study reveals its therapeutic potential is limited because jasmonates attain therapeutic benefit at higher doses (in mM range concentrations). Pharmacokinetic evaluation of jasmonates supports its drug candidature.

ACKNOWLEDGMENTS

The authors are thankful to the P. G. Department of Biosciences for the necessary support. Authors are thankful to Prof. Arabinda Ray, PDPIAS, CHARUSAT, Changa for support in the pharmacokinetic evaluation of the compounds. Mr. Parth Thakor is thankful to DST inspire program for the fellowship (DST INSPIRE FELLOWSHIP 2013/44).

KEYWORDS

- **anticancerous effect**
- **apoptosis**
- **cytotoxic activity**
- **jasmonates**
- **pharmacokinetic parameters**
- **renal cell carcinoma**

REFERENCES

Cragg, G. M., & Newman, D. J., (2000). Antineoplastic agents from natural sources: Achievements and future directions. *Expert Opin. Investig. Drugs.*, 9, 2783–2797.

Cragg, G. M., & Newman, D. J., (2005). Plants as a source of anti-cancer agents. In: Elaine, E., & Nina, L. E., (eds.), *Ethno Pharmacology*. In the encyclopedia of life support systems (EOLSS), developed under the Auspices of the UNESCO, Eolss Publishers, Oxford, UK. http://www.eolss.net (accessed on 22 February 2020).

Fingrut, O., & Flescher, E., (2002). Plant stress hormones suppress the proliferation and induce apoptosis in human cancer cells. *Leukemia*, 16, 608–616.

Flescher, E., (2005). Jasmonates—A new family of anti-cancer agents. *Anti-Cancer Drugs*, 16, 911–916.

Flescher, E., (2007). Jasmonates in cancer therapy. *Cancer Lett.*, 245(1/2), 1–10.

Karad, S. C., Purohit, V. B., Thummar, R. P., Vaghasiya, B. K., Kamani, R. D., Thakor, P., Thakkar, V. R., Thakkar, S. S., Ray, A., & Raval, D. K., (2017). Synthesis and biological screening of novel 2-morpholinoquinoline nucleus clubbed with 1,2,4-oxadiazole motifs. *Eur. J. Med. Chem., 126*, 894–909.

Lopez-Beltrana, A., Scarpelli, M., Montironi, R., & Kirkali, Z., (2006). WHO classification of the renal tumors of the adults. *Eur. Urol., 49*(5), 798–805.

Malani, K., Thakkar, S. S., Thakur, M. C., Ray, A., & Doshi, H., (2016). Synthesis, characterization and *in silico* designing of diethyl-3-methyl-5-(6-methyl-2-thioxo-4-phenyl-1,2,3,4-tetrahydropyrimidine-5-carboxamido)thiophene-2,4-dicarboxylate derivative as anti-proliferative and anti-microbial agents. *Bioorg. Chem., 68*, 265–274.

Patel, P. B., & Thakkar, V., (2013). Cell proliferation and DNA damage study by SCGE in fission yeast exposed to curcumin and 5-fluorouracil. *Asian Journal of Cell Biology, 8*(1), 22–32.

Sembdner, G., & Parthier, B., (1993). The biochemistry and the physiological and molecular actions of jasmonates. *Annu. Rev. Plant Physiol. Plant Mol. Bio., 144*, 569–589. doi: 10.1146/annurev.pp.44.060193.003033.

Sulkowska-Ziaja, K., Muszynska, B., & Konska, G., (2005). Biologically active compounds of fungal origin displaying antitumor activity. *Acta Pol. Pharm., 62*, 153–159.

Tan, G., Gyllenhaal, C., & Soejarto, D. D., (2006). Biodiversity as a source of anticancer drugs. *Curr. Drug Targets, 7*, 265–277.

Thakkar, S. S., Thakor, P., Doshi, H., & Ray, A., (2017a). 1,2,4-triazole and 1,3,4-oxadiazole analogs: Synthesis, MO studies, *in silico* molecular docking studies, antimalarial as DHFR inhibitor and antimicrobial activities. *Bio. Org. Med. Chem., 25*(15), 4064–4075.

Thakkar, S. S., Thakor, P., Ray, A., Doshi, H., & Thakkar, V. R., (2017b). Benzothiazole analogs: Synthesis, characterization, MO calculations with PM6 and DFT, *in silico* studies and *in vitro* antimalarial as DHFR inhibitors and antimicrobial activities. *Bioorg. Med. Chem.*

Thakor, P., (2017). *Investigation of Novel Apoptosis-Inducing Substances and Their Mode of Action.*

Thakor, P., Mehta, J. B., Patel, R. R., Patel, D. D., Subramanian, R. B., & Thakkar, V. R., (2016). Extraction and purification of phytol from Abutilon indicum: Cytotoxic and apoptotic activity. *RSC Adv., 6*(54), 48336–48345.

Thakor, P., Song, W., Subramanian, R. B., Thakkar, V. R., Vesey, D. A., & Gobe, G. C., (2017a). Maslinic acid inhibits proliferation of renal cell carcinoma cell lines and suppresses angiogenesis of endothelial cells. *J. of Kidney cancer and VHL, 4*(1), 16–24.

Thakor, P., Subramanian, R. B., Thakkar, S. S., Ray, A., & Thakkar, V. R., (2017b). Phytol induces ROS mediated apoptosis by induction of caspase 9 and 3 through activation of TRAIL, FAS, and TNF receptors and inhibits tumor progression factor glucose 6 phosphate dehydrogenase in lung carcinoma cell line (A549). *Biomed. Pharmacother., 92*, 491–500.

CHAPTER 10

Apoptosis Activity of 1,2,-Benzene Dicarboxylic Acid Isolated from *Andrographis paniculata* on KB, SiHa and IMR Cancer Cell Line

P. KRISHNAMOORTHY and D. KALAISELVAN

Department of Zoology, Periyar E.V.R College (Autonomous), Thiruchirappalli – 620 023, Tamil Nadu, India, E-mail: pkmoorthy68@rediffmail.com (P. Krishnamoorthy)

ABSTRACT

Plastizicer compound of 1,2-benzene dicarboxylic acid fractionated from methanol extract of leaves of *Andrographis paniculata* which exhibits an *in vitro* antiproliferative effect, cell viability, reactive oxygen species, mitochondrial membrane potential and apoptotic pathway on KB, SiHa, IMR-32 cancer cells. The results showed that the plastizcer compound 1,2-benzene dicarboxylic acid has the potentiality to induce membrane blebbing, chromatin condensation, and fragment of nuclear DNA into oligonucleosome-sized DNA fragments. Therefore, the plastizicer compound 1,2-benzene dicarboxylic acid can be used after further experiments studies in therapies to improve the efficacy and decrease the side effects.

10.1 INTRODUCTION

Of different causes of death in the world, cancer occupies the second place next to the cardiovascular disorders (Reddy et al., 2003). Though there are several advanced therapeutics are available, the cancer mortality rate cannot be checked and there are higher amount of side effects which result from available cancer treatments. Drugs prepared from plants and microorganisms play a significant role in treating the cancer (Mann, 2002). It should be studied that inhibition of proliferation and apoptosis induction in cancer cells for profiling the activities of anticancer drugs. The interesting and new therapeutic approach to cancer treatment is direct influence of biochemistry machinery which regulates apoptosis (Jianze, 2006). Recent researches revealed that apoptosis-based cancer treatment is future promising treatment option (Reddy et al., 2003; Guy, 2001). Many plant-based apoptosis inducers were developed and they are successfully used in breast cancer treatment (Senaratne et al., 2000).

The main compounds of *Andrographis paniculata* are andrographolide and 1,2-Benzne dicarboxylic acid. Andrographolide possesses anticancer activity in several cancer cell lines under *in vitro* conditions (Satyanarayana et al., 2004). It was reported that andrographolide inhibited cell cycle by activating the protein p27 and decreasing expression of cyclin-dependent kinase (Satyanarayana et al., 2004). It suggests the inhibiting impact of andrographolide could contribute to its cytotoxicity both by DNA fragmentation and induction of apoptosis (Sukardiman et al., 2005). The earlier studies used the only andrographolide in treating cancer and it was found that no studies which previously carried out treating the cancer by using 1,2-Benzne dicarboxylic acid. Therefore, an attempt is made here to study the cancer treatment using 1,2-Benzne dicarboxylic acid isolated from *Andrographis paniculata* (Acanthaceae) on KB, SiHa, and IMR cancer cell lines.

10.2 MATERIALS AND METHODS

10.2.1 *APOPTOTIC MORPHOLOGICAL CHANGES BY ACRIDINE ORANGE (AO)/ETHIDIUM BROMIDE (EB) DUAL STAINING*

KB (oral cancer), SiHa (Cervical cancer), and IMR-32 (neuroblastoma) cell lines were purchased from NCCS, Pune.

Distinctive characteristics of apoptotic morphology in KB, SiHa, and IMR-32 cells treated with Plasticizer compound of 1,2-benzene dicarboxylic acid was detected by AO/EtBr staining (Kasibhatla et al., 2006). AO and EB staining of DNA allows visualization of the condensed chromatin of dead apoptotic cells. AO is a cell-permeable dye and it stains all the cells, dead or alive. DNA of cancer cells becomes condensed and gets fragmented during apoptosis. It could be noted when several visible spots of DNA in the cells. Later, the cell membrane becomes permeable to EtBr. The DNA spots turn orange-red after intercalation of EtBr with DNA.

- **Procedure:** Totally nine groups were maintained for each cell (including control, standard Paclitaxel and seven different concentrations of 1,2-Benzne dicarboxylic acid). KB, SiHa, and IMR-32 cells of each group were treated with 100 µg/mL of AO and EtBr dye mixture. Cells of all groups were incubated for 20 minutes at 36°C in dark conditions. Then the excess unbinding dye was removed by washing with PBS. The cells were observed under a fluorescence microscope. Shrunken Cells, fragmented nuclei, and condensed chromatin were noted. The number of apoptotic cells and live cancer cells in the field were counted to calculate the percentage of apoptosis.

10.3 RESULTS

In the present investigation, the effect of plasticizer compound of 1,2-benzene dicarboxylic acid on morphological changes in KB, SiHa, IMR-32 cancer cells were analyzed by the method of AO-EB dual staining assay. Untreated control cells appeared green in color (AO stained) where as the 1,2-benzene dicarboxylic acid-treated cells appeared orange in color (EB stained), (Figure 10.1). AO is a cationic dye that enters only live cells and stain DNA and hence the live cells observed as green under blue emission. On the other hand, EB stains DNA in the cells undergoing apoptosis and hence apoptotic cells were appeared orange in color. Effect of plasticizer compound 1,2-benzenedicarboxylic acid on morphology of KB cells were observed. The present results showed significant morphological changes in the plasticizer compound incubated cells. The initial structural changes were the condensation of the cytoplasm and nucleus, loss of microvilli and disruption of intracellular junctions at the KB cells (Figure 10.1). The incubation of plasticizer compound 1,2-benzene dicarboxylic acid 5 µg/ml, 10 µg/ml, 15 µg/ml, 25 µg/ml, 50 µg/ml, 75 µg/ml and 100 µg/ml induced the apoptosis in cancer cells. This was observed after the cells were stained with acridine orange-EB and incubated in CO_2 incubator for at room temperature (Table 10.1, Figure 10.1). 1,2-benzene dicarboxylic acid 5 µg/ml, 10 µg/ml, 15 µg/ml, concentration incubated cells showed 1.35%, 17%, 26.13% of apoptosis and in 50 µg/ml plasticizer compound treated KB cells were clearly indicating the induced apoptosis 48.08%, but the compound 1,2-benzene dicarboxylic acid 75 µg/ml and 100 µg/ml concentration treated cells were resulted a transparent more effective expression of induced apoptosis 70% and 70.16%. When the apoptosis activity was 99% in standard (25 µg/ml) incubation and 0.00% was in control. When the IMR-32 cell lines incubated with 1,2-benzene dicarboxylic acid which induces apoptosis, and leads to morphological changes in the cells. The initial structural changes were condensation of the cytoplasm and nucleus, loss of microvilli, and disruption of intracellular junctions (Figure 10.2). The 1,2-benzene dicarboxylic acid 10 µg/ml, and 20 µg/ml incubated cells resulted 13.16%, 27.5% of changes, and 30 µg/ml incubated cells showed a clear designated apoptosis 62.33% of changes, 40 µg/ml of 1,2-benzene dicarboxylic acid incubated cells resulted a defined most crystalline visible apoptosis 75.6% and 50 µg/ml treated cancer cells were shown sharp defined induced apoptosis 86% in control the apoptosis was 4.5% and standard (30 µg/ml) treated cells shown 98.8% of changes (Table 10.2 and Figure 10.2). 10 µg/ml and 20 µg/ml concentration of plasticizer compound incubated cells showed the 12.6% and 27% of changes and 30 µg/ml treated cells showed a clear induced apoptosis 65% and 40 µg/ml of the compound incubated cells sharply defined the induced apoptosis 78.6% of changes and 50 µg/ml treated cancer cells were shown glassy most precise visible induced apoptosis 89% in control 4% and standard (30 µg/ml), of incubated cells revealed the changes 98% (Table 10.3 and Figure 10.3).

TABLE 10.1 Percentage of Apoptosis Changes in 1,2-Benzenedicarboxylic Acid Treated KB Cells

S. No	Drug Concentration	Percentage of Apoptosis
1.	0 (Control)	0
2.	25 µg/ml (Std. Paclitaxel)	99.7 ± 0.15
3.	5 µg/ml	1.35 ± 0.10
4.	10 µg/ml	17.92 ± 0.19
5.	15 µg/ml	26.13 ± 0.10
6.	25 µg/ml	38.12 ± 0.07
7.	50 µg/ml	48.08 ± 0.06
8	75 µg/ml	70.33 ± 0.33
9	100 µg/ml	70.16 ± 0.16

Control = without treatment, Std = Standard drug Paclitaxel.
Values are expressed as mean ± Standard Error of Mean (S.E) of six experiments in each group.

Apoptosis Activity of 1,2,-Benzene Dicarboxylic Acid Isolated

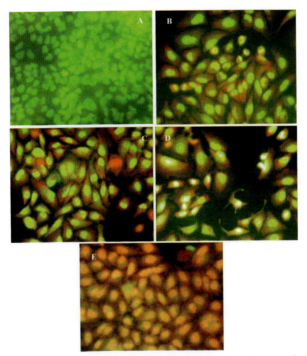

FIGURE 10.1 Fluorescence microphotograph of KB cells showed apoptotic morphological changes in different treatment groups: (A) control, (B and C) 1,2-benzene dicarboxylic acid (25 and 50 µg/ml) treated cells (D) 1,2-benzene dicarboxylic acid (75 and 100 µg/ml) and (E) Paclitaxel (25 µg/ml) treated cells. Orange-red color indicates the occurrence of apoptosis, while green color indicates the cell not undergoing apoptosis in KB cells.

TABLE 10.2 Percentage of Apoptosis Changes in 1,2-Benzenedicarboxylic Acid Treated IMR-32 Cells

S. No	Drug Concentration	Percentage of Apoptosis
1.	0 (Control)	0
2.	25 µg/ml (Std. Paclitaxel)	98.83 ± 0.30
3.	5 µg/ml	13.16 ± 0.74
4.	10 µg/ml	27.5 ± 0.42
5.	15 µg/ml	62.33 ± 0.95
6.	25 µg/ml	75.66 ± 0.49
7.	50 µg/ml	86.83 ± 0.60

Control = Without treatment, Std = Standard drug Paclitaxel.
Values are expressed as mean ± Standard Error of Mean (S.E) of six experiments in each group.

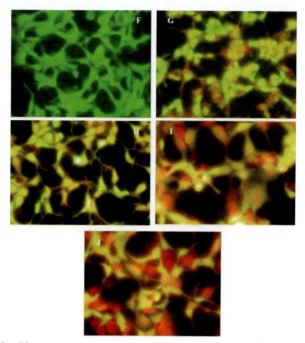

FIGURE 10.2 Fluorescence microphotograph of IMR-32 cells showed apoptotic morphological changes in different treatment groups: (F) control, (G and H) plasticizer compound of 1,2-benzene dicarboxylic acid (30 and 40 μg/ml) treated cells (I) plasticizer compound of 1,2-benzene dicarboxylic acid (50 μg/ml) and (J) Paclitaxel (30 μg/ml) treated cells. Orange-red color indicates the occurrence of apoptosis, while green color indicates the cell not undergoing apoptosis in IMR-32 cells.

TABLE 10.3 Percentage of Apoptosis Changes in 1,2-Benzenedicarboxylic Acid Treated SiHa Cancer Cells

S. No	Drug Concentration	Percentage of Apoptosis
1.	0 (Control)	0
2.	25 μg/ml (Std. Paclitaxel)	98.93 ± 0.20
3.	5 μg/ml	12.61 ± 0.74
4.	10 μg/ml	27 ± 0.64
5.	15 μg/ml	65.33 ± 0.95
6.	25 μg/ml	78.61 ± 0.49
7.	50 μg/ml	89.68 ± 0.60

Control = Without treatment, Std = Standard drug Paclitaxel.
Values are expressed as mean ± Standard Error of Mean (S.E) of six experiments in each group.

Apoptosis Activity of 1,2,-Benzene Dicarboxylic Acid Isolated 135

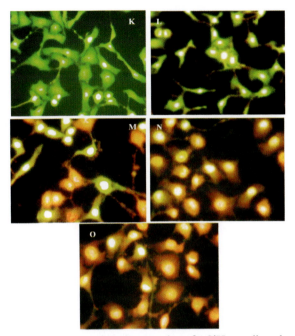

FIGURE 10.3 Fluorescence microphotograph of SiHa cells showed apoptotic morphological changes in different treatment groups: (K) control, (L and M) plasticizer compound of 1,2-benzene dicarboxylic acid (30 and 40 µg/ml) treated cells (N) plasticizer compound of 1,2-benzene dicarboxylic acid (50 µg/ml) and (O) Paclitaxel (30 µg/ml) treated cells. Orange-red color indicates the occurrence of apoptosis, while green color indicates the cell not undergoing apoptosis in SiHa cells.

10.4 DISCUSSION

The present results revealed that the plasticizer compound 1,2-benzene dicarboxylic acid induced the cytotoxicity and produced apoptosis in the cells which were incubated with different concentrations. Similarly, apoptosis is an important phenomenon in cancer chemotherapy, because anticancer drugs exert their antitumor effect against cancer cells by inducing apoptosis (Salomons et al., 1999). Therefore, it is suggested that Plasticizer compound 1,2-benzene dicarboxylic acid may exert its cytotoxic activity on different cancer cells by inducing apoptosis. In this present study, the cytotoxic effect of Plasticizer compound 1,2-benzene dicarboxylic acid is found to be associated with apoptosis. In view of these finding, the investigation was performed AO/EB staining and it was observed that cell shrinkage in Plasticizer compound 1,2-benzene dicarboxylic acid incubated cells, a major characteristic of nuclear fragmentation due to which the dying of cells were taking place in comparison to untreated control cells. AO/EB staining demonstrated that 1,2-benzene dicarboxylic acid extract induced in nuclear morphology. Syed Jafar Mehdi (2011) stated that the apoptosis is a physiological process of cell elimination and DNA fragmentation is one of the hallmarks of cell apoptosis, the apoptosis proportion of cells was increased by treatment of carvacrol in HeLa and SiHa cervical cancer cell lines. The increasing apoptotic cells during Plasticizer compound 1,2-benzene dicarboxylic acid treatments clearly indicated that this combination would be able to induce apoptosis. Plasticizer compound 1,2-benzene dicarboxylic acid (75 and 100 µg/ml) treatment significantly increased apoptotic morphological changes in KB cells. The increased ROS levels and subsequent reduced matrix metalloproteinase levels might be the reason for the increased apoptotic morphological changes in the Plasticizer compound 1,2-benzene dicarboxylic acid-treated cells. DNA is an important molecular target for tumor cell killing (McMillan and Steel, 1997). A significant DNA damage was observed in 1,2-benzene dicarboxylic acid-treated KB cells. Similarly, Chendil et al. (2004) reported that the Polyphenols are known to enhance oxidative DNA damage in cancer cells. Flavonoids auto-oxidize in an aqueous medium and may form highly reactive free radicals. Polyphenols may act as a substrate for peroxidases and other metalloenzymes, yielding quinone-type prooxidants (Metodiewa et al., 1999). This might be the reason for the increased oxidative DNA damage. The 1,2-benzene dicarboxylic acids inhibit KB, SiHa, and IMR-32 cancer cell viability by causing apoptosis that is characterized by the appearance of cytoplasmic histone-associated DNA fragmentation, subdiploid cells (Kalaiselvan, 2017). In contrast, KB, SiHa, and IMR-32 are more resistant to apoptosis induction by Plasticizer compound 1,2-benzene dicarboxylic acid. These results clearly indicate that antitumor activity of Plasticizer compounds against KB, SiHa, and IMR-32 cancer cells is associated with apoptosis induction.

The apoptotic bodies are engulfed by macrophages and thus are removed from the tissue without causing an inflammatory response. These morphological changes are a consequence of characteristic molecular and biochemical events occurring in an apoptotic cell, most notably the activation of proteolytic enzymes which eventually mediate the cleavage DNA into oligonucleosomal fragments as well as the cleavage of a multitude of specific protein substrates which usually determine the integrity and shape of the cytoplasm or organelles (Saraste and Pulkki, 2000). KB, SiHa, and IMR-32 human cancer cells are treated with Plasticizer compound 1,2-benzene dicarboxylic acid at different concentration, 30 µg/ml, 40 µg/ml, 50 µg/ml and 75 µg/ml, 100 µg/ml, and the nuclear structure using AO/EB staining, under the fluorescent microscopy, which exhibited condensation and fragmentation of some nuclei. Sukardiman et al. (2007) stated that the andrographolide isolated from *A. paniculata* induced apoptosis in TD-47 human breast cancer cell line in a time and concentration-dependent manner by increase expression of p53 bax, caspase-3, and decrease expression of bcl-2. The AO-EB dual staining cells exhibited the morphological and biochemical feature that characterize of apoptosis, as shown by membrane blebbing oligonucleosomal DNA fragmentation, chromatin condensation, and fragmentation of the cell into apoptotic bodies. The present study was identify that the active components existing in the 1,2-benzene dicarboxylic acid for its antiproliferative activity. Therefore, induction of apoptosis by 1,2-benzene dicarboxylic acid was due to single or combined effects of multiple agents contained in the extract. Although *A. paniculata* are much less toxic than most chemotherapeutic agents used to fight cancer, it may be seeing natural remedies combined with synthetic chemotherapeutic compounds that might improve efficacy and decrease side effects. The *A. paniculata* is an inexpensive and easily obtained, so it would benefit many, especially people in developing countries where cancer is almost catastrophic.

10.5 CONCLUSION

In conclusion, these *A. paniculata* extracts have an anti-proliferative effect in the cervical cancer cell line of human origin by inducing apoptosis. Therefore, our findings open up the possibility that natural compound found in these spices may be used to develop new treatment modality for KB, SiHa, IMR-32 cancer.

ACKNOWLEDGMENT

Authors thank the Principal and Head of the Department of Zoology, Periyar EVR Govt. Arts College (Autonomous), Tiruchirappalli for providing necessary facilities.

KEYWORDS

- antiproliferative activity
- apoptosis
- cytotoxic activity
- Gc-MS
- phytochemistry

REFERENCES

Chendil, D., Ranga, R. S., Meigooni, D., Sathishkumar, S., & Ahmed, M. M., (2004). Curcumin confers radiosensitizing effect in prostate cancer cell line PC-3. *Oncogene, 23,* 1599–1607.

Fesik, S. W., (2005). Promoting apoptosis as a strategy for cancer drug discovery. *Nat. Rev. Cancer, 5,* 876–885.

Fischer, S. F., Belz, G. T., & Strasser, A., (2007). BH3-only protein Puma contributes to death of antigen-specific T cells during shutdown of an immune response to acute viral infection. *Proc. Natl. Acad. Sci. USA., 105,* 3035–3040.

Guy, M., & Croline, D., (2001). Apoptosis and cancer chemotherapy. *Trends in Cell Biology, 1,* 22–26.

Jianze, L., Brenda, L., & Amy, S. L., (2006). Multiple pathways and activation of p53-upregulated modulator of apoptosis (puma) and noxa by p53. *The J. of Biological Che., 281,* 7260–7270.

Kalaiselvan, D. (2017). Anticancer efficacy of plasticizer compound 1,2-benzene dicarboxylic acid from leaves extract of *Andrographis paniculata* (NEES) on selected cell lines. *PhD Thesis.* Submitted to Bharathidasan University, Trichy, India.

Kasibhatla, S., Gustavo, P., Amaranate-Mendes, Bossy-Wetzed, E., Brunner, T., & Green, D. R., (2006). Acridine organe/ethidium bromide (A0/EB) staining to detect apoptosis. *Cold Spring Harb. Protoc.*

Mann, J., (2002). Natural products in cancer chemotherapy: Past, present and future. *Nature Review Cancer, 2,* 143–148.

McMillan, T. J., & Steel, G. G., (1997). Genetic control of the cellular response to ionizing radiation. In: Steel, G. G., (ed.), *Basic Clinical Radiobiolgoy* (2nd edn., pp. 72–77). London: Arnold.

Metodiewa, D., Jaiswal, A. K., Cenas, N., Dickancaite, E., & Segura-Aguilar, J., (1999). Quercetin may act as a cytotoxic prooxidant after its metabolic activation to semiquinone and quinoidal product. *Free Radical Biology Medicine, 26*(12), 107–116.

Reddy, L., Odhav, B., & Bhoola, K. D., (2003). Natural products for cancer prevention: A global perspective. *Pharma. and Thera., 99,* 1–13.

Salomons, G. S., Smets, L. A., Verwijs-Janssen, M., Hart, A. A., Haarman, E. G., & Kaspers, G. J., (1999). Bcl-2 family members in childhood acute lymphoblastic leukemia: Relationships with features at presentation, *in vitro* and *in vivo* drug response, and long-term clinical outcome. *Leukemia, 13,* 1574–1580.

Saraste, A., & Pulkki, K., (2000). Morphologic and biochemical hallmarks of apoptosis. *Cardiovascular Research, 45*(3), 528–537.

Satyanarayana, C., Dhanavanthri, S. D., Rajagopalan, R., Nanduri, S., & Sriram, R., (2004). DCFR 3188 a novel semi-synthetic analog of andrapholide: Cellular response to MCF-7 breast cancer cells. *BMC Cancer, 4*(26), 1–8.

Senaratne, S. G., Pirianov, G., Mansi, J. L., Arnett, T. R., & Colston, K. W., (2000). Bisphosphonates induce apoptosis in human breast cancer cell lines. *Br. J. Cancer, 82*(8), 1459–1468.

Sukardiman, H., Aty, W., Sismindari., & Noor, C. Z., (2007). Apoptosis-inducing effect of andrographolide on TD-47 human breast cancer cell line. *Afr. J. of Tradit., CAM, 4*(3), 345–351.

Sukardiman, S., & Noor-Cholies, Z., (2005). Anticancer activity of pinostrobin and andrographolide. *Proceeding of Congress of Pharmaceutical Future*. Tokyo, Japan.

Syed, J. M., Aijaz, A., Irshad, M., Nikhat, M., & Moshahid, A. R. M., (2011). Cytotoxic effect of carvacrol on human cervical cancer cells. *Biology and Medicine, 3*(2), 307–312.

CHAPTER 11

Anticancer Drug Discovery from the Indian Spice Box

M. V. N. L. CHAITANYA,[1] SANTHIVARDHAN CHINNI,[2]
P. RAMALINGAM,[3] and Y. PADMANABHA REDDY[4]

[1]*Department of Pharmaceutical Quality Assurance, Phytomedicine and Phytochemistry Division, Raghavendra Institute of Pharmaceutical Education and Research (Autonomous), Saigram, Anantapuramu – 515721, Andhra Pradesh, India,*
E-mails: chaitanya.phyto@gmail.com, drchaitanya@riper.ac.in

[2]*Department of Pharmacology, Pharmacology and Toxicology Division, Raghavendra Institute of Pharmaceutical Education and Research (Autonomous), Saigram, Anantapuramu – 515721, Andhra Pradesh, India*

[3]*RERDS-CPR, Raghavendra Institute of Pharmaceutical Education and Research (Autonomous), Saigram, Anantapuramu – 515721, Andhra Pradesh, India*

[4]*Department of Pharmaceutical Analysis, Raghavendra Institute of Pharmaceutical Education and Research (Autonomous), Saigram, Anantapuramu – 515721, Andhra Pradesh, India*

ABSTRACT

Cancer was defined as something evil or malignant that spreads destructively by Merriam-Webster and as per the National Library of Medicine. It can also be defined as Malignant Neoplasm or as a disease with the uncontrollable division of abnormal cells and spreads rapidly. The modern medicine has failed in preventing cancer but can control cancer to some extent. The anticancer drug discovery became the interesting hub for many multinational pharmaceutical companies and investing huge amount of money. However, it is a failure of synthetic chemistry to deliver a single potent anticancer molecule. The main problem with cancer is its resistance and there is a need of complex new molecules to fight cancer, this is possible only from fruits, spices, and greens. The current chapter focuses on the importance of spice box in everyone kitchen globally and how these superhero spices can hit the villain cancer and also explains the anticancer drug discovery from Indian ancestral spice box which contain anticancer super spices like *Allium sativum (Amaryllidaceae), Capsicum annum (Solanaceae), Carum carvi (Apiaceae), Cinnamomum zeylanicum (Lauraceae), Coriandrum sativum (Apiaceae), Crocus sativus (Iridiaceae), Curcuma longa (Zingiberaceae), Cuminum cyminum (Apiaceae), Elettaria cardamomum (Zingiberaceae), Foeniculum vulgare (Apiaceae), Myristica fragrans (Myristicaceae), Piper nigrum (piperaceae), Syzygium aromaticum (Myrtaceae)* and *Trigonella foenum-graecum (Fabaceae).*

11.1 INTRODUCTION

Cancer can be defined as an ultimate challenging deadly disease of the 21st century in which modern sciences and synthetic molecules have failed to treat or restore the cancer cells to its original existence. The main problem with cancer is resistance and there is a very urgent need for the discovery of new anticancer phytochemical leads having multi-target actions (Singh et al., 2002; Balachandran and Govindrajan et al., 2005). The modern sciences have failed to unreveal the real mechanism of cancer and which was explained earlier by the ancient Vedic scripture known as Ayurveda (Thatte et al., 1999; Smit et al., 1995). According to Charaka and Sushruta Samhitas of Ayurveda, cancer is defined as a 'Granthi' (minor neoplasm) or 'Arbuda' (major neoplasm). The malignant tumors are a result of imbalance of tridoshas namely, vatham, pitham, and kapham can causes tissue damage may result in tissue damage and proliferation (Cravotto et al., 2010). Commonly used Ayurvedic elixirs known as Asavas, Aristas, and rasayanas proved to have great potential in cancer cure. Further, these elixirs proved to have multiple target acting mechanisms which is an important essential criteria for the anticancer drug (Patel et al., 2010).

11.1.1 NUTRACEUTICALS IN KITCHEN IS AN EVOLVING ALTERNATIVE APPROACH

In India the art of healing have been started from the traditional kitchen, the modern sciences proved that Indian food or ingredients or spices that uses in this kitchen proved to have anticancer properties and can be a preventive measure. This make the modern sciences to think and the anticancer drug discovery was started in the Indian kitchen mainly from the Indian Spice box proved that every spice in this is box is a superhero to combat cancer and many anticancer phytochemical leads came into existence through this spice-box, starting from turmeric ending with garlic or the curry leaf (Sengupta et al., 2004).

11.1.2 ANTICANCER DRUG DISCOVERY FROM THE SPICE BOX

The health secret of every healthy Indian is the spice box in every kitchen. It is a medicinal box in modern definitions, the spice box have compartments and the following important spices are present in this box, i.e., *Allium sativum (Amaryllidaceae), Capsicum annum (Solanaceae), Carum carvi (Apiaceae), Cinnamomum zeylanicum (Lauraceae), Coriandrum sativum (Apiaceae), Crocus sativus (Iridiaceae), Curcuma longa (Zingiberaceae), Cuminum cyminum (Apiaceae), Elettaria cardamomum (Zingiberaceae), Foeniculum vulgare (Apiaceae), Myristica fragrans (Myristicaceae), Piper nigrum (piperaceae), Syzygium aromaticum (Myrtaceae)* and *Trigonella foenum-graecum (Fabaceae).* Many recipes like biryani, sambar, rasam can be considered as the cooked herbal decoctions in pulses with tamarind. The modern sciences proved that these traditional kitchen spices possess to have many pharmacological activities like anti-inflammatory, immunomodulatory, and may also be useful in treatment and prevention of various types of cancers. Hence, this spice box became the cotton candy of many big health care sectors to discover new anticancer phytochemical leads from these spices. The modern sciences proved that these spices have complex mixture of synergistic phytochemicals leads which can prevent cancer in person whose daily diet with mixture of spices. The Ayurveda long back documented that these spices have the superpower to restore the cells into natural system of harmony or balance that prevents the cancer or makes the cancer cell to change back to its natural system of balance and can help in becoming the normal cell (Jie et al., 2016). However, not even a single anticancer molecule or product came into market till 2018 even from the turmeric or the garlic, hence this current chapter is stressing to explain in-depth, the importance of spices in the spice box for the future anticancer drug products or new phytochemical drugs or leads and also trying to create awareness how this spice box can prevent cancer. Phytochemical leads for cancer are illustrated in Figure 11.1 and the important anticancer spices are shown in Figure 11.2.

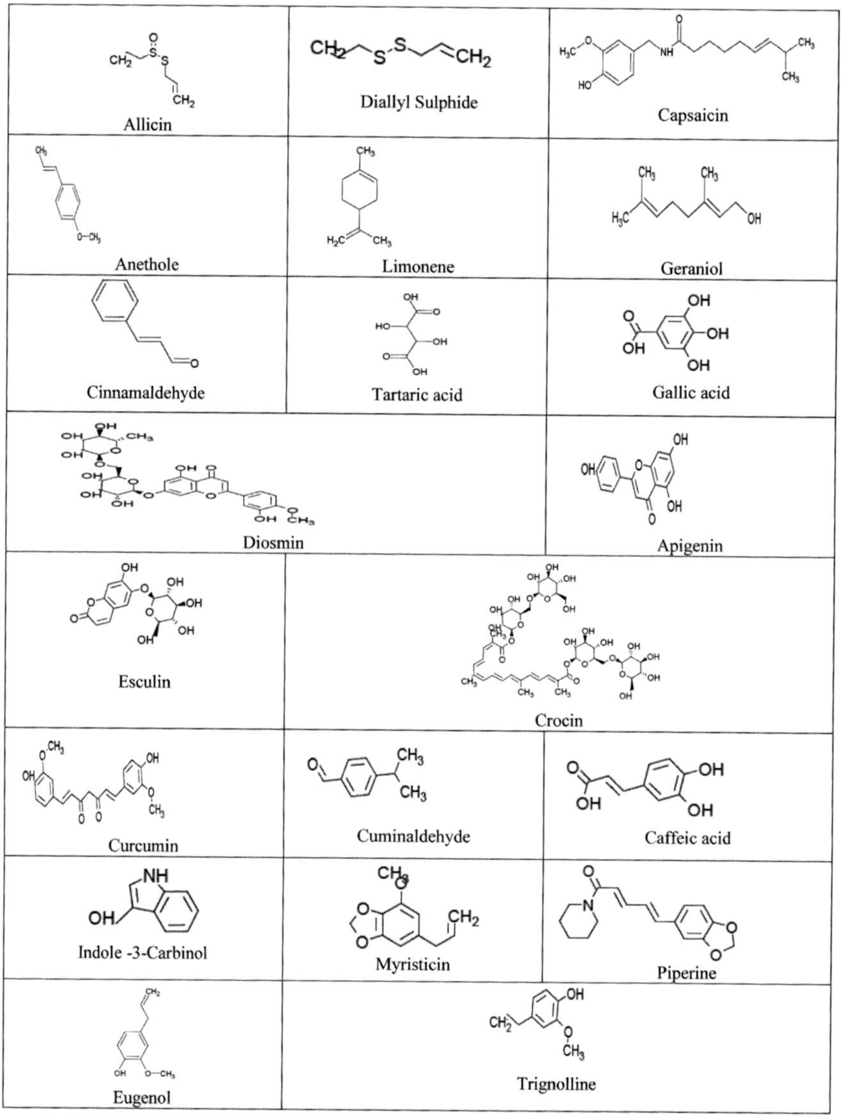

FIGURE 11.1 Phytochemical leads from the super spices of India in spice box.

Anticancer Drug Discovery from the Indian Spice Box

FIGURE 11.2   Anticancer spices.

11.2   ANTICANCER SPICES FROM THE INDIAN KITCHEN BOX

11.2.1   ALLIUM SATIVUM

The dried bulbs of *Allium sativum* is belonging to the family Amaryllidaceae. The common name of garlic in different languages are Naharu,

Lahsun, Lassan, Belluli, Vellulli, Purunvar, Acanam, Hanam, and Velluli. The compounds from garlic are allicin, diallyl disulfide (DADS), diallyl trisulfide used as lead for anticancer drug (Omar and Alwabel, 2010) which activate cytochrome P_{450} enzyme system and inhibition of cancer DNA adducts. Garlic boosts the immune system and protects the cells from external toxicants. The Allicin and quercetin in garlic helps to prevent various types of cancers. The clinical studies proved that individuals who consumes garlic as regular diet proved to have 30% lower risk of stomach and colorectal cancers. Garlic helps in prevention of various other infections in cancer patients. That will allow the enzyme allinase to convert the compound allin to its active form, allicin, which fights cancer (Omar and Alwabel, 2010). Finely sliced garlic should be stable for a few (4–6) hours at room temperature, and 2–3 days in the fridge. pre-chopped garlic in pockets lost their potential. By adding garlic raw at the end of recipe, instead of cooking, this preserved all the anti-cancer properties.

11.2.2 CAPSICUM ANNUUM

It belongs to the family solanaceae. Common names are capsicum, sweet pepper, chilli pepper, cayenne pepper, paprika, and shimla mirchi. Capsaicinoids from capsicum used as a lead for cancer (Amruth Raj et al., 2014). Further, they stated that it modulates free radicals release and it had high anti-inflammatory potential. It lowers the risk of cancer due to its anti-inflammatory and antioxidant properties. The enzyme pectolyse inside the capsicum prevents gastric and lung cancer (Amruth et al., 2014).

11.2.3 CARUM CARVI

Carum carvi belongs to the family apiaceae. Common names are caraway, jangi dhania, shahajire, asitajiraka, bahugandha, seema jeeraka, seemai sompu, shimaisapu. The compound anethole from *Carum carvi* used as lead for cancer (Soodabeh and Mohammad, 2014). It prevents cancer by DNA protecting activity. Daily intake of caraway seeds from soups or curries or salads may prevent cancer.

11.2.4　CINNAMOMUM ZEYLANICUM

Cinnamomum zeylanicum consists of the dried inner bark of the shoots of coppiced trees belongs to the family lauraceae. Common names are karruwa, karuva, dalichina chekka, dalchini. Limonene and geraniol and 20-benzyloxycinnamaldehyde from this spice used as lead for treating cancer (Gelb et al., 1995) by the action of Caspase 3 activation (Herdwaini et al., 2016). Poly-phenolic compounds in cinnamon acts on inflammatory gene expression and help in the prevention of cancer. The cinnamon extracts proved to act against VEGF (vascular endothelial growth factor) without major side effects. The polyphenols in cinnamon extract enhance the expression of TNF (a group of cytokines), which causes the apoptosis in cancer cells (Herdwaini et al., 2016).

11.2.5　CORIANDRUM SATIVUM

The dried fruits of *Coriandrum* sativum belong to the family Apiaceae. Common names are coriander, dhaniya, dhanayak, dhana, dhanyalu. The compounds from spice coriandrum are tartaric acid, gallic acid, diosmin, dicoumarin, 4-hydroxycoumarin apigenin, esculin used as lead for anticancer (Ganesan et al., 2013). The vitamins, polyphenols, flavones in coriander leaves and fruits proved to protect the cells from various mutations helps in prevention of chronic cancers.

11.2.6　CROCUS SATIVUS

Saffron is dried stigma and styletops of *Crocus sativus* belong to the family Iridiaceae. Common names are saffron flower, saffron crocus, zafran, kashmirajanman, kesar, kashmiram, crocin from crocus used as a lead for treating cancer by the action of inhibition of synthesis of DNA and RNA, interaction with cellular topoisomerase, suppression of the telomerase activity and active STAT3, and targeting of microtubules (Reyhane et al., 2017). Some studies showed that saffron has antibacterial properties and possible anti-tumor effects and UAE proved that saffron extract prevents the liver cancer in rats by inhibiting cell proliferation, induces apoptosis and suppresses inflammation. The Chittaranjan National Cancer Institute (NCI) in India proved that the saffron extract proved to have significant anticancer on skin cancer cells in mice.

11.2.7 CURCUMA LONGA

Curcuma longa is a perennial flowering herb that belongs to the ginger family, Zingiberaceae. Common names are turmeric, halodhi, halud, haldar, arishina, arisina, manjal, halad, haldi, haladi, haridra, marmarii, manjal, yaingang, haridra, haldi. Curcumin from C. *longa* used as a lead for anticancer which inhibits the STAT3, Sp-1 and NF-κB signaling pathways, which play key roles in cancer development and progression. The curcumin proved to be active against various types of cancer and reduces angiogenesis and metastasis, in a clinical study the curcumin proved to reduce the number of lesions by 40% (Goel et al., 2008).

11.2.8 CUMINUM CYMINUM

Cuminum cyminum consists of the dried, ripe fruit of belongs to the family Apiaceae. Vernacular names are cumin, jeera, jira, zira, jeerige, jirak, and jilakarra. Cuminaldehyde from this spice used as lead for cancer drug. It works on cancer by topoisomerase inhibition (Kuendaw et al., 2016). Cumin due to its antioxidant, chemopreventive, and anti-carcinogenic properties prevent various cancer, especially colon and breast cancer.

11.2.9 ELETTARIA CARDAMOMUM

Elettaria cardamomum consists of the dried nearly ripe fruits belong to Zingiberaceae. Common name are cardamom, malabar cardamom, ceylon cardamom, elaichi, elam ancha, elatarri, elaki, elakki, elaichi, trutih. Anti-cancer leads from cardamomum are diindolylmethane, indole-3-carbinol, limonene, caffeic acid which are worked against cancer by blocking of cyclooxygenase-2, cytochrome P_{450} and NF-kB (Neha and Sita, 2018). In vivo studies proved that the cardamom extract significantly reduces the cancer. Cineole and limonene in cardamom, proved to have a protective role against cancer progression.

11.2.10 FOENICULUM VULGARE

Foeniculum vulgare consists of dried fruits belongs to the family Apiaceae. Vernacular names are fennel, sweet fennel, Florence fennel, finocchio, moti saunf, hop, sompu, preumjirakam, peddajilakarra, dodda sompu, mauri, misreya, madhurika. An anticancer lead of *F. vulgare* is Anethole, which inhibits TNF-induced NF-kappa B (Chanchal et al., 2009). Anethole shuts down or prevents the activation of NF-kappaB, a gene-altering, inflammation-triggering molecule.

11.2.11 MYRISTICA FRAGRANS

Myristica fragrans consists of dried kernels of the seeds and belongs to the family Myristicaceae. Vernacular names are nutmeg, jaiphol, jatiphala, jayaphala, jati-phal, jayaphal, jakayi, Konkani: cati-k-kay, jayaphal. myristicin from this spice act as lead for cancer medicine (Ekta and Dwijendra et al., 2013). Due to the presence of omega-6 fatty acids, nutmeg is having anticancer properties on leukemia cells.

11.2.12 PIPER NIGRUM

Piper nigrum consists of the dried fruits and belongs to piperaceae. The common names are kalimirch, miryalu, maricha. Anticancer lead of this spice is piperine. It works on cancer prevention by inhibiting activity and expression of multidrug resistance transporters such as P-gp and MRP-1 (Manayi et al., 2017). The piperine in black pepper can be credited with the prevention of cancer due to presence of vitamins, flavanoids, and carotenes. The best way to enjoy the health benefits from pepper is by eating it as freshly grounded powder.

11.2.13 SYZYGIUM AROMATICUM

Syzygium aromaticum consists of dried flower buds and belongs to myrtaceae. Vernacular names are laung, luvanga, lavanga, kirambu, lavangam, lavangalu. Eugenol acts as an anticancer agent by which enhanced expression of Bcl-2 protein (Fadilah et al., 2017). The clinical trials proved that the clove extract is effective against in prevention of breast cancer, colon cancer, cervical cancer, and ovarian cancers.

11.2.14 TRIGONELLA FOENUM-GRAECUM

Trigonella foenum-graecum seeds and leaves are used as spice and belong to the family fabaceae. Vernacular names are fenugreek, sickle fruit fenugre, methika, methi. Trignolline isolated from this spice act against cancer prevention. Fenugreek, proved to be safe on normal cells but very effective in prostrate and pancreatic cancers at a dose of 20 µg/ml (Shabana et al., 2009).

11.3 CONCLUSION

This chapter is very important to a common reader and motivates how to prevent, protect, and combat the dreadful villain cancer with this super kitchen box containing super spices, even if every individual adopts this at least one spice daily in their regular dietary style, definitely they will be protect from cancer and these superfoods or spices helps to restore the nature balance in every cell. However, billions, and trillions of dollars are being invested in cancer drug discovery but till now not even single phytochemical lead proved to be useful against dreadful resistant cancer. Even lot of documentation proved that these super spices are the sources of new phytochemical anticancer leads, though they have been neglected and not even single preventive product against cancer have not been released into market. Hence, the authors are requesting the common people to read and gain knowledge how to prevent cancer by implementing these super spices in everyone's daily dietary style and also this chapter giving small hint there is much scope for the phytochemists to discover many leads from these spices.

KEYWORDS

- anticancer drug discovery
- cancer
- phytochemical anticancer
- phytochemicals
- spices
- *Trigonella foenum-graecum*

REFERENCES

Amruth, R. N. J., (2014). *In vitro* studies on anticancer activity of capsaicinoids from capsicum chinense against human hepatocellular carcinoma cells. *Int. J. Pharm. Pharm. Sci., 6,* 254–258.

Balachandran, P., & Govindaraja, R., (2005). Cancer: An ayurvedic perspective. *Pharmacol. Res., 51,* 19–30.

Bhishagratha, K. L., (1991). Sushruta Samhita, Sushruta samhita. *Varanasi: Choukhamba Orientalia.*

Chanchal, G., Khan, S. A., Ansari, S. H., Suman, A., & Munish, G., (2009). Chemical composition, therapeutic potential, and perspectives of *Foeniculum vulgare*. *Pharmacogn Rev., 3,* 346–352.

Cravotto, G., Boffa, L., Genzini, L., & Garella, D., (2010). Phytotherapeutics: An evaluation of the potential of 1000 plants. *J. Clin. Pharm. Ther., 35,* 11–48.

Ganesan, P., Phaiphan, A., Murugan, Y., & Baharin, B. S., (2013). Comparative study of bioactive compounds in curry and coriander leaves: An update. *J. Chem. Pharm. Res., 5,* 590–594.

Gelb, M. H., Tamanoi, F., Yokoyama, K., Ghomashchi, F., Esson, K., & Gould, M. N., (1995). The inhibition of protein prenyltransferases by oxygenated metabolites of limonene and perillyl alcohol. *Cancer Lett., 91*(2), 169–175.

Herdwiani, W., Soemardji, A., & Elfahmi, T. M., (2016). A review of cinnamon as a potent anticancer drug. *Asian J. Pharm. Clin. Res., 9,* 8–13.

Kuen-daw, T., (2016). Cuminaldehyde from *Cinnamomum verum* Induces cell death through targeting topoisomerase 1 and 2 in human colorectal adenocarcinoma COLO 205 Cells. *Nutrients, 8,* 318–321.

Manayi, A., Nabavi, S. M., Setzer, W. N., & Jafari, S., (2017). Piperine as a potential anti-cancer agent: A review on preclinical studies. *Curr. Med. Chem.*, doi: 10.2174/092986732 4666170523120656.

Neha, V., & Sita, S., (2018). Natural and herbal medicine for breast cancer using *Elettaria cardamomum* (L.) Maton. *IJHM, 6,* 91–96.

Omar, S. H., & Al-Wabel, N. A., (2010). Organosulfur compounds and possible mechanism of garlic in cancer. *Saudi Pharm. J., 18,* 51–58.

Patel, B., Das, S., Prakash, R., & Yasir, M., (2010). Natural bioactive compound with anticancer potential. *Int. J. Advan. Pharmaceut. Sci., 1,* 32–41.

Sengupta, A., Ghosh, S., Bhattacharjee, S., & Das, S., (2004). Indian food ingredients and cancer prevention: An experimental evaluation of anticarcinogenic effects of garlic in rat colon. *Asian Pac. J Cancer Prev., 5,* 126–132.

Shabana, S., (2009). Fenugreek: A naturally occurring edible spice as an anticancer agent. *Cancer Biology and Therapy, 8,* 272–278. doi: 10.4161/cbt.8.3.7443.

Smit, H. F., Woerdenbag, H. J., Singh, R. H., Meulenbeld, G. J., Labadie, R. P., & Zwaving, J. H., (1995). Ayurvedic herbal drugs with possible cytostatic activity. *J. Ethnopharmacol., 47,* 75–84.

Soodabeh, S., & Mohammad, A., (2014). Perspective studies on novel anticancer drugs from natural origin: A Comprehensive Review. *IJP, 10,* 90–108.

Thatte, U., & Dhahanukar, S., (1991). Ayurveda, the natural alternative. *Sci. Today, 2001,* 12–8.

Zheng, J., Zhou, Y., Li, Dong-Ping, X., Sha, Li, & Hua-Bin Li., (2016). Spices for prevention and treatment of cancers. *Nutrients, 8,* 495–498.

WEB INFORMATION

http://2beingfit.com/nutmeg-nutrition-facts-health-reasons-benefits-uses-for-good-health/# (accessed on 20 February 2020).

http://ar.iiarjournals.org/content/35/2/645.full (accessed on 20 February 2020).

http://www.cookingwithbooks.net/2014/06/coconut-cardamom-summer-cocktail.html (accessed on 20 February 2020).

http://www.hrpub.org/download/201309/ujfns.2013.010101.pdf (accessed on 20 February 2020).

http://www.softschools.com/facts/plants/crocus_facts/502/ (accessed on 20 February 2020).

https://anticancerclub.com/recipes-and-eating-healthy/cilantro-and-coriander-one-and-the-same/21. https://doi.org/10.1111/jphp.12776 (accessed on 20 February 2020).

https://anticancerclub.com/recipes-and-eating-healthy/cinnamon-and-anti-cancer/ (accessed on 20 February 2020).

https://eatandbeatcancer.com/2012/11/10/anti-cancer-recipes-how-to-handle-garlic/ (accessed on 20 February 2020).

https://food.ndtv.com/food-drinks/fenugreek-water-benefits-5-reasons-to-drink-this-up-every-morning-1755913 (accessed on 20 February 2020).

https://food.ndtv.com/opinions/black-pepper-benefits-more-than-just-a-spice-1238993 (accessed on 20 February 2020).

https://foodfacts.mercola.com/fennel.html (accessed on 20 February 2020).

https://pdfs.semanticscholar.org/63b4/02879d63fd6a6e876e8bfec9405785219f7d.pdf (accessed on 20 February 2020).

https://thetruthaboutcancer.com/turmeric-tea-recipe/ (accessed on 20 February 2020).

https://www.geniuskitchen.com/recipe/nutmeg-and-ginger-tea-15491 (accessed on 20 February 2020).

https://www.healthline.com/nutrition/top-10-evidence-based-health-benefits-of-turmeric#section6 (accessed on 20 February 2020).

https://www.medicalnewstoday.com/articles/319651.php (accessed on 20 February 2020).

https://www.naturalfoodseries.com/13-benefits-capsicum/ (accessed on 20 February 2020).

https://www.naturalhealthnews.uk/food/2013/12/study-reveals-the-anti-cancer-properties-of-spices/ (accessed on 20 February 2020).

https://www.natureword.com/tag/coriander-anticancer/ (accessed on 20 February 2020).

https://www.organicfacts.net/health-benefits/seed-and-nut/health-benefits-of-cumin.html (accessed on 20 February 2020).

https://www.realnatural.org/cloves-stop-the-growth-of-several-cancers/ (accessed on 20 February 2020).

https://www.webmd.com/vitamins/ai/ingredientmono-844/saffron (accessed on 20 February 2020).

https://www.wikihow.com/Prepare-Cumin-Tea (accessed on 20 February 2020).

CHAPTER 12

Phytochemical Investigation and Evaluation of Anti-Breast Cancer Activity of Chloroform Extract of *Tagetes erecta*

R. ARUL PRIYA,[1] K. SARAVANAN,[1] and CHUKWUEBUKA EGBUNA[2]

[1]Department of Zoology, Nehru Memorial College (Autonomous), Puthanampatti – 621 007, Tiruchirappalli, Tamil Nadu, India, E-mail: arulpriyaphd@gmail.com (R. A. Priya)

[2]Department of Biochemistry, Faculty of Natural Sciences, Chukwuemeka Odumegwu Ojukwu University, Nigeria

ABSTRACT

The present research work was implemented to detect the phytochemicals present in the chloroform extract of *Tagetes erecta* flowers using GC-MS analysis. Further, the cytotoxic effects of this extract were evaluated on human breast adenocarcinoma (Michigan cancer foundation-7: MCF-7) cell lines by MTT assay and AO/EB staining fluorescent assay. Preliminary phytochemical tests indicated the presence of flavonoids, phenols, phytosterols, tannins, and terpenoids. GC-MS profile of *T. erecta* flower extract showed the presence of 11 compounds. The anticancer activity studies confess the *T. erecta* flower extract is a potent anticancer agent for the treatment of breast cancer. Thus, the Marie gold flower may be used as a lead for the preparation cancer drug

12.1 INTRODUCTION

Medicinal plants have played an important role in the treatment of cancer and, indeed, most new clinical applications of plant metabolites and their derivatives have been applied towards fighting cancer (Balunas and

Kinghorn, 2005). Cancer is a multistep disease incorporating physical, environmental, metabolic, chemical, and genetic factors, which play a key role in the induction and deterioration of cancers. The limited success in available cancer treatment options including radiation, chemotherapy, immune-modulation, and surgery indicates that there is urgent need of alternative drugs for cancer treatment (Dai and Mumper, 2010). *Tagetes erecta* (Marigold) is an ornamental plant widely cultivated for the cut-flower trade. The flowers are carminitive, diuretic, and vermifuge (Manandhar, 2002). Moreover, decoction of this flower and its paste are used for treatment of many diseases including cold, mumps (Duke and Ayensu, 1985), skin diseases, and conjunctivitis (Duke and Ayensu, 1985; Manandhar, 2002). Flavonoids are also a kind of natural product with antioxidant properties capable of scavenging free superoxide radicals, having anti-aging properties, as well as reducing the cancer risk (Middleton et al., 2000; Yao et al., 2004). Several authors reported that flavonoids are able to control cancer cell growth in the human body (Davis et al., 2000; Arts et al., 2002; Shukla et al., 2007; Park et al., 2008). Flavonoids are found in almost all plants especially rich in colorful flowering plants. Based on the above-said points, it was hypothesized that flavonoid might rich in marigold flower (*T. erecta*) and the flower can cure cancer. To test this hypothesis, the marigold flower was selected to identify the phytochemical compounds and was studied its anticancer activity against breast cancer (MCF-7) cell lines.

12.2 MATERIALS AND METHODS

12.2.1 COLLECTION OF PLANT MATERIAL

In the present study, *T. erecta* flowers were purchased from local flower market, Trichy, and Namakkal. They were thoroughly washed with tap water and dried under shadow. Then they were ground well using domestic grinder.

12.2.2 PREPARATION OF EXTRACTS

About 250 gm of dried, fine flower powder of *T. erecta* was loaded into the Soxhlet apparatus with 500 ml of chloroform solvent and extracted at the temperature of 61.2°C. The extraction was made upto two cycles for 48 hrs. Then, the extract was concentrated by distillation and the obtained residues were stored in the desiccator until use.

12.2.3 PHYTOCHEMICAL SCREENING OF T. ERECTA FLOWER EXTRACT

The phytochemical screening of *T. erecta* flower extract was made by the standard biochemical procedure (Kokate, 1997).

12.2.4 GC-MS ANALYSIS

Gas chromatography-mass spectrum (GC-MS) analyses were carried out to detect the bioactive compounds found in the chloroform extract of *T. erecta*. GC analysis was done by JEOL GC MATE II instrument employing the following conditions: Front inept temperature 220°C; Column HP 5Ms; Helium gas (99.99%) was utilized as carrier gas at a constant flow rate of 1 ml/min. Temperature was maintained from 50 to 250°C at 10°C/min. The ion chamber temperature and GC interface temperature was maintained at 250°C. Mass analysis was done with the help of Quadruple Double Focusing Analyzer. The Photon Multiplier tube was used for detection. At 70 eV, mass spectra were taken. The necessary data were gathered by the full-scan spectra within the scan range of 50–600 amu. Percentage peak area (PA) (i.e., percent composition of constituents of the extract) was also obtained.

12.2.4.1 IDENTIFICATION OF COMPONENTS

On the basis of GC RT (retention time), chemical compounds in chloroform extract of the *T. erecta* were identified. Further, the mass spectra were matched with the standard mass spectra available in National Institute of Standard and Technology (NIST) library. The compound name, molecular weight (MW), molecular formula (MF), and chemical structure of the phytochemicals were ascertained from NIST and Pubchem libraries.

12.2.5 ANTICANCER ACTIVITY

12.2.5.1 PROCUREMENT OF CELL LINES

Cell lines of human breast cancer (MCF-7) were obtained from National Center for Cell Science (NCCS), Pune, India. These cells were stored in DMEM medium supplemented with 10% FBS (Sigma-Aldrich, St. Louis, Mo, USA). Penicillin at 100 U/mL and streptomycin at 100 µg/mL were used as antibiotics (Himedia, Mumbai, India). At humidified atmosphere, the culture was kept at 5% CO_2 level in a CO_2 incubator at 37°C (Forma, Thermo Scientific, USA).

12.2.5.2 CELL VIABILITY ASSAY

The MTT tetrazolium salt colorimetric assay, described by Mosmann (1983) was performed to measure the cytotoxicity of chloroform extract of *T. erecta*. The plant extract were dissolved in 100% DMSO to prepare a stock. The stock solution was diluted separately with fresh medium to get various concentrations from 0 to 0.3 mg. Exactly 100 µL of sample was added to well of 96-well plate. DMSO (0.02%) was used as the solvent control. After 24 hrs, 20 µL of MTT solution (5 mg/mL in PBS) was added to each well and the plate was covered with aluminum foil and incubated for 3 hrs at 37°C. The purple Formozan product was dissolved by addition of 100 µL of DMSO to each well. The absorbance was observed at 570 nm (measurement) and 630 nm (reference) using a 96-well plate reader (Bio-Rad, iMark, USA). In order to calculate the respective mean values, data were collected for triplicates each. The following formula was used to calculate the percentage of growth inhibition.

$$\text{Growth inhibition} = \frac{\text{Mean OD of control cells} - \text{Mean OD of treated cells}}{\text{Mean OD of Control}} \times 100$$

From the obtained values, IC_{50} for 24 hrs for MCF-7 cells were computed by probit analysis using statistical package for the social sciences (SPSS) windows based software.

12.2.5.3 ACRIDINE ORANGE (AO)/ETHIDIUM BROMIDE (EB) (AO/EB) STAINING FLUORESCENT ASSAY FOR CELL DEATH

Apoptotic morphology was investigated by AO/EB double staining technique as described by (Spector et al., 1998). MCF-7 cells were cultured separately in 6-well plates and treated with IC_{50} concentration of the plant extracts for 24 hrs. DMSO (0.02%) was used as solvent control. The treated and untreated cells (25 µL of suspension containing 5000 cells) were incubated with the solution of acridine orange (AO) and ethidium bromide (EB) (1 part of 100 µg/mL each of AO and EB in PBS) and examined in a fluorescent microscope (Carl Zeiss, Jena, Germany) using a UV filter (450–490 nm). For each sample, three hundred cells were counted, in triplicate. In every time, point, and scored as viable or dead cells by apoptosis or necrosis as judged from the nuclear morphology and cytoplasmic organization. The percentage of apoptotic (dead cells) and necrotic cells was calculated from the data. Morphological features of dead cells were photographed under fluorescent Microscope.

12.2.6 STATISTICAL ANALYSIS

Data from each of the three experiments are expressed as mean ± standard deviation (SD). IC$_{50}$ values were calculated by probit analysis using windows based SPSS statistical software.

12.3 RESULTS AND DISCUSSION

12.3.1 PRELIMINARY PHYTOCHEMICAL SCREENING

The phytochemical study of marigold flower showed the presence of flavonoids, alkaloids, phenols, phytosterols, tannins, and terpenoids (Table 12.1). Flavonoids are potent water-soluble antioxidants and free radical scavengers, which prevent oxidative cell damage and have strong anticancer activity (Del-Rio et al., 1887; Okwu, 2004). They can also inhibit enzymes such as prostaglandin synthase, lypoxygenase, and cyclooxygenase, closely related to tumorogenesis (Laughton et al., 1991; Smith and Yang, 1994). Plants which constitute a high content of polyphenols and flavanoids exhibited good anticancer, inflammatory, and antifungal activities (Attarde et al., 2011). Thus, presence of flavonoids in the study plant indicates towards the potential anticancer activity of the plants.

12.3.2 GC-MS ANALYSIS OF CHLOROFORM EXTRACT OF T. ERECTA FLOWER

GC-MS study was employed for the identification of plant materials which constitute high complexes of phytocompound (Robards et al., 1999; Nikolic, 2006). The spectrum profile of GC-MS (Figure 12.1) confirmed the presence of eleven major compounds. Identified compounds in chloroform extract of *T. erecta* were 4-[2,6,6-Trimethyl-cyclohex-1-enyl]-but-3-en-2-one oxime, Flavone, Methyl tetradecanoate, 4H-1-Benzopyran-4-one, 5,7-dihydroxy-3-phenyl-, 4'5,7-Trihydroxy isoflavone, 4'-Methoxy-5,7-dihydroxy isoflavone, 14,17-Octadecadienoic acid, methyl ester, 4H-1-Benzopyran-4-one, 2-[3,4-dimethoxyphenyl]-7-hydroxy, 4H-1-Benzopyran-4-one, 2[3,4-dimethoxyphenyl]-7-methoxy-, 4H-1-Benzopyran-4-one, 3,5,7-Trihydroxy-2-[4-hydroxyphenyl]-6-methoxy and 4H-1-Benzopyran-4-one,

2-[2,6-dimethoxyphenyl]-5,6-dimethoxy with their RT of 5.28, 5.46, 7.29, 5.69, 25.14, 5.32, 14.68, 12.64, 5.91, 6.54 and 6.05% respectively. The compound 4'5,7-Trihydroxy isoflavone showed highest PA (25.14%). The molecular structure (MS), MF, and MW of compounds are illustrated in Table 12.2. Many of these compounds are possessed different therapeutic values including cancer treatment. Genistein is also called as 4'5,7-Trihydroxy isoflavone and 4H-1-Benzopyran-4-one, 6,7-dimethoxy-3-phenyl are reported to have anticancer activity (Li et al., 2012). They stated that Genistein and its synthetic derivatives may be an emerging new type of anticancer agent.

TABLE 12.1 Phytochemical Screening of Chloroform Extract of *T. Erecta*

S. No.	Phytochemical	Test Name	Results
1.	Alkaloids	a. Hagers test	Positive
		b. Mayers test	Positive
		c. Wagners test	Positive
2.	Carbohydrates	a. Barfoeds test	Negative
		b. Benedict test	Negative
		c. Fehlingsi test	Negative
3.	Cardiac glycosides	a. Bromine water test	Negative
4.	Flavonoids	a. Alkaline test	Positive
		b. Ferric chloride test	Positive
		c. Lead acetate test	Positive
		d. Shinoda test	Positive
		e. Mg turning test	Positive
		f. Zinc test	Positive
5.	Phenols	a. Alkaline test	Positive
		b. Ferric chloride test	Positive
		c. Lead acetate test	Positive
6.	Protein	a. Millions test	Negative
		b. Xanthoproteic test	Negative
7.	Saponins	a. Bubble test	Negative
		b. Emulsion test	Negative
		c. Foam test	Negative
8.	Phytosterols	a. Liberman-Burchards test	Positive
9.	Tannins	a. Ferric chloride test	Positive
		b. Lead acetate test	Positive
10.	Terpenoids	a. Con H_2SO_4 test	Positive
		b. Liberman-Burchards test	Positive

Phytochemical Investigation and Evaluation of Anti-Breast Cancer

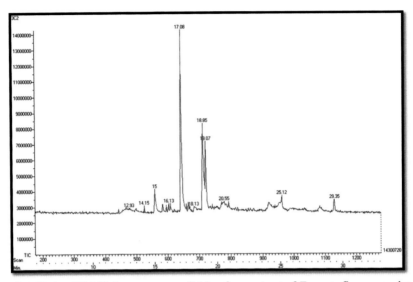

FIGURE 12.1 GC-MS chromatogram of chloroform extract of *T. erecta* flower powder.

TABLE 12.2 Bioactive Compounds Identified in Chloroform Extract of *T. Erecta* Flower Powder by GC-MS Analysis

S. No	RT	Compound Name	MF	MS	MW	PA (%)
1.	13	4-[2,6,6-Trimethyl-cyclohex-1-enyl]-but-3-en-2-one oxime	$C_{13}H_{20}O$		192	5.28
2.	14	Flavone	$C_{15}H_{10}O_2$		222	5.46
3.	15	Methyl tetradecanoate	$C_{15}H_{30}O_2$		242	7.29
4.	16	4H-1-Benzopyran-4-one, 5,7-dihydroxy-3-phenyl-	$C_{15}H_{10}O_4$		254	5.69
5.	17	4',5,7-Trihydroxy isoflavone	$C_{15}H_{10}O_5$		270	25.14

TABLE 12.2 *(Continued)*

S. No	RT	Compound Name	MF	MS	MW	PA (%)
6.	18	4'-Methoxy-5,7-dihydroxy isoflavone	$C_{16}H_{12}O_5$		284	5.32
7.	19	14,17-Octadecadienoic acid, methyl ester	$C_{19}H_{34}O_2$		294	14.68
8.	19	4H-1-Benzopyran-4-one, 2-[3,4-dimethoxyphenyl]-7-hydroxy	$C_{17}H_{14}O_5$		298	12.64
9.	21	4H-1-Benzopyran-4-one, 2[3,4-dimethoxyphenyl]-7-methoxy-	$C_{18}H_{16}O_5$		312	5.91
10.	25	4H-1-Benzopyran-4-one, 3,5,7-Trihydroxy-2-[4-hydroxyphenyl]-6-methoxy	$C_{16}H_{17}O_7$		320	6.54
11.	29	4H-1-Benzopyran-4-one, 2-[2,6-dimethoxyphenyl]-5,6-dimethoxy	$C_{19}H_{18}O_6$		342	6.05

MF: Molecular Formula; MS: Molecular Structure; MW: Molecular Weight; PA: Peak Area.

12.3.3 EFFECT OF CHLOROFORM EXTRACT OF T. ERECTA TREATMENT ON THE VIABILITY OF BREAST CANCER CELL LINE

Most of the currently used anticancer drugs are highly toxic, expensive, and resistance mechanisms pose a significant problem (Lippert et al., 2008). Extensive reported have documented on medicinal plant extract induced cytotoxicity to cancer cells (Sivalokanathan et al., 2000). Experiments on cell lines and in animals demonstrated that the role of herbal drugs anticancer activity by inducing apoptosis and differentiation, enhancing the immune system, inhibiting angiogenesis and reversing multidrug resistance

(Romero et al., 2005). Cytotoxicity assays are an important approach for drug discovery from natural products. In the present study, cytotoxicity of the chloroform extract of *T. erecta* at different concentrations for 24 hrs on MCF-7 cells was investigated by MTT assay. The chloroform extract of *T. erecta* possessed a higher cytotoxic effect on MCF-7 cell lines with the IC$_{50}$ values of 2.464mg/ml (Figure 12.2). The maximum growth inhibition (66.2 ± 0.52%) was obtained at the concentration of 3.0mg/ml) (Table 12.3). The cytotoxic effect of marigold may be due to the presence of flavonoids and other phytochemical compounds. Flavonoids proved their anticancer activity by arresting the cell cycle. Cyclin-dependent kinases (CDKs) are recognized as key regulators of cell cycle progression. Alteration and deregulation of CDK activity resulted in a non-stop cycle. They are pathogenic hallmarks of neoplasia. Various types of cancers are associated with hyperactivation of CDKs due to the mutation of CDK genes or CDK inhibitor genes. Therefore, inhibitors or modulators are of great interest as novel therapeutic agents in cancer (Senderowicz, 2001). Checkpoints at both G1/S and G2/M of the cell cycle in cultured cancer cell lines have been found to be perturbed by flavonoids, such as silymarin, genistein, quercetin, daidzein, luteolin, kaempferol, apigenin, and epigallocatechin 3-gallate (Choi et al., 2001). Many studies revealed that flavonoids could arrest cell cycle during either G1 or G2/M by inhibiting all CDKs (Sakagami, 2001; Wang, 2000).

TABLE 12.3 Cytotoxic Effect of Chloroform Extract of *T. Erecta* Flower on MCF-7 Cell Lines

Concentration	Cell Inhibition (%) (Mean ± SD)	Cell Viability (%) (Mean ± SD)
Control	0.0 ± 0.00	100.0 ± 0.00
	(0.0–0.0)	(100.0–100.0)
0.3	3.8 ± 5.41	96.2 ± 5.41
	(0.0–7.7)	(92.4–100.0)
0.6	12.0 ± 7.22	88.0 ± 7.22
	(6.9–17.1)	(82.9–93.1)
0.9	20.1 ± 2.93	79.9 ± 2.93
	(18.0–22.2)	(77.8–82.0)
1.2	24.7 ± 17.88	75.3 ± 17.88
	(12.0–37.3)	(62.7–88.0)
1.5	25.2 ± 5.52	74.8 ± 5.52
	(21.3–29.1)	(70.9–78.7)
1.8	29.0 ± 17.12	71.0 ± 17.12
	(16.9–41.1)	(58.9–83.1)

TABLE 12.3 *(Continued)*

Concentration	Cell Inhibition (%) (Mean ± SD)	Cell Viability (%) (Mean ± SD)
2.1	36.5 ± 13.27 (27.1–45.9)	63.5 ± 13.27 (54.1–72.9)
2.4	48.1 ± 7.40 (42.9–53.4)	51.9 ± 7.40 (46.6–57.1)
2.7	56.4 ± 7.64 (51.0–61.8)	43.6 ± 7.64 (38.2–49.0)
3.0	66.2 ± 0.52 (65.8–66.5)	33.8 ± 0.52 (33.5–34.2)

Values within the parentheses are the range of respective mean.

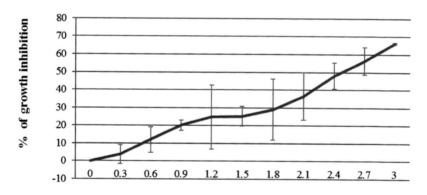

FIGURE 12.2 Cytotoxic activity of chloroform extract of *T. erecta* flower against MCF-7 breast cancer cell line by MTT assay.

12.4 MORPHOLOGY OF NORMAL AND EXTRACT TREATED MCF-7 CELLS BY AO/EB STAINING

AO/EB staining method was adopted for studying morphological changes in treated cells. MCF-7 cells treated with chloroform extract of *T. erecta* exhibited notable morphological changes related to apoptosis-like blebbing formation, nuclear formation, and apoptotic bodies. The treatment with chloroform extract apoptotic cells showed early apoptotic cells (perinuclear chromatin condensation is shown as bright green patches.). A fluorescent image reveals that *T. erecta* extracts triggered morphological characters that related to apoptosis. In several such cases, the fluorescent bright green color (chromatin condensation), apoptotic body formation (cell membrane cleavage, cell shrinkage, scattered cytoplasm, and cellular content release) were observed. However, early apoptosis bright green nuclei with perinuclear condensation were mostly observed in chloroform extract of *T. erecta* treated MCF-7 cell lines. The mean percentage of apoptosis was found to be (59.0%) and necrosis was found to be 8.3% (Table 12.4) in the MCF-7 cells treated with chloroform extract of *T. erecta*. The chloroform extract of *T. erecta* induced more necrosis in the MCF-7 cells than other extracts (Figure 12.3).

TABLE 12.4 Percentage of Normal, Apoptotic, and Necrotic cells on Human Breast Cancer Cell Line (MCF-7) After the Treatment with Chloroform Extract of *T. Erecta* Flower Powder

S. No.	Type of Extract	Normal Cells (%)	Mode of Cell Death	
			Apoptosis (%)	Necrotic (%)
1	Control	90.3 ± 2.08	5.3 ± 2.52	4.3 ± 1.53
		(88.0–92.0)	(3.0–8.0)	(3.0–6.0)
2	Chloroform extract of *T. erecta*	32.7 ± 1.53	59.0 ± 1.73	8.3 ± 0.58
		(31.0–34.0)	(58.0–61.0)	(8.0–9.0)

Value within the parentheses are ranges of respective mean.

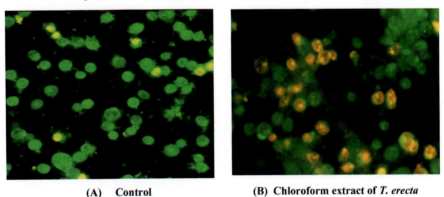

(A) Control (B) Chloroform extract of *T. erecta*

FIGURE 12.3 Morphological changes observed for control and chloroform extract of *T. erecta* treated (24 hrs) MCF-7 cells stained with acridine orange and ethidium bromide. Viable cells have uniform bright green nuclei with organized structure. Early apoptotic cells have green nuclei, but perinuclear chromatin condensation is visible as bright green patches or fragments. Late apoptotic cells have orange to red nuclei with condensed or fragmented chromatin.

12.5 CONCLUSION

Phytochemical and GC-MS analysis of marigold flower exhibited the presence of flavonids and other anticancer compounds. The results obtained from the *in vitro* studies performed using the MCF-7 cell lines revealed chloroform extract of *Tagetes erecta* flower has potent anticancer activity. Thus, the marigold is a potential resource for the discovery of novel leads for the preparation of effective and safe anticancer drugs.

ACKNOWLEDGMENT

The authors thank the Management, the Principal, and Head of the Department of Zoology, Nehru Memorial College, Puthanampatti, Tiruchirappalli district, Tamil Nadu, India for providing necessary facilities to do this research work successfully. Authors thank Dr. M. Akbarsha, Director, Mahatma Gandhi Doerenkamp Center for alternatives to use of animals in life science education, Bharathidasan University, Tiruchirappalli for giving permission to do cell line work and the first author thank UGC, New Delhi, for awarding Rajiv Gandhi National Fellowship (RGNF-SRF) and financial support.

KEYWORDS

- apoptosis
- cytotoxic activity
- flavonoids
- gas chromatography-mass spectrometry
- phytochemistry
- *Tagetes erecta*

REFERENCES

Arts, I. C., Jacobs, D. R., J., Gross, M., Harnack, L. J., & Folsom, A. R., (2002). Dietary catechins and cancer incidence among postmenopausal women: The low a women's health study (United States). *Cancer Cause and Control, 13*(4), 373–382.

Attarde, D. L., Chaudhari, A. J. M., & Murali, A., (2011). Phytochemical investigation and *in vitro* antioxidant activity of extracts from leaves of *Limonia acidissima* L. (Rutaceae). *J. Pharm. Res., 33*(2), 225–227.

Balunas, M. J., & Kinghorn, A. D., (2005). Drug discovery from medicinal plants. *Life Science, 78*(5), 431–441.

Choi, J. A., Kim, J. Y., Lee, J. Y., Kang, C. M., Kwon, H. J., & Yoo, Y. D., (2001). Induction of cell cycle arrest and apoptosis in human breast cancer cells by quercetin. *Int. J. Oncol., 19*(4), 837–844.

Dai, J., & Mumper, R. J., (2010). Plant phenolics: Extraction, analysis and their antioxidant and anticancer properties. *Molecules, 15*(10), 7313–7352.

Davis, W., Lamson, M. S., Matthew, S., & Brignall, N. D., (2000). Antioxidants and cancer III: Quercetin. *Altern. Med. Rev., 5*(3), 196–208.

Del-Rio, A., Obdululio, B. G., Casfillo, J., Marin, F. G., & Ortuno, A., (1887). Uses and properties of citrus flavonoids. *J. Agric. Food. Chem., 45,* 4505–4515.

Duke, J. A., & Ayensu, E. S., (1985). *Medicinal Plants of China*. Reference Publication, Inc. ISBN 0-917256-20-4.

Kokate, A. P., (1997). *Purohit and Ghokhale*. Pharmacognosy. Nirali Prakashan, Pune, India.

Laughton, J. M., Evans, P. J., Moroney, M. A., Hoult, J. R., & Halliwell, B., (1991). Constitutive over expression of 89 kDa heat shock protein gene in the HBL100 mammary cell line converted to a tumorigenic phenotype by the EJ/ T24 Harvey-rats oncogene. *Oncogene., 6*(7), 1125–1132.

Li, H. Q., Luo, Y., & Qiao, C. H., (2012). The mechanisms of anticancer agents by genistein and synthetic derivatives of isoflavone. *Mini Rev Med Chem., 12*(4), 359–362.

Lipppert, T. H., Ruoff, H. J., & Volm, M., (2008). Instrinsic and acquired drug resistance in malignant tumors. The main reason for therapeutic failure. *Arzneimittelforchung, 58*(6), 261–264.

Manandhar, N. P., (2002). *Excellent Book, Covering Over 1500 Species of Useful Plants from Nepal Together with Information on the Geography and Peoples of Nepal*. Plants and people of Nepal Timber Press. *Oregon*. ISBN 0-88192-527-6.

Middleton, V., Vukovi, N., Niforovi, N., Soluji, S., Mladenovi, M., & Maskovi, P., (2011). The effects of plant flavonoids on mammalian cells: Implications for inflammation heart disease and cancer. *Pharmcol. Rev., 52*(4), 673–751.

Mosmann, T., (1983). Rapid colorimetric assay for cellular growth and survival: Application to proliferation and cytotoxicity assays. *J. Immunol. Methods., 65*(1), 55–63.

Nikolic, K. M., (2006). Theoretical study of phenolic antioxidants properties in reaction with oxygen centered radicals. *J. Mol. Struct. Theochem., 774*(3), 95–105.

Okwu, D. E., (2004). Phytochemical and vitamin content of indigenous spices of Southeastern. *Nigeria. J. Sustain. Agric. Environ., 6*(1), 30–37.

Park, S. J., Myoung, H., Kim, Y. Y., Paeng, J. Y., Park, J. W., Kim, M. J., & Hong, S. M., (2008). Anticancer effects of *Annona squamosa* seed extracts is through the generation of free radicals and induction of apoptosis. *Indian J. Biochem. Biophys., 41*(4), 167–172.

Robarts, K., Pernzler, P. D., Tucker, G., Swatsitang, P., & Glover, W., (1999). Phenolic compounds and their role in oxidative processes in fruits. *Food. Chem., 66*(4), 401–436.

Romero, M. R., Efferth, T., Serrano, M. A., Castano, B., Macias, R. I. R., Briz, O., & Marin, J. J. G., (2005). Effect of artemisinin/artesunate as inhibitors of hepatitis B virus production in an "*in vitro*" replicative system. *Antiviral. Res., 68*(2), 75–83.

Sakagami, H., Jiang, Y., Kusama, K., Atsumi, T., Ueha, T., Toguchi, M., Iwakura, I., Satoh, K., Fukai, T., & Nomura, T., (2000). Induction of apoptosis by flavones, flavonols (3-hydroxyflavones) and isoprenoid-substituted flavonoids in human oral tumor cell lines. *Anticancer. Res., 20*(1A), 271–277.

Senderowicz, A. M., (2001). Development of cyclin-dependent kinase modulators as novel therapeutic approaches for hematological malignancies. *Leukemia, 15*(1), 1–9.

Shukla, Y., Prasad, S., Tripathi, C., Singh, M., George, J., & Kalra, N., (2007). *In vitro* and *in vivo* modulation of testosterone mediated alteration in apoptosis-related protein by [6]-gingerol. *Mol. Nutr. Food. Res., 51*(12), 1492–1502.

Sivalokanathan, S., Ilayaraja, M., & Balasubramanium, M. P., (2000). Efficacy of Terminalia Arjuna (Roxb) on N-nitrosodiethylamine induced hepatocellular carcinoma in rats. *Indian J. Exp. Biol., 43,* 264–267.

Smith, T. J., & Yang, C. S., (1994). Effect of food phytochemical on metabolism and tumorigenesis. In: Huang, M. T., (ed.), *Food Phytochemicals for Cancer Prevention 1* (p. 48). American Chemical Society, Washington, Dc.

Spector, D. L., Goldman, R. D., & Leiwand, L. A., (1998). A simple technique for quantitation of low levels of DNA damage in individual cells. *Exp. Cell. Res., 175*(1), 184–191.

Wang, H. K., (2000). The therapeutic potential of flavonoids. *Expert. Opin. Investig. Drugs, 9*(9), 2103–2119.

Yao, L. H., Jiang, Y. M., Shi, J., Tomas-Barberan, F. A., Datta, N., Singanusong, R., & Chen, S. S., (2004). Flavonoids in food and their health benefits. *Plant Food Hum. Nut., 59*(3), 113–122.

CHAPTER 13

Isolation, Identification and *In Silico* Evaluation of Anticancer Activity of Flavone from *Pisum sativum*

R. ARUL PRIYA, K. SARAVANAN, and P. KARUPPANNAN

P.G. and Research Department of Zoology, Nehru Memorial College (Autonomous), Puthanampatti, Trichy District, Tamil Nadu, India, E-mail: kaliyaperumalsaravanan72@gmail.com (K. Saravanan)

ABSTRACT

The present investigation was aimed to discover the lead compound from *P. sativum* seed for cancer treatment. Molecular docking studies were performed for the isolated compound against cancer target, Heat Shock Protein (Hsp 90). Preliminary phytochemical tests confirm the presence of flavonoids. GC-MS profile showed the occurrence of 11 compounds in the seeds of *P. sativum*. Among these, flavone was selected as ligand for docking studies. Docking study showed that the ligand flavone was strongly bound with target Hsp 90 with low binding energy. Thus, the flavone isolated from *P. sativum* is an emerging lead for the process of bringing a new cancer drug.

13.1 INTRODUCTION

Cancer is the second leading cause of death, after cardiovascular disease (Jemel et al., 2007) and the most progressive and destructive disease causing mortality to all over the world. In spite of good advancements in diagnosis and treatment, cancer is still a big threat to human society (Kotnis et al., 2005). Of late, the paradigm of cancer treatment has advanced to use a combination approach, which targets an individual protein. Inhibition of heat-shock protein 90 (Hsp90) is one of the novel key cancer targets. Because of its ability to target several signaling pathways, Hsp90 inhibition emerged as a useful strategy to cure different types of cancers (Kumalo et al., 2015).

Flavonoids are important bioactive compounds of plants (Karimi et al., 2011). They had potent antioxidant and anti-aging properties (Middleton et al., 2000). They are also having the property of reducing the risk of cancer (Yao et al., 2004) and ability to inhibit cancer cell growth (Davis et al., 2000; Arts et al., 2002; Shukla et al., 2007; Mavundza et al., 2010). Flavonoids are found in almost all plants especially rich in plants of the Leguminosae family. Thus, this research work was attempted to screen flavonoid content in pea seeds *Pisum sativum* and evaluated the molecular docking efficiency of flavone against a cancer target protein (Hsp90).

13.2 MATERIALS AND METHODS

13.2.1 COLLECTION OF PLANT MATERIALS

In the present study, *Pisum sativum* (Pea) seeds were collected from local area Namakkal district. They were dried under shadow condition and powdered using a mechanical grinder.

13.2.2 EXTRACTION PREPARATION

Plant extract was prepared using ethyl acetate solvent by Soxhlet extraction method. Then the extract was concentrated by distillation unit and stored at 4°C in the desiccator.

13.2.3 PHYTOCHEMICAL AND GC-MS ANALYSIS

Phytochemical screening was done by the method of Harborne (1998) and Kokate (2005). The bioactive compounds present in the extract were identified by GC-MS analysis. The *P. sativum* extract was analyzed for identification of flavonoid compounds by using JEOL GC MATE II instrument. Interpretation of Mass spectrum of GC-MS was done by comparing with the database at the National Institute of Standard and Technology (NIST). The name of the compound, molecular weight (MW) and the molecular formula (MF) of the compounds of the plant extracts were also retrieved from NIST, Guidechem, Chemspider, and Pubchem Libraries.

13.2.4 MOLECULAR DOCKING

The target protein, Heat Shock Protein 90 (Hsp 90) was retrieved from protein data bank (PDB). The structure of ligand was sketched using Chemsketch software. The molecular docking was done by the Autodock ver. 4.0.

13.3 RESULTS AND DISCUSSION

13.3.1 PHYTOCHEMICAL AND GC-MS ANALYSIS

P. sativum is a legume plant its phytochemical screening confirmed the occurrence of flavonoids (Table 13.1). Plant-derived products and traditional Indian medicine in cancer treatment reduce adverse side effects and exhibited low or almost no toxicity to normal tissues (Dai et al., 2011; Shynu et al., 2011). Flavonoids had potent anti-mutagenic and anti-malignant effect (Fotsis et al., 1997). Chemopreventive role of flavonoids in cancer by signal transduction in cell proliferation and angiogenesis (Wagner et al., 1986). GC-MS profile revealed the presence of 11 bioactive compounds (Figure 13.1 and Table 13.2). Flavone is one among the 11 compounds and it was reported as an anticancer compound (Batra and Sharma, 2013).

TABLE 13.1 Phytochemical Screening of Ethyl Acetate Extract of *P. sativum*

S. No	Phytocompounds	Status
1	Alkaloids	Present
2	Carbohydrates	Present
3	Cardiac glycosides/Glycosides	Present
4	Flavonoids	Present
5	Phenols	Present
6	Protein	Present
7	Saponin	Present
8	Phytosterols	Present
9	Tannins	Present
10	Triterpenoids	Present

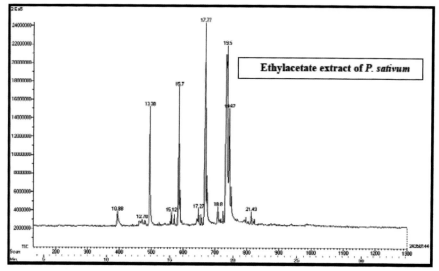

FIGURE 13.1 GC-MS chromatogram of ethylacetate extract of *P. sativum* seed.

TABLE 13.2 GC-MS Results of Ethyl Acetate Extract of *P. sativum* Seeds

S. No	RT Value	Compound Name	Molecular Formula	Molecular Weight (g/mol)	Compound Structure	% Peak Area
1.	10.88	Tetradecane	$C_{14}H_{30}$	198		3.33
2.	15.12	Hexadecane, 2 methyl-	$C_{17}H_{36}$	240		3.28
3.	17.78	n-Hexadecanoic acid	$C_{16}H_{32}O_2$	256		20.37
4.	15.7	Octadecane	$C_{18}H_{38}$	254		14.98
5.	19.67	Docosane	$C_{22}H_{46}$	311		12.95
6.	13.38	Hexadecane	$C_{16}H_{34}$	226		13.10
7.	17.27	4H-1-Benzopyran-4-one, 7-hydroxy-3-[4-methoxyphenyl]-				3.58
8.	21.43	Tetracosane	$C_{24}H_{50}$	339		3.35

TABLE 13.2 *(Continued)*

9.	19.5	4'-Methoxy-5,7-dihydroxy isoflavone	$C_{16}H_{12}O_5$	284		18.84
10.	12.82	Flavone	$C_{15}H_{10}O_2$	222		2.44
11.	18.8	4H-1-Benzopyran-4-one, 5,7-dihydroxy-2-[4-hydroxyphenyl]-6-methoxy-	$C_{16}H_{12}O_6$	300		3.79

13.3.2 MOLECULAR DOCKING STUDIES

The flavone (Figure 13.3) was selected as a ligand for docking against Heat shock protein (Hsp 90) (Figure 13.2). Molecular docking study is very important to know the interaction properties like binding energy, geometry complementarity, electron distribution, hydrogen donor-acceptor, hydrophobicity, and polarizability.

So, it contributed a vital role in the process of making the new drug in the identification of innovative small molecular scaffolds, exhibiting the important properties with selectivity for the target (Krovat, 2005). Drug-likeliness and docking efficiency of flavone with HsP 90 protein were studied through *in silico* docking, and results were presented in Table 13.3 and illustrated in Figures 13.3–13.5. Molecular properties of the flavone were observed to be decent score and drug properties. The Molecular properties such as MF: $C_{12}H_{10}O_2$, MW: 222.07, Number of HBA:2, Number of HBD:0, Mol log P: 3.96, Mol Logs: –4.56, MolPSA:21.79 A^2, Mol Vol.: 228.78 A^3, The drug-likeness score of the flavone was observed to be 0.58.

The activity of Hsp90 in cancerous cells appears to be closely associated with the proliferation of malignant cells and it allows tumor cells to escape from apoptotic death (Takayama et al., 2003). Further, HsP 90 plays a vital part in controlling cell growth, survival, and development of tissue during tumorigenesis by facilitating an escape from normal proteolytic turnover (Pratt, 1998). Hsp90 substrate and its active site include mixed hydrophobic, polar, and charged amino acids. The residues include Leu48, Asn51, Asp54, Ala55, Lys58, Ile91, Asp93, Ile96, Gly97, Met98, Asn106, Leu107, Lys112, Gly135, Phe138, Val150, Thr184, and Val186 were found as the potential catalytic sites to bind with the ligand molecule.

From the docking study, the bioactive compound flavone was effectively bound with target protein Hsp 90 with minimum binding energy (–6.99 Kcal/mol) (Figure 13.5). When the pocket is increasingly hydrophobic toward the bottom, one charged residue and one polar residue are retained as Asp93 and Thr184 respectively (Stebbins et al., 1997). The adenine ring in ATP sites at the bottom of the pocket and its N_6 group form direct hydrogen bonds with Asp93, and water-mediated hydrogen bonds with the Ser52 and Leu48. Also, Glu47 gets involved in Hsp90 ATPase activity (Prodromou et al., 1997; Obermann et al., 1998). The high docking frequency and score were obtained in the HsP 90. Hence the *In silico* study recommends that the flavone as a cancer drug candidate.

TABLE 13.3 Molecular Properties and Drug Likeness Score of Flavone Isolated from Ethylacetate Extract of *P. sativum*

Molecular Formula	Molecular Weight	No. of HBA	No. of HBD	Mol. LogP	Mol Logs moles/L	Mol PSA (A^2)	Mol Vol (A^3)	Drug Score
$C_{12}H_{10}O_2$	222.07	2	0	3.96(<5)	–4.56 (6.06)	21.79	228.78	0.58

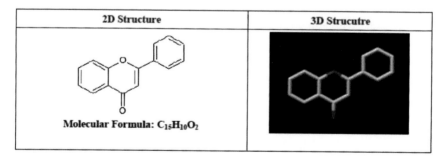

FIGURE 13.2 2D and 3D structure of ligand (flavone).

Isolation, Identification and *In Silico* Evaluation of Anticancer Activity   173

Molecular Properties and Drug-likeness.

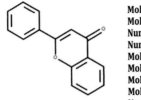

Molecular formula: C15 H10 O2
Molecular weight: 222.07
Number of HBA: 2
Number of HBD: 0
MolLogP: 3.96
MolLogS: -4.56 (in Log(moles/L)) 6.06 (in mg/L)
MolPSA: 21.79 Å^2
MolVol: 228.78 Å^3
Number of stereo centers: 0

Drug-likeness model score: 0.58

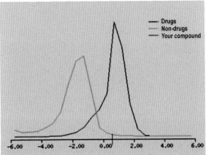

FIGURE 13.3 Prediction of molecular properties and drug-likeness of ligand flavone.

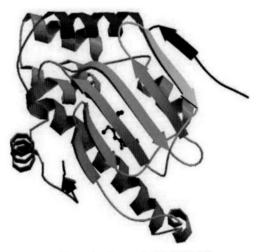

FIGURE 13.4 3D structure of heat shock protein 90 (HSP 90).

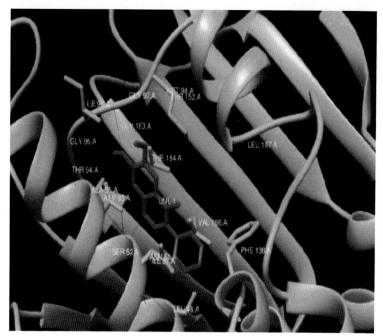

FIGURE 13.5 Docking result of Hsp 90 with ligand flavone.

13.4 CONCLUSION

Ethyl acetate extract of *P. sativum* seeds possesses several bioactive compounds including flavonoids. GC-MS analysis confesses the presence of Flavone in the ethyl acetate extract of *P. sativum* seeds. Docking results exhibited that flavone isolated from pea plant was effectively bound with target protein Hsp 90 with minimum binding energy. Henceforth the flavone lead was proposed and prescribed for successful designing of a drug for HsP90 targeted cancer therapies.

ACKNOWLEDGMENT

The authors thank the Authority and Head of the institution of Nehru Memorial College (Autonomous), Puthanampatti for providing necessary facilities. The first author acknowledges the UGC, New Delhi for providing financial assistance by awarding UGC-RGNF.

KEYWORDS

- flavone
- GC-MS
- Hsp 90
- molecular docking
- phytochemistry
- *Pisum sativum*

REFERENCES

Arts, I. C., Jacobs, D. R. J., Gross, M., Harnack, L. J., & Folsom, A. R., (2002). Dietary catechins and cancer incidence among postmenopausal women: The low a women's health study (United States). *Cancer Cause and Control., 13*(4), 373–382.

Batra, P., & Sharma, A. K., (2013). Anti-cancer potential of flavonoids: Recent trends and future perspectives. *Biotech., 3*(6), 439–459.

Dai, Z. J., Gao, J., Li, Z. F., Ji, Z. Z., Kang, H. F., Guan, H. T., Diao, Y., Wang, B. F., & Wang, X. J., (2011). In vitro and in vivo antitumor activity of *scutellaria barbate* extract on murine liver cancer. *Molecules, 16,* 4389–4400.

Davis, W., Lamson, M. S., Matthew, S., & Brignall, N. D., (2000). Antioxidants and cancer III: Quercetin. *Altern. Med. Rev., 5*(3), 196–208.

Fotsis, T., Pepper, M. S., Aktas, E., Breit, S., Rasku, S., & Adlercreutz, H., (1997). Flavonoid, dietary-derived inhibitors of cell proliferation and *in vitro* angiogenesis. *Cancer Research, 57,* 2916–2921.

Harborne, J. B., (1998). *Phytochemical Methods: A Guide to Modern Techniques of Plant Analysis* (2nd edn., pp. 54–84). London: Chapman and Hall.

Jemel, A., Siegel, R., Ward, E., Murray, T., Xu, J., & Thun, M. J., (2007). Cancer statistics, CA. *Cancer. J. Clin., 57*(1), 43–66.

Karimi, E., Jaafar, H. Z. E., & Ahmed, S., (2011). Phytochemical analysis and antimicrobial activities of methanolic extracts of leaf stem and root from different varieties of *Labsia Pumila* benth. *Molecules, 16*(6), 4438–4450.

Kokate, C. K., (2005). *A Textbook of Practical Pharmacognosy* (5th edn., pp. 107–111). Vallabh Prakashan, New Delhi.

Kotnis, A., Sarin, R., & Mulherkar, R., (2005). Genotype, phenotype, and cancer: Role of low penetrance genes and environment in tumor susceptibility. *J. Biosci., 30,* 93–102.

Krovat, E. M., Stenidl, T., & Lange, T., (2005). Recent advance in docking and scoring. *Curr. Computer-Aided Drug Des., 1,* 93–102.

Kumalo, H. M., Bhakat, S., & Soliman, M. E., (2015). Heat-shock protein 90 (Hsp90) as anticancer target for drug discovery: An ample computational perspective. *Chem. Biol. Drug Des., 86,* 1131–1160.

Mavundza, E. J., Tshikalange, T. E., Lall, N., Hussein, A. A., Mudau, F. N., & Meyer, J. J. M., (2010). Antioxidant activity and cytotoxicity effect of flavonoids isolated from *Athrixia phylicoides*. *J. Med. Plant. Res., 4*(23), 2584–2587.

Middleton, E., Kandaswami, C., & Theoharides, T. C., (2000). The effects of plant flavonoids on mammalian cells: Implications for inflammation, heart disease, and cancer. *Pharmacol. Rev., 52*(4), 673–751.

Obermann, W. M., Sondermann, H., Russo, A. A., Pavletich, N. P., & Hartl, F. U., (1988). *In vivo* function of Hsp90 is dependent on ATP binding and ATP hydrolysis. *J. Cell. Biol., 143*(4), 901–910.

Pratt, W. B., (1998). The hsp90-based chaperone system: Involvement in signal transduction from a variety of hormone and growth factor receptors. *Proc. Soc. Exp. Biol. Med., 217*, 420–434.

Prodromou, C., Roe, S. M., O'Brien, R., Ladbury, J. E., Piper, P. W., & Pearl, L. H., (1997). Identification and structural characterization of the ATP/ADP-binding site in the Hsp90 molecular chaperone. *Cell, 90*(1), 65–75.

Shukla, Y., Prasad, S., Tripathi, C., Singh, M., George, J., & Kalra, N., (2007). *In vitro* and *in vivo* modulation of testosterone mediated alteration in apoptosis related protein by [6]-gingerol. *Mol. Nutr. Food. Res., 51*(12), 1492–1502.

Shynu, M., Gupta, P. K., & Saini, M., (2011). Antineoplastic potential of medicinal plants. *Recent Patents on Biotechnology, 5*, 85–94.

Stebbins, C. E., Russo, A. A., Schneider, C., Rosen, N., Hartl, F. U., & Pavletich, N. P., (1977). Crystal structure of an Hsp90-geldanamycin complex: Targeting of a protein chaperone by an antitumor agent. *Cell, 89*(2), 239–250.

Takayama, S., Reed, J. C., & Homma, S., (2003). Heat-shock proteins as regulators of apoptosis. *Oncogene, 22*, 9041–9047.

Wagner, H., Geyer, B., Yoshinobu, K., & Govind, S. R., (1986). Coumestan as the main active principles of liver drugs *Eclipta alba* and *Wedelica calendulaceae*. *Planta Medica., 5*, 370–372.

Yao, L. H., Jiang, Y. M., Shi, J., Tomas-Barberan, F. A., Datta, N., Singanusong, R., & Chen, S. S., (2004). Flavonoids in food and their health benefits. *Plant Food Hum. Nut., 59*(3), 113–122.

CHAPTER 14

Functional Lead Phytochemicals from the Rutaceae and Zingiberaceae Plants for Development of Anticancer Drugs

K. SARAVANAN,[1] B. UMARANI,[1] V. BALAN,[1] P. PREMALATHA,[1] and BIR BAHADUR[2]

[1]P.G. and Research Department of Zoology, Nehru Memorial College (Autonomous), Puthanampatti – 621 007, Tiruchirappalli, Tamil Nadu, India, E-mail: umamiraa86@gmail.com (B. Umarani)

[2]Department of Botany, Kakatiya University, Warangal, India

ABSTRACT

Currently, phytochemicals are an important source of drug discovery in recent research for curing various forms of metabolic disorders and diseases including cancer. From this study, data were analyzed to explore the secondary metabolites from herbs with potent cancer treatment property. Functional lead compounds were manually mined from the phytochemical database. We selected two families' *viz.*, Rutaceae, and Zingiberaceae to identify functional lead compounds for the development of anti-cancer drugs. The three plant species from Rutaceae *(Aegle marmelous, Citrus sinensis,* and *Citrus limon),* three plant species from Zingiberaceae family *(Amomum xanthioides, Curcuma longa,* and *Zingiber officinale)* are possessed many bioactive compounds with curative effects of cancer. From these 6 plants, many phytocompounds were identified as anticancer, anticarcinogenic antitumor, and cancer preventive. The present study focused on the leads for the development of anticancer drugs from herbs.

14.1 INTRODUCTION

Cancer is characterized by loss of cell control and uncontrolled growth. As a result of cancer, tumors are appeared with the potential of invading surrounding tissues and organs (Ochwang et al., 2014). Recent days various therapies are available like radiotherapy, chemotherapy, and immunotherapy. And also using drugs synthesized by chemicals. However, these treatments not given as much as satisfaction due to their severances of affecting normal tissues instead of cancer cells and damaging the patient health (Cancer Research, 2014). Therefore, have a need of alternative treatments and therapies against different forms of cancer. Herbs have been used for treatments of various diseases from ancient times. It is the primary agents for treatment of many diseases in developing countries. Thus, researchers are interested to develop to investigate the potential activity behind the medicinal herbs (Sivaraj et al., 2014). This article aims to identify the functional lead compounds for the development of anticancer drugs and their availability in the plants of Rutaceae and Zingiberaceae families.

14.2 MATERIALS AND METHODS

14.2.1 DATA-MINING OF PHYTOCHEMICALS

Phytochemicals with effectiveness of cancer-treating were manually mined from Dr. Duke's phytochemistry and ethnobotanical database (www.ars-grin.gov/duke). Cancer curing activity of phytochemical was taken from research articles. The journals were choose to capture the considerable quantity and quality data in a cost, and time-consuming manner. From research articles, tested compounds details, assays performed information and any others related information for the compounds were abstracted.

14.3 RESULTS AND DISCUSSION

In Rutaceae family, three plants are possessed bioactive compounds with curative effect of cancer. The plants and number of phytocompounds are *Aegle marmelous* (9), *Citrus sinensis* (44), and *Citrus limon* (42). These compounds possess different cancer-curing nature that is anticancer, anticarcinogenic antitumor and cancer-preventive effects (Table 14.1). In Zingiberaceae family, 3 plants have found to possess bioactive compounds with efficiency of treating cancer. The selected species are *Curcuma longa, Amomum xanthioides, and Zingiber officinale* and they included 1, 19 and 41 phytochemicals with cancer-treating effects. A list of bio-actives found in plants species of Rutaceae and Zingiberaceae are given in Tables 14.1–14.3).

TABLE 14.1 List of Compounds with Cancer Treating Efficiency and Their Plant Sources

S. No	Compound Name	Activity	Plant Species	Compound Found in Number of Plant Species
1.	Alanine	Cancer preventive	*Citrus sinensis, Zingiber officinale*	2
2.	Alpha-carotene	Anticancer	*Citrus sinensis*	1
3.	Alpha-tocopherol	Anticancer, cancer preventive, anti-tumor	*Citrus sinensis*	1
4.	Alpha-curcumene	Antitumor activity	*Zingiber officinale*	1
5.	Alpha-humulene	Antitumor	*Citrus lemon*	1
6.	Alpha-linoleic-acid	Cancer preventive	*Citrus sinensis, Zingiber officinale*	2
7.	Alpha-pinene	Cancer preventive	*Citrus sinensis, Citrus lemon, Amomum xanthioides, Curcuma longa, Zingiber officinale*	5
8.	Alpha-terpineol	Anticancer	*Citrus sinensis, Citrus lemon, Curcuma longa, Zingiber officinale*	4
9.	Ar-tumerone	Anti tumor, anticancer	*Curcuma longa*	2
10.	Ascorbic acid	Cancer preventive	*Citrus lemon, Zingiber officinale, Aegle marmelous, Citrus sinensis, Curcuma longa*	5
11.	Bergapten	Cancer preventive, anti-tumor	*Citrus lemon*	1

TABLE 14.1 *(Continued)*

S. No	Compound Name	Activity	Plant Species	Compound Found in Number of Plant Species
12.	Beta carotene	Anticancer, cancer preventive, anti-tumor	*Aegle marmelous, Citrus sinensis, Citrus lemon, Curcuma longa, Zingiber officinale*	5
13.	Beta-ionone	Cancer preventive, anti tumor	*Zingiber officinale*	1
14.	Beta-myrcene	Cancer preventive	*Zingiber officinale*	1
15.	Beta-sitosterol	Cancer preventive	*Citrus sinensis, Citrus lemon, Zingiber officinale*	3
16.	Butyric-acid	Anticancer, anti-tumor	*Citrus sinensis*	1
17.	Caffeic acid	Anticancer, cancer preventive, anti-tumor, anticarcinogenic	*Citrus lemon, Curcuma longa, Zingiber officinale, Citrus sinensis*	3
18.	Caffeine	Cancer-preventive, anti-tumor, anticarcinogenic	*Citrus sinensis, Citrus lemon*	2
19.	Camphor	Cancer preventive	*Zingiber officinale*	1
20.	Capsaicin	Cancer preventive	*Zingiber officinale*	1
21.	Carvone	Cancer preventive	*Citrus sinensis, Citrus lemon*	2
22.	Caryophyllene	Anti-tumor	*Citrus sinensis, Citrus lemon, Curcuma longa*	3
23.	Caryophyllene-oxide	Anti-tumor	*Citrus lemon*	1
24.	Chlorogenic acid	Cancer preventive, anti-tumor, anticarcinogenic	*Zingiber officinale*	1
25.	Cinnamic acid	Cancer preventive	*Curcuma longa*	1
26.	Citral	Anticancer, cancer preventive, anti-tumor	*Aegle marmelous, Citrus lemon, Zingiber officinale*	3

TABLE 14.1 *(Continued)*

S. No	Compound Name	Activity	Plant Species	Compound Found in Number of Plant Species
27.	Citric acid	Antitumor	*Citrus lemon, Citrus sinensis*	2
28.	Curcumenol	Anticancer, anti tumor	*Curcuma longa*	2
29.	Curcumin	Anticancer, cancer preventive, anti tumor, anticarcinogenic	*Curcuma longa, Zingiber officinale*	2
30.	Curdione	Anti tumor	*Curcuma longa*	1
31.	D-limonene	Cancer preventive	*Aegle marmelous*	1
32.	Decon-1-al	Cancer preventive	*Citrus lemon*	1
33.	Delphinidin	Cancer preventive	*Zingiber officinale*	1
34.	Diosmetin	Cancer preventive	*Citrus lemon*	1
35.	Diosmin	Anti-carcinogenic, anticancer	*Citrus sinensis, Citrus lemon*	2
36.	Eugenol	Cancer preventive, anti tumor	*Curcuma longa*	1
37.	Ferulic acid	Cancer preventive, anti-tumor anticarcinogenic	*Zingiber officinale, Citrus sinensis, Citrus lemon*	3
38.	Fiber	Cancer preventive, antitumor	*Aegle marmelous, Citrus sinensis, Citrus lemon, Curcuma longa, Zingiber officinale*	5
39.	Geraniol	Cancer preventive, anti-tumor	*Citrus sinensis, Citrus lemon, Zingiber officinale*	3
40.	Glycine	Cancer preventive	*Citrus sinensis, Zingiber officinale*	2
41.	Imperatorin	Cancer preventive	*Citrus lemon*	1
42.	Isopimpinellin	Cancer preventive	*Citrus lemon*	1
43.	Isorhamnetin	Cancer preventive	*Citrus lemon*	1

TABLE 14.1 *(Continued)*

S. No	Compound Name	Activity	Plant Species	Compound Found in Number of Plant Species
44.	Isovitexin	Cancer preventive	*Citrus lemon*	1
45.	Kamfperol	Anticancer, cancer preventive, anti tumor	*Zingiber officinale*	1
46.	Limonene	Anticancer, cancer preventive, anti-tumor	*Citrus sinensis, Citrus lemon, Curcuma longa, Zingiber officinale*	4
47.	Linoleic-acid	Cancer preventive	*Citrus sinensis, Zingiber officinale*	2
48.	Luteolin	Cancer preventive, anti-tumor and anticarcinogenic	*Citrus lemon*	1
49.	Malic-acid	Anti-tumor	*Citrus sinensis*	1
50.	Methionine	Cancer preventive	*Citrus sinensis, Zingiber officinale*	2
51.	Mucilage	Cancer preventive	*Citrus lemon*	1
52.	Mufa	Cancer preventive	*Zingiber officinale*	1
53.	Myricetin	Cancer preventive	*Zingiber officinale*	1
54.	Myristic acid	Cancer preventive	*Zingiber officinale*	1
55.	Naringenin	Anticancer, cancer preventive, anti-tumor	*Citrus sinensis*	1
56.	Naringin	Cancer preventive	*Citrus lemon, Citrus sinensis*	2
57.	Neohesperidin	Anticancer, cancer preventive	*Citrus sinensis, Citrus lemon*	3
58.	Niacin	Cancer preventive,	*Aegle marmelous, Citrus sinensis, Citrus lemon, Curcuma longa, Zingiber officinale*	5

TABLE 14.1 *(Continued)*

S. No	Compound Name	Activity	Plant Species	Compound Found in Number of Plant Species
59.	N-Methyl tyramine	Cancer preventive	*Citrus lemon, Citrus sinensis*	2
60.	Oleic acid	Cancer preventive	*Zingiber officinale, Citrus sinensis*	2
61.	P-coumaric acid	Cancer preventive, anti tumor	*Citrus lemon, Curcuma longa*	2
62.	P-hydroxyl-benzoic acid	Cancer preventive	*Zingiber officinale*	1
63.	Pantothenic-acid	Cancer preventive	*Zingiber officinale, Citrus sinensis*	2
64.	P-coumaric acid	Cancer preventive, anti-tumor	*Zingiber officinale, Citrus sinensis*	2
65.	Pectin	Cancer preventive	*Citrus sinensis, Citrus lemon*	2
66.	Psoralen	Cancer preventive, antitumor	*Aegle marmelous*	1
67.	Quercetin	Cancer preventive, anti tumor	*Citrus lemon, Zingiber officinale*	2
68.	Riboflavin	Cancer preventive,	*Aegle marmelous, Citrus sinensis, Citrus lemon, Curcuma longa, Zingiber officinale*	5
69.	Rutin	Anticancer, cancer preventive, anti-tumor	*Citrus sinensis, Citrus lemon*	2
70.	Scopoletin	Cancer preventive, anti-tumor	*Citrus lemon*	1
71.	Scutellarein	Cancer preventive	*Citrus sinensis*	1
72.	Selenium	Cancer preventive, anti-tumor	*Citrus sinensis, Zingiber officinale*	2

TABLE 14.1 *(Continued)*

S. No	Compound Name	Activity	Plant Species	Compound Found in Number of Plant Species
73.	Serine	Cancer preventive	Citrus sinensis, Zingiber officinale	2
74.	Shikimic-acid	Anticancer, cancer preventive, anti-tumor	Zingiber officinale	1
75.	Sinapic-acid	Cancer preventive	Citrus lemon, Citrus sinensis	2
76.	Stigmasterol	Cancer preventive	Citrus sinensis	1
77.	Succinic-acid	Cancer preventive	Citrus sinensis	1
78.	Synephrine	Cancer preventive	Citrus sinensis, Citrus lemon	2
79.	Tangeretin	Cancer preventive	Citrus sinensis	1
80.	Tannin	Anticancer, cancer preventive, anti-tumor	Aegle marmelous	1
81.	Terpineol	Anticancer	Curcuma longa	1
82.	Tyrosine	Cancer preventive	Citrus sinensis, Zingiber officinale	2
83.	Umbelliferone	Cancer preventive	Citrus lemon	1
84.	Vanillic acid	Anticancer, cancer preventive, anti tumor	Curcuma longa, Zingiber officinale	4
85.	vanillin	Anticancer, cancer preventive, anti tumor	Zingiber officinale	1
86.	Vicenin-2	Anticancer	Citrus lemon	1

TABLE 14.2 Phytochemical Compounds with Anticancer Activity in Plants of Rutaceae and Zingiberaceae Families

S. No	Plant Name	Family	No. of Cancer Treatment Compound
1.	Aegle marmelous	Rutaceae	9
2.	Citrus sinensis	Rutaceae	44
3.	Citrus limon	Rutaceae	42
4.	Amomum xanthioides	Zingberaceae	1
5.	Curcuma longa	Zingberaceae	19
6.	Zingiber officinale	Zingberaceae	41

TABLE 14.3 Number of Compounds with Cancer Treatment Properties in Rutaceae and Zingiberaceae Families

S. No	Cancer Treatment Properties	#Total Number of Compounds	
		Rutaceae	Zingiberaceae
1.	Anticancer activity	12	14
2.	Cancer preventive	45	33
3.	Antitumor activity	07	06
4.	Anticarcinogenic activity	01	–

14.3.1 AEGLE MARMELOUS (RUTACEAE)

Aegle marmelous is a tree which is deciduous, glabrous, and armed. It is commonly called Bael fruit. In this plant, nine bio-actives with cancer treatment properties were identified. Among nine bioactives, three compounds having anticancer nature, all nine compounds possessing cancer-preventive nature and five compounds possessing antitumor activity. Beta-carotene, Citral, and Tannin having three combined nature *viz.* anticancer, antitumor, and cancer preventive. Fiber and Psoralen possessed nature of cancer preventive and antitumor. According to Jegetia et al. (2005), leaves of *A. marmelous* have shown *in vivo* antitumor and anticancer effect in Ehrlich ascites carcinoma (EAC). According to Costa-lotulo et al. (2005), extract of *A. marmelous* exhibits cytotoxicity against tumor by MTT and BSLA assay.

14.3.2 CITRUS SINENSIS (RUTACEAE)

Citrus sinensis is a tree or shrub with small, shallow-rooted. It is commonly called as sweet orange. From this plant, 44 bioactives are mined related with cancer-curing nature. Among 44 bioactives, 11 bioactive compounds are anticancer nature, 36 bioactive compounds are cancer preventive nature, 16 bioactives are antitumor nature, and five phytocompounds are anticarcinogenic nature. Out of 44 bioactives, five compounds having three combined nature viz., anticancer, cancer preventive, and antitumor. Four compounds having two combined nature of cancer preventive and antitumor. Caffeic acid blessed with having all four-cancer treatment potential. Ferulic acid having activity of cancer-preventive, antitumor, and anticarcinogenic. Neohesperidin having nature of cancer preventive and anticancer. Caffeine has both effect of antitumor and anticarcinogenic and butyric acid has two combined effect of anticancer and antitumor. According to Jie et al. (2013), D-Limonene from citrus plant which showed apoptosis against human colon cancer cell by the mechanism of mitochondrial death pathway and suppressing PI3K/Akt pathway. Perillyl alcohol from citrus has a good effect in malignant brain tumor patients (Cen et al., 2015). According to Murthy et al. (2012), essential oils from orange could inhibit metastasis, angiogenesis in human colon cancer cells.

14.3.3   CITRUS LIMON (L.) (RUTACEAE)

Citrus limon is a tree which originated from Asia. It is commonly called as lemon. Totally 42 bioactive compounds are mined from this plant. Among 42 bioactives of *C. lemon*, 17 bioactives having anticancer property, 34 phytocompounds having cancer-preventive activity, 18 phytocompounds possessed antitumor activity and five bioactive compounds possessed anticarcinogenic nature. From the above compounds, 4 possessed anti-cancer, cancer preventive and antitumor effect. Caffeic acid has blessed with all four combined nature of anticancer, cancer preventive, antitumor, and anticarcinogenic. Among them, six compounds having both effect of cancer preventive and antitumor. Three compounds possessed activity of cancer-preventive, antitumor, and anticarcinogenic. Citrus fruits showed various health beneficial effects that related with human health, including anti-inflammatory, anti-oxidant, anti-microbial, and anticancer effects, as well as protection against cardiovascular and nervous system injury (Lu et al., 2015; Mandalari et al., 2017).

14.3.4   AMOMUM XANTHIOIDES (ZINGIBERACEAE)

It is a large herb of *Amomum villosum* var. It is from Thailand and Myanmar. It is known as Malabar cardamom. It has only one compound that is Alpha-pinene which contains cancer preventive activity. *Amomum xanthiodes* is useful for treatment of gastrointestinal (GI) gland related cancer (Yong et al., 2007).

14.3.5   CURCUMA LONGA (ZINGIBERACEAE)

It is a rhizomatous herbaceous perennial plant. It is native to tropical south Asia. It is commonly called Turmeric. From this plant, 19 phytochemicals are mined. Among them, nine compounds having anticancer nature, nine phytochemicals having cancer-preventive nature and five compounds having antitumor nature. Caffeic acid and curcumin possessed all four cancer treatment properties *viz.* cancer-preventive, anticancer, anticarcinogenic, and antitumor. The compounds beta-carotene, limonene has combined the nature of antitumor and anticancer. The compounds Eugenol, fiber, P-coumaric acid, and vanillic acid possessed both antitumor and cancer preventive nature. The rhizome extract of *Curcuma longa* possesses anticancer activities (Susma et al., 2012). *Curcuma longa* is commonly called turmeric. It is used as a flavor of foods and has high medicinal value. It is used as medicine in various medicinal systems such as Ayurveda, Unani, and Siddha. It is used as a home remedy for many diseases and is called 'wound-healer.' It contains turmeric or curcumin which act as a potent anti-carcinogenic compound and was given by the National Cancer Institute (NCI) (Rathaur et al., 1987). The curcumin from *Curcuma longa* as a good anticancer agent which induces apoptosis and prevents cancerous cell growth (Collett et al., 2001).

14.3.6 ZINGIBER OFFICINALE (ZINGIBERACEAE)

It is herb and perennial, native to India. It is commonly called ginger. This plant totally contains 41 compounds with nature of cancer treatment. Among them, 10 compounds have anticancer potential, 30 phytocompounds having cancer preventive activities and Alpha-curcumene has antitumor activity. Ferulic acid and chlorogenic acid have three effects viz., cancer preventive, antitumor, and anticarcinogenic. Six phytochemicals having both natures of antitumor and cancer preventive. Seven phytochemicals having three effects of anticancer, cancer preventive and antitumor activity. Caffeic acid and curcumin are blessed with all four effects viz., anticancer, cancer preventive, antitumor, and anticarcinogenic activity. The rhizome of *Zingiber officinale* is widely for preparation of various foods and beverages. Ginger leaves lead to transcriptional activation of activating transcription factor 3 (ATF3) which may relate with the reduction of viable cells and induced apoptosis in human colorectal cancer cells (Gwang et al., 2014).

Among the 86 bioactive compounds with nature of cancer-treating, alpha-pinene, ascorbic acid, β carotene, fiber, niacin, and riboflavin are common phytochemicals which present in five species out of six plant species under this review. *A. marmelous, C. sinensis, C. lemon, C. longa, Z. officinale*. These herbs are potent with anticancer nature by many researchers. Thus, research to be needed on these bioactives for the development of anticancer drugs.

14.4 CONCLUSION

This study was done to explore bioactive phytochemicals with a nature of cancer prevention. In present days, researchers are directed towards identification of biologically active compounds from medicinal herbs. Herbs play an essential role in medicinal field in developing countries. The medicinal plants, *C. sinensis,* and *C. lemon* of Rutaceae and *C. longa* and *Z. officinale* of Zingiberaceae had large number of lead compounds for the development of anticancer. However, alpha pinene, ascorbic acid, β carotene, fiber, niacin, and riboflavin are present in almost all plants in these two family. Thus, these plants and these phytochemicals may call as drug candidates for the cancer treatment. Further, these all six plants are advised to add daily diet to reduce cancer risk.

ACKNOWLEDGMENT

The authors thank the Management, the Principal and Head of the Department of Zoology, Nehru Memorial College, Puthanampatti, Trichy for providing necessary facilities to do this research work successfully. The third author thanks the University Grants Commission, New Delhi for awarding Rajiv Gandhi National Fellowship (RGNF-SRF) for financial support.

KEYWORDS

- **anticancer**
- **anticarcinogenic**
- **Citrus limon**
- *Curcuma longa*
- **medicinal plants**
- **phytochemicals**
- **Zingiberaceae**

REFERENCES

Cancer Research. UK, (2014). *What is Cancer?* Available at: http://www.cancerresearchuk.org/about-cancer/what-is-cancer (accessed on 20 February 2020).

Chen, T. C., Fonseca, C. O., & Schönthal, A. H., (2015). Preclinical development and clinical use of perillyl alcohol for chemoprevention and cancer therapy. *Am. J. Cancer Res., 5*, 1580–1593.

Collett, G. P., Robson, C. N., Mathers, J. C., & Campbell, F. C., (2001). Curcumin modifies Apc^{min} apoptosis and inhibits 2-amino-1-methyl-6-phenylimidazo [4,5-b] Pyridine (PhIp) induced tumor formation of Apc ^{min} Mice. *Carcinogenesis, 22*(5), 821–825.

Costa-Lotulo, L. V., Khan, M. T., Ather, A., Wilke, D. K., Jimenez, P. C., Pessoa, C., De Moraes, M. E., De & Morales, M. O., (2005). Studies of the anticancer potential of plants used in Bangladeshi folk medicine. *J. Ethnopharmacol., 21*, 99.

Gwang, H., Jae, H. P., Hun, M. S., Hyun, J. E., Mi, K. K., Jin, W. L., Man, H. L., Kiu, H. C., Jeong, R. L., Hyeon, J. C., & Jin, B. J., (2014). Anti-cancer activity of Ginger (*Zingiber officinale*) leaf through the expression of activating transcription factor three in human colorectal cancer cells. *BMC Complement Altern. Med., 14*, 408.

Jagetia, G. C., Venkatesh, P., & Baliga, M. S., (2005). *Aegle marmelos* (L.) Correa inhibits the proliferation of transplanted Ehrlich ascites carcinoma in mice. *Biol. Pharm. Bull., 28*, 58.

Jia, S. S., Xi, G. P., Zhang, M., Chen, Y. B., Lei, B., Dong, X. S., & Yang, Y. M., (2013). Induction of apoptosis by D-limonene is mediated by inactivation of Akt in LS174T human colon cancer cells. *Oncol. Rep., 29*, 349–354.

Lv, X., Zhao, S., Ning, Z., Zeng, H., Shu, Y., & Tao, O., (2015). Citrus fruits as a treasure trove of active natural metabolites that potentially provide benefits for human health. *Chem. Cent. J., 9*, 68. 10.1186/s13065-015-0145-9.

Mandalari, G., Bisignano, C., Cirmi, S., & Navarra, M., (2017). Effectiveness of *Citrus* Fruits on *Helicobacter pylori*. *Evid. Based Complement Altern. Med.*, 8379262. 10.1155/2017/8379262.

Murthy, K. N. C., Jayaprakasha, G. K., & Patil, B. S., (2012). D-Limonene rich volatile oil from blood oranges inhibits angiogenesis, metastasis, and cell death in human colon cancer cells. *Life Sci., 91*, 429–439.

Ochwang, I. D. O., Kimwele, C. N., Oduma, J. A., Gathumbi, P. K., Mbaria, J. M., & Kiama, S. G., (2014). Medicinal plants used in treatment and management of cancer in Kakamega County Kenya. *Journal of Ethnopharmacology, 151*, 1040–1055.

Rathaur, P., Raja, W., Ramteke, P. W., & Suchit, A., (2012). Turmeric: The golden spice of life. *Int. J. Pharm. Sci. Res., 3*(7), 1987–1994.

Sivaraj, R., Rahman, P. K. S. M., Rajiv, P., Vanathi, P., & Venckatesh, R., (2014). Biosynthesis and characterization of *Acalypha indica* mediated copper oxide nanoparticles and evaluation of its antimicrobial and anticancer activity. *Spectrochimica Acta Part A: Molecular and Biomolecular Spectroscopy, 129*, 255–258.

Susma, K., Rani, P., & Praveen, K., (2012). Medicinal plants of Asian origin having anticancer potential: Short review. *Asian Journal of Biomedical and Pharmaceutical Sciences, 2*(10), 1–7.

Yong, S. L., Min, H. K., So, Y. C., & Choon, S. J., (2007). Effects of constituents of *Ammomum xanthioides* on gastritis in Rats and on growth of gastric cancer cells. *Archpharm Res., 30*(4), 436–443.

CHAPTER 15

In Vitro Anticancer Activity of *Biophytum sensitivum* on Liver Cancer Lines (HEPG2)

M. P. SANTHI,[1] K. SARAVANAN,[1] and P. KARUPPANNAN[2]

[1] P.G. and Research Department of Zoology, Nehru Memorial College (Autonomous), Puthanampatti, Trichy District – 621 007, Tamil Nadu, India, E-mail: kaliyaperumalsaravanan72@gmail.com (K. Saravanan)

[2] P.G. and Research Department of Zoology, Holy Cross College (Autonomous), Tiruchirappalli, Tamil Nadu, India

ABSTRACT

Biophytum sensitivum is a traditionally used medicinal plant for the treatment of different ailments including cancer. However, there were no scientific studies on anticancer activity of this plant. Thus, this study focuses anticancer activity of *B. sensitivum* against HEP G2 (liver cancer cell line). *B. sensitivum* whole plant extracts possessed cytotoxic activity against HEP G2 cancer cell lines. However, water extract of *B. sensitivum* showed good cytotoxic effects compared to other extracts. These results support the traditionally used of *B. sensitivum* as anticancer agent and suggest that they could provide a possible source for the preparation of a new anticancer drug.

15.1 INTRODUCTION

Cancer is the most progressive and destructive disease causing mortality to all over the world. A theoretical state that 20 million new cancer cases will diagnose in 2020, particularly the proportion of new cases will increase up to 70% in developing countries like India (Rao and Ganesh, 1998). Among different cancer types, liver cancer (hepatocellular carcinoma (HCC)) is the sixth risky cancer in worldwide. Approximately 24 individuals per 100,000 people were affected by HCC in a year and it is one of the predominant diseases in the Asian region (Satija, 2014). The available treatment methods for liver cancer (chemotherapy, surgery, and radiation) cause unwanted side effects and to kill normal cells and tissues around the cancer cells. Thus, there is an urgent need for an alternative treatment option for cancer treatment. The secondary metabolites, especially the phytocompounds present in the medicinal plants, provided the basic platform for several sophisticated traditional medicine systems like Ayurveda, Unani, Folk, and Chinese (Bosch et al., 2004). Herbal plants still remain the main resource for treating majority of health problems. *Biophytum sensitivum* is a small, sensitive annual herb belongs to the family Oxalidaceae. It is traditionally used for the cure of various ailments including diabetes, tuberculosis, and cancer for long time (Gambarin-Gelwan et al., 2000; Wong et al., 2000; Bosch et al., 2004). But no scientific validation for *B. sensitivum* with regard to treatment of cancer. So, the present study was focused to explore the anticancer activity of *B. sensitivum* against liver cancer cell line.

15.2 METHODOLOGY

15.2.1 PLANT MATERIAL

A large quantity of whole plant of *B. sensitivum* were collected from Puthanampatti village, Thuraiyur taluk, Trichy district, Tamil Nadu, India. The whole plant was washed with running water, dried under shadow, finely powdered and stored.

15.2.2 PREPARATION OF WHOLE PLANT EXTRACTS OF B. SENSITIVUM

Extraction was made by Soxhlet apparatus. About 10 g of *B. sensitivum* whole plant powder and 200 ml of solvent were taken in the thimble and round bottom flask of Soxhlet apparatus, respectively. Ethanol, Chloroform, Petroleum ether, and water were used as solvent. Extraction was continued for 48 hrs. After the completion of extraction, the liquid portion of extract was distillate by vacuum evaporator. About 100 μg of extract was dissolved using DMSO (1 ml) and then diluted in the ratio of 1:3 for *in vitro* test.

15.2.3 EVALUATION OF ANTICANCER ACTIVITY

15.2.3.1 PROCUREMENT OF LIVER CANCER CELL LINE

HEPG2 (Human liver cancer) cell lines were purchased from NCCS (National Center for Cell Sciences), Pune. After procurement, the cells were kept in Minimal Essential Media added with FBS (10%), penicillin (100 µg/ml), and streptomycin (100 µg/ml) in a CO_2 incubator with humidified condition of 50 µg/ml CO_2 at 37°C.

15.2.3.2 MTT (3-[4,5-DIMETHYLTHIAZOL-2-YL]-2,5-DIPHENYLTETRAZOLIUM BROMIDE) ASSAY

HEPG2 cells which maintained in the CO_2 incubator were trypsinized and 5×10^4 cells/well were seeded in the microplate with 100 µL of culture medium and kept at CO_2 incubator with temperature of 37°C and 5% of CO_2 for 24 hrs. After forming monolayer, the supernatant was drained out and added 100 µL of various concentrations of *B. sensitivum* extracts with diluted media and incubated at CO_2 incubator for 72 hours. Incubated HEPG2 cells were periodically observed for their morphological changes such as granularity, shrinkage, and swelling. After completion of 72 hours, removed media with extracts in the wells and added 10 µL of MTT dye to each well. The plates were gently shaken and incubated for 4 hours at 37°C in 5% CO_2 incubator. The supernatant was removed and 100 µL of DMSO was added and the microplates were shaken to solubilize the formed Formosan. Then, the OD was measured at 540 nm using a microplate reader with a reference filter of 620 nm (Skehan, 2003). The percentage of cell growth inhibition or percentage cytotoxicity was calculated by the following formula:

$$\% \text{ Growth inhibition} = \frac{100 - \text{Mean OD of individual Test Group}}{\text{Mean OD of individual Control Group}} \times 100$$

15.3 OBSERVATION AND RESULTS

Antiproliferative activity of whole plant extracts of *B. sensitivum* against HEPG2 cancer cells were determined by MTT assay. Results of cytotoxicity study of *B. sensitivum* whole plant extracts are illustrated in Figure 15.1 and photomicrography of extracts treated cells and untreated cells of liver are shown in Figure 15.2. *B. sensitivum* extracts showed a strong cytotoxic effect in the HEPG2 cancer cell lines. Ethanolic, chloroform, petroleum ether, and aqueous extracts of *B. sensitivum* exhibited potent activity at the concentration of 30 µg, 40 µg, 65 µg, and 80 µg, respectively. The ethanolic extract showed higher cytotoxicity followed by chloroform and peteroleum ether extracts. All the extracts showed gradual increase in antiproliferation activity with increase of concentration.

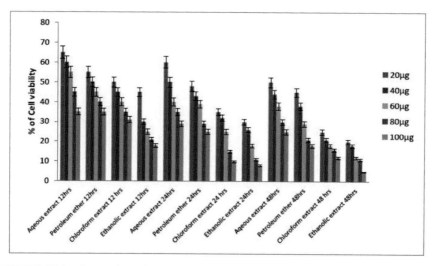

FIGURE 15.1 Cytotoxicty of *B. sensitivum* extracts on HEPG2 cells.

15.4 DISCUSSION

HCC is the most abundant malignancies and causes 10 lakhs death annually (Fecht and Befeler, 2004). The medicinal plants have been clinically proven effective treatment for various cancers and demonstrated their ability to inhibit tumor growth without any side effects (Niu and He, 1992). Whole plant *B. sensitivum* extracts were used to clinically prove their anticancer activities against liver cancer. MTT assay revealed the ethanol and chloroform extracts of *B. sensitivum* possess efficient antiproliferation activity against HEPG2 cells. Inhibition of cell growth and an initiation of cell death are two major processes of cancer treatment. Ethanol and chloroform extracts exhibited the strong inhibition of cancer growth. The 50% inhibition concentration (IC50) value of the extracts showed significant

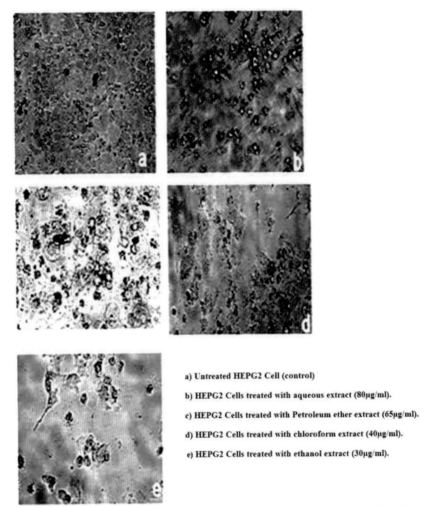

FIGURE 15.2 Anti-proliferative activity of *B. sensitivum* extracts against HEPG2 cell lines.

reduction in cancer cell growth at 40 μg and 60 μg concentration, respectively. It may be due to the occurrence of secondary metabolites such as biflavones (cupressuflavone and amentoflavone), flavonoids, (luteolin 7-Methyl ether, isoorinentin, and 3-methoxyluteolin 7-O-glucoside) and acids (4-caffeoylquinic acid and 5-caffeoylquinic acid) (Lin and Wang, 2000).

15.5 CONCLUSION

The medicinal plant, *B. sensitivum* had efficient cytotoxic effect against the liver cancer cell lines and was remarkably inhibited the abnormal growth of liver cells. Hence, the herbal plant *B. sensitivum* may be used as an emerging drug candidate for the treatment of liver cancer.

ACKNOWLEDGMENT

The authors thank the President, Secretary, and the Principal Nehru Memorial College (Autonomous) for providing necessary facilities to do this research work.

KEYWORDS

- **anticancer activity**
- **antiproliferation**
- ***Biophytum sensitivum***
- **cytotoxic effect**
- **herbal medicine**
- **liver cancer**

REFERENCES

Alison, M. R., (2005). Liver stem cells: implications for Hepatocarcinogenesis. *Stem Cell Rev., 1*(3), 253–60.

Bosch, F. X., Ribes, J., & Diaz, M., (2004). Primary liver cancer: Worldwide incidence and trends. *Gastroenterology, 127,* S5–S16.

Fecht, W. J., & Befeler, A. S., (2004). Hepatocellular carcinoma: Updates in primary prevention. *Curr. Gastroenterol. Rep., 6,* 37–43.

Gambarin-Gelwan, M., Wolf, D. C., Shapiro, R., Schwartz, M. E., & Min, A. D., (2000). Sensitivity of commonly available screening tests in detecting hepatocellular carcinoma in cirrhotic patients undergoing liver transplantation. *Am. J. Gastroenterol., 95*(6), 1535–1538.

Knick, V. C., Eberwein, D. J., & Miller, C. G., (1995). Vinorelbine tartrate and paclitaxel combinations: Enhanced activity against *in vivo* P388 murine leukemia cells. *Journal of the National Cancer Institute, 87,* (1071).

Lin, Y., & Wang, W., (2000). Chemical constituents of *Biophytum sensitivum. Chin. Pharm. J., 55,* 71–75.

Mosmann, T., (1983). Rapid colorimetric assay for cellular growth and survival: Application to proliferation and cytotoxicity assays. *Journal of Immunological Methods, 65,* 55–63.

Niu, C. Q., & He, L. Y., (1992). [HPTLC separation and fluorodensitometric determination of isoquinoline alkaloids in *Chelidonine majus* L]. *Yao Xue Xue Bao., 27*(1), 69–73.

Okpako, D. T., (1999). Traditional African medicine: Theory and pharmacology explored. *Trends Pharmacol. Sci., 20,* 482–485.

Rao, D. N., & Ganesh, B., (1998). Estimate of cancer incidence in India in (1991). *Indian J. Cancer, 35,* 10–18.

Satija, A., (2014). *Cervical Cancer in India.* South Asia center for chronic disease.

Skehan, P., (1990). New colorimetric cytotoxicity assay for anti-cancer drug screening. *Journal of the National Cancer Institute, 82,* 1107–1112.

Skehan, P., Storeng, R., Scudiero, D., Monks, A., Mc Mahon, J., Vistica, D., Warren, J. T., Bokesch, H., Kenny, S., & Boyd, M. R., (2003). New colorimetric cytotoxicity assay for anti-cancer drug screening. *Journal of the National Cancer Institute, 82,* 1107–1112.

Wong, J. B., McQuillan, G. M., McHutchison, J. G., & Poynard, T., (2000). Estimating future hepatitis C morbidity, mortality, and costs in the United States. *Am. J. Public Health, 90*(10), 1562–1569.

CHAPTER 16

Saffron and Its Active Ingredients: A Natural Product with Potent Anticancer Property

ASHUTOSH GUPTA and ABHAY K. PANDEY

Department of Biochemistry, University of Allahabad, Allahabad – 211002, India,
E-mail: akpandey23@rediffmail.com (A. K. Pandey)

ABSTRACT

Cancer defined as an abnormal growth of cells that feast and metastasize to different parts from the origin with the help of uncontrolled cell proliferation. Even though there are developing therapeutic alternatives for cancer patients, their efficiency is time-dependent and non-remedial. Thus to overwhelmed these problems, continuous selection for potent and safer remedies has been consistently discovered for several years, resulting in the exploration of cancer preventive efficacy of various natural products. Chemoprevention using traditional therapeutic knowledge, i.e., an active ingredient from fruits, spices, herbs, and vegetables are appropriate approaches in the current scenario for cancer prevention. *Crocus sativus L.* (saffron) has gained attention because it's pharmacological studies including radical scavenging, anti-mutagenic, and immuno-modulating activities, etc. The major active compounds of saffron are crocetin, picrocrocin, crocin, and safranal. Broad spectrum studies on animal models and humanized cancer cell lines revealed the antitumor activities of saffron and its bioactive constituent. This chapter delivers a comprehensive understanding of the anticancer properties and molecular mechanism of saffron along with its potent bioactive compounds.

16.1 INTRODUCTION

Cancer is defined as an abnormal growth of cells that feast and metastasize to different parts from the origin with the help of uncontrolled cell proliferation.

Presently, it is a significant health concern for living beings, challenging millions of lives every year. The adverse effect of cancer has on the fitness of humans primarily depends on the age, response to medication and invasiveness (Samarghandian and Borji, 2014). In most of the cases, routine chemotherapy is incapable to attain satisfactory effects due to its serious side effects and dose-limiting toxicity. Administration of bioactive compounds or man-made formulations or their combinations is extensively used as a therapeutic possibility to inhibit cancer (Chermahini et al., 2010) apart from additional evolving techniques *viz.* catalytic therapy, radiotherapy, photo-dynamic therapy and sometimes surgical intervention in serious conditions (Samarghandian and Borji, 2014). Various studies are conducted to explore the molecular signaling functioning which governed the advancement of cancer as a feasible target for therapy (Bhandari, 2015). Now researchers have been shifted their interest towards the natural remedies which include spices, fruits, vegetables, and herbs to identify the potent compounds which possess cytotoxic efficacy and anti-cancer activity (Gupta et al., 2017). A large number of plant species have studied till now whose fractions possess anticancer activity, like *Curcuma longa* (curcumin), *Terminalia bellerica* (ellagic acid), *Taxus brevifolia* (Taxanes), *Digitalis purpurea (digitalin)*, *Catharanthus roseus* (vincristine, vinblastine), *Nigella sativa* (Thymoquinone), *Solanum nigrum* (solanine) and *Zingiber officinale* (curcumin).

Crocus sativus L. (saffron) is a key representative member of family Iridaceae. It is a stemless small perennial plant which is extensively cultivated throughout the world in dry and mild climates, including Spain, France, Italy, Greece, United Arab Emirates, Turkey, Egypt, India, Pakistan, Japan, Australia, China, and especially Iran (Zarinkamar et al., 2011). The total annual saffron production is assessed about 205 tons world widely, in which more than 80% produced by Iran and thus pronounced as 'Red Gold.' In Europe, Spain is the main source of cultivated saffron, which exports around 60 tons of total saffron production annually. Consequently, Greece produces 4.5 tons of saffron per year. The therapeutic potential of saffron could be related to its bioactive compounds *viz.,* safranal, crocins, picrocrocin, and crocetin (Abdullaev and Espinosa-Aguirre, 2004). Chemical analysis of saffron suggested that it contains approximately 150 volatile and nonvolatile compounds in which nearly 50 compounds have been identified (Winterhalter and Straubinger, 2000). Safranal is present as major volatiles compounds whereas, non-volatile compounds include crocins, crocetin, and picrocrocin (Liakopoulou-Kyriakides and Kyriakidis, 2002). Saffron possesses broad-spectrum medicinal properties such as antispasmodic, expectorant, stomachic, aphrodisiac, cardiotonic, carminative, and stimulant.

Saffron and Its Active Ingredients

The modern pharmacological finding indicated that saffron extract has antitumor, free radical scavenging, anticonvulsant, antidepressant, antioxidant, anti-inflammatory, and anti-hyperlipidemic properties. This chapter summarizes the chemical properties of saffron and its major bioactive compounds along with their anticancer properties.

16.2 ACTIVE CONSTITUENT AND THEIR CHEMICAL PROPERTIES

Saffron is the dried stigma of *C. sativus* which is red and approximately 2 mg in weight. Various chemical investigation suggested that it comprises water (14–16%), sugar (12–15%), volatile oil (1%), nitrogenous matters (11–13%), fiber (4–5%), extract soluble (41–44%), protein (12%), mineral (5%) and carbohydrates such as dextrins, starch, gum, reducing sugar and pectin. It also contains vitamins (riboflavin and thiamine) with a small amount of β-carotene. The concentration of thiamine is 0.7–4 μg/g, which is commonly found in different fruits and vegetables. Riboflavin concentration is 56–130 μg/g and is the highest amount present in any food (Bhat et al., 1953; Shahi et al., 2016). Reports on petroleum ether extract of saffron bulbs have shown the presence of essential fatty acids (linolenic and linoleic), sterol (stigmasterol, β-sitosterol, and campesterol), ursolic, palmitoleic, oleanolic, oleic, and palmitic acids (Loukis et al., 1983). Crocetin, its glucosidic derivatives, safranal, crocins, picrocrocin, and flavonoids (quercetin and kaempferol) are the major active compounds found in saffron (Figure 16.1) (Shahi et al., 2016).

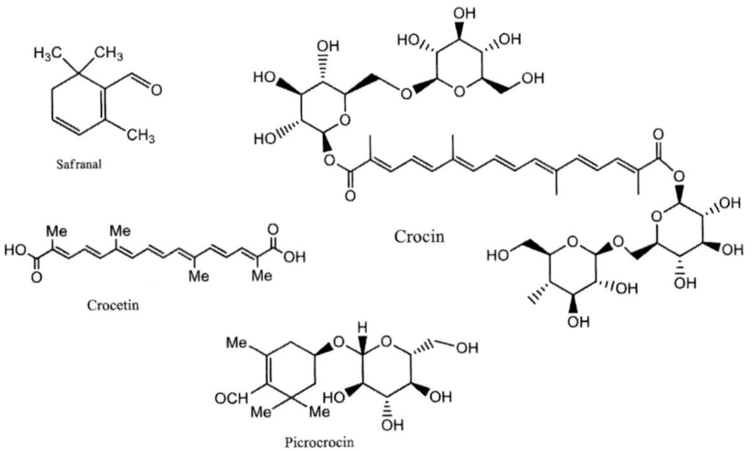

FIGURE 16.1 Structure of some important bioactive compounds present in saffron.

16.2.1 PICROCROCIN

Picrocrocin (4-hydroxy-2,6,6-trimethyl-1-cyclohexen-1-carboxaldehyde) is an oxidative degradative product of the carotenoid, zeaxanthin. This molecule has a chemical formula of $C_{16}H_{26}O_7$ with a molecular weight (MW) of 330.37 g/mol and boiling point of 520.4°C (Shahi et al., 2016). Picrocrocin is the second major component (1–13%) of the saffron. It is bitter and acts as a precursor molecule for safranal, an aroma component. Picrocrocin shows maximum absorbance at 254 nm (Lage and Cantrell, 2009).

16.2.2 SAFRANAL

Safranal (2,6,6-trimethyl-1,3-cyclohexadiene-1-carboxaldehyde), is a cyclic monoterpene aldehyde and aglycon of picrocrocin with a MW of 150.21 g/mol. It is a prime component of the saffron's distilled essential oil which develops aroma. In some species, saffron may comprise up to 70% of the total volatile fraction in the form of safranal (Curro et al., 1986). In the past few years, the aroma of saffron gained a lot of attraction because it is not present in fresh stigma due to the lack of safranal and is formed by the action of enzyme β-glucosidase. This enzyme dehydrates picrocrocin (heating associated with an enzymatic action) during post-harvest drying and storage process (Carmona et al., 2007; Maggi et al., 2010). The maximum absorbance for safranal is obtained at 330 nm (Lage and Cantrell, 2009).

16.2.3 CROCIN

Crocin (8,8-diapocarotene-8,8-dioic acid), is the essential glycoside carotenoid that provides its specific color and has a chemical formula $C_{44}H_{70}O_{28}$ with a MW of 976.96 g/mol (Figure 16.1) (Samarghandian and Borji, 2014). It is the diester derived from the disaccharide of gentiobiose and the dicarboxylic acid crocetin. Crocin is water-soluble, but most of the other carotenoids are insoluble in an aqueous medium. It is a special carotenoid used more frequently as a food additive and medicine. Crocin showed maximum absorbance at 440 nm (Lage and Cantrell, 2009). The other form of crocin includes trans-crocin 2, trans-crocin 3, trans-crocin 4, cis-crocin 2 and cis-crocin 4 (Caballero-Ortega et al., 2007).

16.2.4 CROCETIN

Crocetin (8, 80-diapo-8, 80-carotenoic acid) is a lipophilic carotenoid with several unsaturated conjugated olefin acid structures. Its common chemical formula is $C_{20}H_{24}O_4$ while MW is 328.4 g/mol and having a melting point of 285°C (Xi et al., 2007). It develops yellow color in soluble condition and forms red needles like crystal in melting point. Several studies suggested that nearly 94% of the total crocetin existing in the form of glycosides while the remaining 6% is present in free form. It is less soluble in basic solutions (20 µM at pH 8) but completely soluble in organic bases, like pyridine. Crocetin is well known for their antioxidant attributes and is one the most studied constituent of saffron. The maximum absorbance of crocetin is 464 nm.

16.3 CURRENT BIOMEDICAL STATUS OF SAFFRON AND ITS INGREDIENTS

Various studies have been reported that saffron and its bioactive compounds may be useful for management of neurodegenerative problems and associated ischemic retinopathy, memory impairment, or age-related macular degeneration, seizure, disorder related to blood pressure, mild to moderate depression, coronary artery disease, acute, and chronic inflammatory abnormalities, and parkinsonism. In addition, antimutagenic, antioxidant, tumoricidal, and antigenotoxic properties have also been established in saffron and its active phytoconstituent (Figure 16.2) (Abdullaev and Espinosa-Aguirre, 2004).

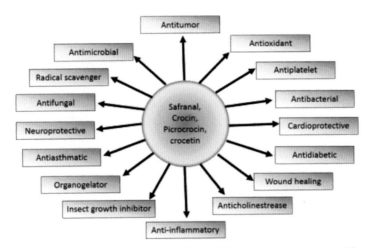

FIGURE 16.2 Major pharmacological activities of bioactive compounds of saffron.

16.4 TOXICITY AND SAFETY LEVEL OF SAFFRON AND ITS BIOACTIVE COMPOUNDS

A study designed by Bahmani et al. (2014) explored the value of lethal dose and suggested that the LD_{50} value in case of saffron was 4.12 ± 0.556 g/kg. Similarly, the LD_{50} value of safranal was 1.48 mL/kg and 1.88 mL/kg in male and female rats respectively (Hosseinzadeh et al., 2013). Another study on stigma and petal extracts of saffron revealed that LD_{50} values were 1.6–6 g/kg (Karimi et al., 2004). Whereas, crocin or aqueous fraction of saffron have shown to possess no side effect on schizophrenia patients (Mousavi et al., 2015). Safranal is less-toxic when administered through acute intraperitoneal route while non-toxic when administered orally. Moreover, in sub-acute toxicity assay, safranal significantly altered the hematological and biochemical parameters (Hosseinzadeh et al., 2013). Saffron tablets have been shown to possess some modifications in hematological and biochemical parameters; however such changes are within normal ranges (Modaghegh et al., 2008). *In vivo*, sub-acute toxicity indicated that 10 mg/kg dose of saffron enhances survival rate with no mortality (Ziaee et al., 2014). Saffron did not exert any toxic sign on the adult liver or in the neonatal kidneys when histopathological alterations were studied. Recent studies have supported that the injectable dose of saffron between 1.2–2 g may lead to some complications including nausea, vomiting, and diarrhea (Schmidt et al., 2007).

16.5 MOLECULAR MECHANISM OF ANTICANCER EFFECT OF SAFFRON AND ITS BIOACTIVE COMPOUNDS

16.5.1 INHIBITION OF SYNTHESIS OF RNA AND DNA

Sun et al. (2011) stated that crocin appreciably decreases both RNA and DNA contents by inhibiting the synthesis of DNA and RNA in human tongue cancer cell line (Tca8113). This drop in DNA and RNA contents was confirmed as one of the major cause of apoptosis in tongue cancer cell line. Moreover, this outcome was supported by several other cancer cell lines *viz*., promyelocytic leukemia HL-60 cells, HeLa, A549 (Lung adenocarcinoma) and VA13 (SV40-transformed fetal lung fibroblasts) (Bhandari, 2015; Gupta et al., 2019). Although, the complete regulatory mechanism of crocin effects on DNA and RNA formation is still unclear.

16.5.2 INTERACTION WITH CELLULAR TOPOISOMERASE AND SUPPRESSION OF TELOMERASE ACTIVITY

A wide range of studies has been revealed the inhibitory property of saffron and its active phytoconstituents on cell division. As mentioned above, replication of DNA is one of the major characteristic events of cell division (Bhandari, 2015). DNA replication itself regulated by the activity of replication complex (DNA polymerases and helicases). Numerous proteins which are involved in DNA replication altered the topology of DNA. Helicases relax DNA duplex, thus promoting the formation of supercoils whereas topoisomerases catalyze addition or removal of supercoils (Lodish et al., 2000). Bajbouj and coworkers (2012) worked on p53 colorectal cancer and found that DNA topoisomerase enzymes could be responsible for antiproliferation activity of saffron.

By down-regulating the telomerase activity, carcinoma cells can avoid cell death through the mechanism that assists cells to maintain lengths of telomeres and potent cancer cells to divide indefinitely. Noureini et al. (2012) suggested the killing mechanism by measuring the telomerase activity of HepG2 cancer cell line after the treatment of crocin. They observed that cancer cell line down-regulate the expression of a catalytic subunit of telomerase up to 60% as compared to normal cells.

16.5.3 INDUCTION OF APOPTOSIS

Apoptosis (programmed cell death) is an important mechanism through which a cell maintains its morphology and physiology. Normal cell performed apoptosis mainly by two pathways namely intrinsic and extrinsic pathways. In the intrinsic pathway, intracellular signals cause modifications in the inner mitochondrial membrane. This further leads to the release of cytochrome-c and formation of apoptosome complex with the help of cytosolic protein APAF-1. APAF-1 in turn leads to the formation of apoptosome complex which further initiates caspase cascade activation. Besides this, the extrinsic pathway performed by receptor-mediated interactions which are mostly members of the tumor necrosis factor (TNF) receptor and transmits death signals from cell surface to the intracellular signaling pathways (Elmore, 2007). Crocin is reported as an inducer of both intrinsic and extrinsic pathways. Amin and coworker (2011) reported that crocin triggers the death of cancer cells via p53-dependent and p53-independent mechanisms in colorectal cancer cells. They suggested that crocin stimulate an autophagy-independent classical programmed cell death in p53-Null colorectal cancer. Another study indicated that 10 µg/l crocin enhanced apoptotic morphology with reduced cell volume in pancreatic cancer (Bakshi et al., 2010). Crocin increased Bax/Bcl-2 ratio and caspase activation which in turn initiated the process of apoptosis in gastric adenocarcinoma cells.

16.5.4 INCREASING ANTIOXIDANT ENZYMES

Extensive studies have shown the evidence that saffron possesses anticancer potential against diethylnitrosamine (DEN) induced hepatic carcinoma in the rat model. Moreover, Saffron plays a significant role in carbon tetrachloride (CCl_4) induced hepatic injury (Bahashwan et al., 2015). It upregulate the levels of phase-II detoxifying enzymes such as glutathione peroxidase (GPx), glutathione-S-transferase (GST) along with catalase and superoxide dismutase (SOD) while it down-regulated the process of lipid peroxidation (LPO) and malondialdehyde (MDA) formation, protein carbonyl formation and myeloperoxidase activity (Amin et al., 2011; Das et al., 2004). LPO and its product MDA initiate the process of cancer and are assumed to be mutagenic and carcinogenic. GST suppresses the binding of cancerous agents to DNA and reduces carcinogenesis. SOD, catalase, and GPx protect the cell from free radical damages (ROS/RNS) (Das et al., 2004). In an immune histochemical study, saffron promotes inhibition of DEN mediated increase of cyclooxygenase 2, nuclear factor-kappa B p-65, phosphorylated TNF receptor (TNFR), and inducible nitric oxide synthase (Amin et al., 2011).

16.5.5 SUPPRESS INFLAMMATORY RESPONSE

In a study on the anticancer potential of crocetin against methylcholanthrene (MCA)-induced cervical cancer, Chen et al. (2015b) observed that crocetin inhibited polymorphonuclear cells (PMNs) and maleic dialdehyde formation and down-regulated various inflammatory cytokines such as interleukin-1 (IL-1) and tumor necrosis factor-α (TNF-α). PMNs leads to constriction of a vessel discharge of free radicals, cytolytic proteases, and pro-inflammatory cytokines. Kim and his co-worker observed that crocin in association with crocetin activated the nuclear factor erythroid 2-related factor 2 (Nrf2) while reduced lactate dehydrogenase-A (LDHA) expressions in HeLa cells (2014). Moreover, saffron extract and its active constituent explore the mechanism of anticancer on prostate cancer cell lines (PC-3 and 22rv1). Outcomes indicated that saffron reduced epithelial-mesenchymal trans-differentiation (EMT) due to enhanced expression of cytokeratin 18 (K18), E-cadherin, and suppressed the expression of mesenchymal markers (N-cadherin, β-catenin), vimentin (Festuccia et al., 2014).

16.6 ANTICANCER ACTIVITY OF SAFFRON AND ITS BIOACTIVE COMPOUNDS

16.6.1 HEPATIC CANCER

Liver malignancy is the third major cause of death throughout the world. The majority of hepatic cancer is associated with chronic hepatitis virus. More than 180 million people throughout the world are suffering from hepatitis C virus. Presently, hepatitis C is a major global health concern, and is the leading cause of cirrhosis and chronic hepatitis, etc. Investigation of hepatic cancer and its management has witnessed various key modifications which were evaluated in the past two decades (Rasool et al., 2014). A number of studies provide a strong background for the anticancerous property of saffron. In the case of DEN, mediated hepatic carcinoma (HepG2), saffron inhibits cell proliferation by initiation of apoptosis, lessening oxidative damage and down-regulating the inflammatory response. Results suggested that saffron inhibit increased cleavage of caspase-3, DNA damage, nuclear factor-kappa B activation, and cell cycle arrest (Samarghandian et al., 2010). Anti-proliferative potentials of crocin were examined against liver cancer cell line (HepG2). The finding indicated that crocin suppressed the telomerase property of HepG2 cells, which was possibly triggered by down-regulating the expression of enzyme's catalytic subunit (Sun et al., 2013). Furthermore, *in vitro* study on crocetin treated C3H10T1/2 fibroblast cells indicated reduction in aflatoxin B1-DNA adduct formation that proposed protective effect of crocetin against aflatoxin B1 induced carcinogenicity because of increase in glutathione S-transferase (GST), reduced glutathione (GSH), and GPx as cellular defense mechanism (Wang et al., 1991a). Aflatoxin mediated hepatic damage exhibited that crocetin appreciably reduced the level of serum AST, ALT, ALP, and GGT and showed a protective effect (Wang et al., 1991b). Crocetin also reduced the formation of MDA, a product of LPO which is produced as a result of free radical-mediated cell damage (Tseng et al., 1995). Therefore, these findings supported that saffron and its bioactive compounds possess therapeutic effect against cancer by modifying various cellular mechanisms.

16.6.2 SKIN CANCER

An occurrence of skin cancer has increased throughout the different regions of the world. During the past few decades, the attraction has been centered toward the understanding of molecular basis of skin oncogenesis and recognizing substance with anticancer potential for skin cancer. Free radicals (ROS/RNS) are produced by carcinogenic agents or UV irradiation, which play a major role in skin carcinogenesis. A wide range of pharmacological studies revealed that cellular defense mechanism or phase-II detoxification enzymes, which together pronounced as cytoprotective proteins, and protect skin from carcinogenesis (Chun et al., 2014). In recent years, chemoprevention is universally accepted as the main powerful or novel method to prevent and revert the process of oncogenesis by the involvement of natural products. Natural products are well documented as a potent antioxidant, anticancer, and carcinogen detoxifying agents and hence considered as potent chemopreventive agents. 200 mg/kg dose of saffron suppressed ascites tumors formation derived from sarcoma-180 (S-180), Ehrlich ascites carcinoma (EAC), Dalton's lymphoma ascites (DLA). Consequently, saffron significantly improved the life span (2–3 folds) of tumor treated rats (Nair et al., 1991). Additionally, consumption of saffron in encapsulated liposome appreciably promotes the anticancer potential against S-180 and EAC. Another *in vitro* study on T-cell mitogen phytohaemagglutinin, saffron showed improved non-specific proliferation of blood cells and revealed that anticancer efficacy of saffron might be assisted immunologically (Nair et al., 1992). Saffron aqueous extract prevent the proliferation of human transitional cell carcinoma (TCC) and non-neoplastic fibroblast cell line (L929) in dose-dependent manner (Feizzadeh et al., 2008). It was found that carotenoids present in saffron could apply its chemo-preventive properties by altering LPO, antioxidant, and detoxification systems. Similar anticancer property of crocetin was also examined in skin tumor which is developed by the administration of DMBA and croton oil (Mathewsroth, 1982). These findings recommended that crocetin effectively enhanced the life span of experimental model against different types of cancer, although the complete mechanism of anticancer property of crocetin has not been fully understood.

16.6.3 LEUKEMIA

Nair and coworker (1992) studied the anticancer efficacy of dimethylcrocetin and crocin against different cancer cell lines (leukemia P388 and

L1210) and observed that dimethylcrocetin possess 50% cytotoxicity at the dose of 7–30 mg/ml while crocetin showed 50% cytotoxicity at the concentration of 11–39 mg/ml. They found remarkable decline in the synthesis of DNA and proposed that dimethylcrocetin may altered the DNA-protein interactions (toposiomerases II), which is essential for cellular DNA synthesis. Kumar et al. (2001) studied the property of aqueous extract of saffron on mice and reported that mice pre-treatment with saffron can dominantly suppress the cyclophosphamide, mitomycin genotoxicity of cisplatin and urethane. Sun and coworker (2013) found that crocin reduce proliferation of HL-60 and triggered the process of apoptosis and block cell cycle progression at G_0/G_1 stage. Meanwhile, crocin decrease the size and weight of HL-60 xenograft induced tumor in nude mice by lowering the expression of Bcl-2 along with increased expression of Bax in xenograft. Safranal in composition with small amount of imatinib mesylate (drug which is used in management of human chronic myelogenous leukemia (CML)) suppress the expression of Bcr-Abl gene which is responsible for tyrosine kinase activity of Bcr-Abl protein (Geromichalos et al., 2014).

16.6.4 LUNG CANCER

Lung cancer has become a major challenge because of large number of mortality throughout the world. The main cause of this mortality is late detection of cancer where effective treatment is no longer feasible. Crocetin, play a significant role in cancer chemoprevention. In a study, lung cancer-induced animal model was used to evaluate the anticancer property of crocetin in both pre and post-initiation periods. Crocetin normalized the elevated level of LPO and other important marker enzymes which is used to detect the carcinogenicity in animal. Moreover, it also regulates the activity of antioxidant enzymes and glutathione metabolizing enzymes. The physiological changes in carcinogenic animals were remarkably regulated by the administration of crocetin (Magesh et al., 2006). In lung cancer, cell lines (A549) saffron enhance the apoptosis and cytotoxic activities. Samarghandian and coworker (2013) showed that the proliferation of the A549 cells was declined after the application of saffron in time and dose-dependent method. Although, there was an increase in the percentage of cells performed apoptosis. Additionally, the anticancer potential of aqueous extract cold is due to its cell proliferation inhibitory property along with initiation of apoptosis in carcinoma cells through a caspase-dependent pathway. Ethanolic fraction of saffron showed positive results were used against human alveolar basal epithelial cancer cell line (A549). Outcomes indicated that saffron suppress the cell viability in a dose-dependent manner in malignant cells. The extract provided pro-apoptotic properties in a lung cancer cell line and hence it was considered as a potential therapeutic agent against lung cancer (Samarghandian et al., 2011).

16.6.5 PANCREATIC CANCER

Pancreatic cancer cell line (BxPC-3) revealed the apoptotic property of crocetin. Chromatin condensation (characteristic of apoptosis) was observed by Hoechest33258 staining, flow cytometry was used to check cell cycle phases whereas gel electrophoresis technique was used to investigate DNA fragmentation. Finding indicated that pancreatic cancer cell lines are highly sensitive towards crocin-mediated growth inhibition and cell death by apoptosis (Bakshi et al., 2010). Broad-spectrum studies revealed that crocetin has appreciable anti-tumorigenic property on both the *in vitro* pancreatic cancer cells and *in vivo* athymic nude mice tumor by stimulation of apoptosis. Cell cycle proteins and epidermal growth factor receptor (EGFR) were modified by the application of crocetin in pancreatic cancer cells (MIA-PaCa-2). *In vivo* analysis explores remarkable deterioration in tumor development with proliferation inhibition in the crocetin-treated model compared with the control model. Bax/Bcl-2 ratio indicated that crocetin triggered the process of apoptosis in cancer cell line and athymic nude mice cancer (Dhar et al., 2009).

16.6.6 BREAST AND CERVICAL CANCER

Crocetin and its analog down-regulate the proliferation of breast cancer cells. Chryssanthi and his coworker suggested that cancer cells (MCF) and MDA-MB-231 growth were inhibited by crocetin in a dose-dependent manner. This effect was not dependent on the functioning of estrogen receptors (2007). Therefore, the anticancer property of crocetin was separate from hormone regulation. The pro-apoptotic property of crocetin was analyzed by using MCF-7 cells. Results are commended that crocetin promotes the expression of BAX protein, and accelerate the caspase-dependent pathway of apoptosis (Mousavi et al., 2015). It has been found that crocetin like compounds which present in saffron may cause a significant reduction in the development of the colony, RNA, and DNA synthesis at the concentration of 1.2 g/

mL in HeLa cells. Analysis of crocetin under UV spectroscopy confirms the interaction between tRNA with crocetin. It also altered the activity of DNA-dependent RNA polymerase II and RNA synthesis. Consequently, crocetin may have binding properties at the molecular level (Tavakkol-Afshari et al., 2008; Kanakis et al., 2007).

16.7 CONCLUSION

Cancer is the major health concern, and its management is frequently unproductive. Scientists, throughout the world, are more concerned to explore the medicinal attributes of saffron that positively associates with a less threat of various forms of cancers they also explored the existence of a large number of bioactive compounds in saffron. Among them, crocetin, safranal, picrocrocin, and crocin are well established as a chief therapeutically active ingredient and are most commonly studied in various *in vitro* and *in vivo* studies. Till now, the complete mechanism of action of the anticancer property of saffron is not fully understood. Moreover, most of the medicinally active compounds of saffron are comes under the category of carotenoids. Carotenoids such as α-carotene, canthaxanthin, lycopene, β-carotene, and zeaxanthin, etc. possess a number of biological activities such as modulate cell growth regulation, antioxidants immune response and modify gene expression. Because of the lipid-soluble property of carotenoids and might act as in association with the membrane, it is used in cancer prevention and its treatment. The antioxidant properties of these compounds could inhibit damage of protein, RNA, and DNA which mediated by free radicals and related chain reactions. This chapter summarized the potentials of *C. sativus* and its phytoconstituents which act as an effective compound for cancer treatment. Furthermore, the current findings have not yet been confirmed by clinical trials in humans and in-depth studies are necessary to elucidate the usefulness of saffron in treatment and anticipation of cancer. Besides this, the insufficiency and expense in gaining huge amounts of saffron may deliver impairments to cancer treatment and human chemoprevention using this saffron and its potent bioactive phytoconstituents.

ACKNOWLEDGMENT

A. Gupta acknowledges financial support from the University Grants Commission, New Delhi in the form of UGC-CRET fellowship. Both authors acknowledge DST-FIST and UGC-SAP assisted Department of Biochemistry, the University of Allahabad for providing infrastructure facilities.

KEYWORDS

- anti-cancer
- bioactive compounds
- crocetin
- crocin
- *Crocus sativus*
- picrocrocin
- saffron
- safranal

REFERENCES

Abdullaev, F. I., & Espinosa-Aguirre, J. J., (2004). Biomedical properties of saffron and its potential use in cancer therapy and chemoprevention trials. *Cancer Detect. Prev., 28*, 426–432.

Amin, A., Hamza, A. A., Bajbouj, K., Ashraf, S. S., & Daoud, S., (2011). Saffron: A potential candidate for a novel anticancer drug against hepatocellular carcinoma. Hepatology, 54, 857-867.

Bahashwan, S., Hassan, M. H., Aly, H., Ghobara, M. M., El-Beshbishy, H. A., & Busati, I., (2015). Crocin mitigates carbon tetrachloride-induced liver toxicity in rats. *J. Taibah Univ. Med. Sci., 10*(2), 140–149.

Bahmani, M., Rafeian, M., Baradaran, A., Rafeian, S., & Rafeian-Kopaei, M., (2014). Nephrotoxicity and hepatotoxicity evaluation of *Crocus sativus* stigmas in neonates of nursing mice. *J. Nephropathol., 3*, 81–85.

Bajbouj, K., Schulze-Luehrmann, J., Diermeier, S., Amin, A., & Schneider-Stock, R., (2012). The anticancer effect of saffron in two p53 isogenic colorectal cancer cell lines. BMC. Complement. Altern. Med., 28, 12-69.

Bakshi, H., Sam, S., Rozati, R., Sultan, P., Islam, T., Rathore, B., Lone, Z., Sharma, M., Triphati, J., & Saxena, R. C., (2010). DNA fragmentation and cell cycle arrest: A hallmark of apoptosis induced by crocin from Kashmiri saffron in a human pancreatic cancer cell line. Asian Pac. J. Cancer Prev., 11, 675-679.

Bhandari, P., (2015). *Crocus sativus* L. (saffron) for cancer chemoprevention: A mini review. *J. Tradit. Complement Med., 5,* 81–87.

Bhat, J. V., & Broker, R., (1953). Riboflavine and thiamine contents of saffron, *Crocus sativus. Nature, 172*, 544–545.

Caballero-Ortega, H., Pereda-Miranda, R., & Abdullaev, F. I., (2007). HPLC quantification of major active components from 11 different saffron (*Crocus sativus* L.) sources. *Food Chemistry*, *100*, 1126–1131.

Carmona, M., Zalacain, A., Salinas, M. R., & Alonso, G. L., (2007). A new approach to saffron aroma. *Crit. Rev. Food Sci. Nutr.*, *47*, 145–159.

Chen, B., Hou, Z. H., Dong, Z., & Li, C. D., (2015b). Crocetin downregulates the proinflammatory cytokines in methylcholanthrene-induced rodent tumor model and inhibits COX-2 expression in cervical cancer cells. *BioMed. Res. Int.*, 829513.

Chryssanthi, D. G., Dedes, P. G., Karamanos, N. K., Cordopatis, P., & Lamari, F. N., (2011). Crocetin inhibits invasiveness of MDA-MB-231 breast cancer cells via down regulation of matrix metalloproteinases. Planta. Med., 77, 146-151.

Chun, K. S., Kundu, J., Kundu, J. K., & Surh, Y. J., (2014). Targeting Nrf2-Keap1 signaling for chemoprevention of skin carcinogenesis with bioactive phytochemicals. *Toxicol. Lett.*, *229*, 73–38.

Curro, P., Francesco, L., & Giuseppe, M., (1986). *Evaluation of the Volatile Fraction of the Saffron by Gas Chromatography of the Headspace*, *38*, 331–334.

Das, I., Chakrabarty, R. N., & Das, S., (2004). Saffron can prevent chemically induced skin carcinogenesis in Swiss albino mice. *Asian Pac. J. Cancer Prev.*, *5*, 70–76.

Dhar, A., Mehta, S., Dhar, G., Dhar, K., Banerjee, S., Van, V. P., Campbell, D. R., & Banerjee, S. K., (2009). Crocetin inhibits pancreatic cancer cell proliferation and tumor progression in a xenograft mouse model. Mol. Cancer Ther., 8, 315–323.

Elmore, S., (2007). Apoptosis: A review of programmed cell death. *Toxicol. Patho.*, *4*, 495–516.

Feizzadeh, B., Afshari, J. T., Rakhshandeh, H., Rahimi, A., Brook, A., & Doosti, H., (2008). Cytotoxic effect of saffron stigma aqueous extract on human transitional cell carcinoma and mouse fibroblast. Urol. J., 5, 161-167.

Festuccia, C., Llorens, S., Mancini, A., Alonso, G. L., Jannini, E. A., Lenzi, A., Gravina, G. L., et al., (2014). Antitumor effects of saffron-derived carotenoids in prostate cancer cell models. *Bio. Med. Res. Int.*, *13*, 48–50.

Geromichalos, G. D., Papadopoulos, T., Sahpazidou, D., & Sinakos, Z., (2014). Safranal, a *Crocus sativus* L constituent suppresses the growth of K-562 cells of chronic myelogenous leukemia. *In silico* and *in vitro* study. *Food Chem. Toxicol.*, *74*, 45–50.

Gupta, A., & Pandey, A. K., (2019). Phytochemicals as oxidative stress mitigators. In: Egbuna, C., Kumar, S., Ifemeje, J. C., & Kurhekar, J. V., (eds.), *Phytochemistry-Pharmacognosy, Nanomedicine, and Contemporary* (Vol. 2). Apple Academic Press Inc., USA.

Gupta, A., Kumar, R., Kumar, S., & Pandey, A. K., (2017). Pharmacological aspects of *Terminalia belerica*. In: Mahdi, A. A., Abid, M., Khan, A. A., Ansari, M. I., & Maheshwari, R. K., (eds.), *Molecular Biology and Pharmacognosy of Beneficial Plants*. Lenin media private limited, Delhi, India.

Hosseinzadeh, H., Shakib, S. S., Sameni, A. K., & Taghiabadi, E., (2013). Acute and subacute toxicity of safranal, a constituent of saffron, in mice and rats. *Iran J. Pharm. Res.*, *12*, 93–99.

Kanakis, C. D., Tarantilis, P. A., Tajmir-Riahi, H. A., & Polissiou, M. G., (2007). Interaction of tRNA with safranal, crocetin, and dimethylcrocetin. *J. Biomol. Struct. Dyn.*, *24*, 537–545.

Karimi, G. H., Taiebi, N., Hosseinzadeh, H., & Shirzad, F., (2004). Evaluation of subacute toxicity of aqueous extract of *Crocus sativus* L. stigma and petal in rats. *J. Med. Plants*, *3*, 29–35.

Kim, S. H., Lee, J. M., Kim, S. C., Park, C. B., & Lee, P. C., (2014). Proposed cytotoxic mechanisms of the saffron carotenoids crocin and crocetin on cancer cell lines. *Biochem. Cell Biol., 92,* 105–111.

Kumar, P., Abraham, S. K., Santhiya, S. T., Gopinath, P. M., & Ramesh, A., (2001). Inhibition of genotoxicity by saffron (*Crocus sativus L.*) in mice. Drug Chem. Toxicol., 24, 421-428.

Lage, M., & Cantrell, C. L., (2009). Quantification of saffron (*Crocus sativus* L.) metabolites crocins, picrocrocin, and safranal for quality determination of the spice grown under different environmental Moroccan conditions. *Scientia Horticulturae, 12,* 366–373.

Liakopoulou-Kyriakide, M., & Kyriakidis, D. A., (2002). *Croscus sativus*-biological active constituents. *Studies in Natural Products Chemist, 26,* 293–312.

Lodish, H., Berk, A., & Zipursky, S. L., (2000). The role of topoisomerases in DNA replication. In: *Section 12.3, Molecular Cell Biology* (4th edn.). New York: W. H. Freeman. Available from: https://www.ncbi.nlm.nih.gov/books/NBK21703/ (accessed on 22 February 2020).

Loukis, A., Al-Kofahi., A., & Philianos, S., (1983). Constituents of *Crocus sativus* L. bulbs. *Plantes Medicinales et Phytotherapie, 17,* 89–91.

Magesh, V., Singh, J. P., Selvendiran, K., Rajendran, P., & Sakthisekaran, D., (2006). Antitumor activity of crocetin in accordance to tumor incidence, antioxidant status, drug metabolizing enzymes and histopathological studies. Mol. Cell Biochem., 287, 127-135.

Maggi, L., Carmona, M., Zalacain, A., Kanakis, C. D., Anastasaki, E., Tarantilis, P. A., Moschos, G., Polissiou, M. G., & Alonso, G. L., (2010). Changes in saffron volatile profile according to its storage time. *Food Res. Inter., 43,* 1329–1334.

Mathews-Roth, M., (1982). Antitumor activity of beta-carotene, canthaxanthin and phytoene. *Oncology, 39,* 33–37.

Modaghegh, M. H., Shahabian, M., Esmaeili, H. A., Rajbai, O., & Hosseinzadeh, H., (2008). Safety evaluation of saffron (*Crocus sativus*) tablets in healthy volunteers. *Phytomedicine, 15,* 1032–1037.

Mousavi, B., Bathaie, S. Z., Fadai, F., Ashtari, Z., Beigi, N. A., Farhang, S., Hashempour, S., Shahhamzei, N., & Heidarzadeh, H., (2015). Safety evaluation of saffron stigma (*Crocus sativus* L.) aqueous extract and crocin in patients with schizophrenia Avicenna. *J. Phytomed., 5,* 413–419.

Nair, S. C., Pannikar, B., & Pannikar, K. R., (1991). Antitumor activity of saffron (*Crocus sativus*). *Cancer Lett., 57,* 109-114.

Nair, S. C., Salomi, M. J., Varghese, C. D., Panikkar, B., & Panikkar, K. R., (1992). Effect of saffron on thymocyte proliferation, intracellular glutathione levels, and its antitumor activity. Biofactors, 4, 51-54.

Noureini, S. K., & Wink, M., (2012). Antiproliferative effects of crocin in HepG2 cells by telomerase inhibition and hTERT down-regulation. Asian Pac. J. Cancer Prev., 13, 2305-2309.

Rasool, M., Rashid, S., Arooj, M., Ansari, S. A., Khan, K. M., Malik, A., et al., (2014). New possibilities in hepatocellular carcinoma treatment. *Anticancer Res., 34,* 1563–1571.

Samarghandian, S., & Borji, A., (2014). Anticarcinogenic effect of saffron (*Crocus sativus* L.) and its ingredients. *Pharmacognosy Res., 6,* 99–107.

Samarghandian, S., Afshari, J. T., & Davoodi, S., (2011). Chrysin reduces proliferation and induces apoptosis in the human prostate cancer cell line pc-3. *Clinics (Sao Paulo), 66,* 1073-1079.

Samarghandian, S., Borji, A., Farahmand, S. K., Afshari, R., & Davoodi, S., (2013). *Crocus sativus* L. (saffron) stigma aqueous extract induces apoptosis in alveolar human lung cancer cells through caspase-dependent pathways activation. *Biomed Res. Int.,* 417–928.

Samarghandian, S., Boskabady, M. H., & Davoodi, S., (2010). Use of *in vitro* assays to assess the potential antiproliferative and cytotoxic effects of saffron (*Crocus sativus L.*) in human lung cancer cell line. Pharmacogn. Mag., 6, 309-314.

Schmidt, M., Betti, G., & Hensel, A., (2007). Saffron in phytotherapy: Pharmacology and clinical uses. *Wien. Med. Wochenschr., 157,* 315–319.

Shahi, T., Assadpour, E., & Jafari, S. M., (2016). Main chemical compounds and pharmacological activities of stigmas and tepals of 'red gold', saffron. *Trends Food Sci. Technol., 58,* 69–78.

Sun, J., Xu, X. M., Ni, C. Z., Zhang, H., Li, X. Y., Zhang, C. L., Liu, Y. R., Li, S. F., Zhou, Q. Z., & Zhou, H. M., (2011). Crocin inhibits proliferation and nucleic acid synthesis and induces apoptosis in the human tongue squamous cell carcinoma cell lineTca8113. *Asian Pac. J. Cancer Prev., 12,* 2679–2683.

Sun, Y., Xu, H., Zhao, Y., Wang, L., Sun, L., Wang, Z., & Sun, X., (2013). Crocin exhibits antitumor effects on human leukemia HL-60 cells *in vitro* and *in vivo*. *Evid. Based Complement Alternat. Med., 69,* 01–64.

Tavakkol-Afshari, J., Brook, A., & Mousavi, S. H., (2008). Study of cytotoxic and apoptogenic properties of saffron extract in human cancer cell lines. Food Chem. Toxicol., 46, 3443-3447.

Tseng, T. H., Chu, C. Y., Huang, J. M., Shiow, S. J., & Wang, C. J., (1995). Crocetin protects against oxidative damage in rat primary hepatocytes. *Cancer Lett., 97,* 61-67.

Wang, C. J., Hsu, J. D., & Lin, J. K., (1991b). Suppression of aflatoxin B1-induced lesions by crocetin (a natural carotenoid). Carcinogenesis, 12, 1807-1810.

Wang, C. J., Shiow, S. J., & Lin, J. K., (1991a). Effects of crocetin on the hepatotoxicity and hepatic DNA binding of aflatoxin B1 in rats. *Carcinogenesis, 12,* 459-462.

Winterhalter, P., & Straubinger, M., (2000). Saffron-renewed interest in an ancient spice. *Food Rev. Int., 16,* 39–59.

Xi, L., Qian, Z., Xu, G., Zheng, S., Sun, S., Wen, N., Sheng, L., Shi, Y., & Zhang, Y., (2007). Beneficial impact of crocetin, a carotenoid from saffron, on insulin sensitivity in fructose-fed rats. *J. Nutr. Biochem., 18,* 64–72.

Zarinkamar, F., Tajik, S., & Soleimanpour, S., (2011). Effects of altitude on anatomy and concentration of crocin, picrocrocin, and safranal in *Crocus sativus* L. *Aust. J. Crop. Sci., 5,* 831-838.

Ziaee, T., Razavi, B. M., & Hosseinzadeh, H., (2014). Saffron reduced toxic effects of its constituent, safranal, in acute and subacute toxicities in rats. *J. Nat. Pharm. Prod., 9,* 3–8.

PART II
Antidiabetic Drug Discovery

CHAPTER 17

Effects of *Pterocarpus marsupium* in the Management of Type 2 Diabetes Mellitus

PRISCILLA SURESH

P. G. and Research Department of Zoology, Bishop Heber College, Tiruchirappalli – 620 017, India, Tel.: 9789164989, E-mail: priscisf@gmail.com

ABSTRACT

Diabetes mellitus is a metabolic disorder, which causes insulin resistance, relative insulin deficiency, and hyperglycemia. Insulin resistance occurs due to the accumulation of fatty acids or fatty acid derivatives in muscle and liver. Monosodium glutamate (MSG) obesity can be induced in newborn mice with the subcutaneous administration of MSG. It results in lesions in hypothalamic arcuate nucleus and impairs leptin and insulin signaling causing hyperleptinemia and hyperinsulinemia. *Pterocarpus marsupium* is distributed in the hilly regions of the Deccan peninsula in India. *P. marsupium* heartwood was collected from the Idukki district in Kerala. Crude and distilled extracts of two types of extracts of the heartwood of *P. marsupium* were used in this study. To the diabetic mouse, 0.25 ml of 10% of the aqueous extract (25 mg/mouse or nearly 1.25 g/kg body weight) was administered orally. The results of this study show that the administration of the extracts of *P. marsupium* is able to control the MSG diabetic condition to a very great extent. They are able to reduce the obesity situation which is evident from the reduced weight and reduced Lee index in the extracts-fed diabetic animals. When treated with the crude and distilled extracts of *P. marsupium* for a period of 60 days the MSG diabetic mice were able to get over the hyperglycemic effect. The distilled fraction of the aqueous extract contains specific chemical compounds with antidiabetic and antilipidemic properties.

17.1 INTRODUCTION

Obesity results from an imbalance between energy intake and energy expenditure in the body which is a risk factor for type 2 diabetes. Over the last decades, its occurrence has been progressively increasing affecting not only adults but also children and adolescents (Malecka-Tendera and Mazur, 2006). Obesity results in the accumulation of fat in liver, leading to fatty liver disorders (Farrel and Larter, 2006) and results in insulin resistance, hyperinsulinemia, and type II diabetes.

The causes for obesity are primarily due to changing food habits and life styles. Monosodium glutamate (MSG), commonly known as Ajinomoto and a popular flavor enhancer, has also found its place among obesity causing agents (Hirata et al., 1997; Jezova et al., 1998; Harris et al., 2001; Matsuki et al., 2003; Youssef et al., 2002). MSG acts as an excitotoxin as it is absorbed very quickly in the gastrointestinal (GI) tract. It causes lesions in the hypothalamic region and impairs leptin and insulin signaling in this region resulting in hyperleptinemia and hyperinsulinemia (Dawson et al.1997; Broberger et al., 1998; Maletinska et al., 2006). Djazayery et al. (1979) found that MSG induced animals had a dramatic increase in body fat due to a lowered metabolic rate.

Different medicinal systems are making use of plant extracts and bioactive compounds by virtue of traditional knowledge as natural antidiabetic agents. The World Health Organization (WHO) Expert Committee on Diabetes has recommended the traditional medicinal herbs and has been proposed in the recent years a multidirectional approach integrating different therapeutic branches to manage the pathophysiological conditions of diabetes.

Medicinal plants have been used as a natural medicine in India. There are many reviews on Indian medicinal plants having the blood sugar-lowering potential (Mukherjee et al., 1981; Grover et al., 2002; Saxena et al., 2004; Mukherjee et al., 2006). Some of the plants used in traditional medicine to reduce obesity include *Camellia sinensis, Chlorella pyrenoidosa, Citrus aurantium, Garcinia cambogia, Lagerstroemia speciosa, Panax ginseng, Salix matsudana, Nelumbo nucifera, Pterocarpus marsupium,* and *Hibiscus sabdariffa* (Calapai et al., 1999; Han et al., 2003; Hidaka et al., 2004; Ono et al., 2006). The hypoglycemic effect of many herbal extracts has been confirmed in the animal model and in human of type 2 diabetes.

The antidiabetic effect of *Pterocarpus marsupium* is exploited in the traditional medical practices in India (Kedar and Chakrabarti, 1981; Satyavati et al., 1987). Different parts of *P. marsupium* are used in the treatment of diabetes by ayurvedic a traditional Indian medicine system in India.

It is a drug with unique features such as beta-cell protective and regenerative properties apart from blood glucose reduction (Manickam et al., 1997). Epicatechin from the heartwood increased the CAMP content of the pancreatic islets associated with increased insulin release, conversion of proinsulin to insulin and cathepsin B activity in rats (Ahmed et al., 1991).

Reports are also available on the clinical trials made in diabetic patients using the heartwood of *P. marsupium* extract which proved successful in controlling hyperglycemia in them (Manickam et al., 1997; Sheehan and Zemaitis, 2004; Mukhtar et al., 2005; Grover et al., 2002; Maruthupandian and Mohan, 2011). Nearly 10 patients were given the water stored in a heartwood container of this plant for a period of 1 month. The blood glucose levels decreased from the second week of the treatment and were maintained at a normal level when the treatment was withdrawn (Kedar and Chakravarthy, 1981). This work was aimed to find out whether *P. Marsupium* has a diabetic management potential in MSG-induced experimental diabetic obese mouse.

- To evaluate the diabetic management potential of two types of extracts the crude and the distilled extract of the heartwood of *P. marsupium* in MSG induced obese mouse with type 2 diabetes.
- To trace the underlying principles and mechanisms of the two types of extracts of *P. marsupium* by tracing the chemical component differences in them.

17.2 MATERIAL AND METHODS

The inbred strain of Swiss albino mouse, *Mus musculus,* was used for this study. The number of animals used was approved by CPCSEA (Committee for the Permission and Control of Supervision on Experimental Animals). MSG is a food additive, marketed as a "flavor enhancer." Commonly known as MSG, Ajinomoto, vetsin or accent, the glutamate of MSG confers the same umami taste of glutamate from other foods, being chemically identical. For this study, the commercial product of MSG (Ajinomoto) is used which is available as a white crystalline powder.

17.2.1 INDUCTION OF OBESITY-BASED DIABETES

Diabetes was induced in the mouse with the subcutaneous injection of MSG. The dosage used was 4 mg/g of body weight as reported by Olney (1969).

The required amount of salt was weighed, dissolved in 0.5 ml saline and was injected subcutaneously into the neonatal mice. Each animal received a total of eight injections, on every alternate day from the 5th day of birth onwards. The mouse was tested for obesity and type II diabetes on the 60th day after the administration of the MSG by the confirmatory test.

17.2.2 COLLECTION OF PLANTS AND EXTRACT PREPARATION

The heartwood of mature *Pterocarpus marsupium* was collected from the Idukki district in Kerala. It was dried, powdered, and stored in glass containers at normal room temperature. Two types of extracts of the heartwood of *P. marsupium* were used in this study, namely crude and distilled extracts. The process of extract preparation, based on the principle of extraction of organic compounds from plant materials (Clarke and Haynes, 1975; Chakravarthy et al., 1980) and the dosage to be used had already been standardized in our laboratory by Farzana (2005). An oral administration of 0.25 ml of 10% of the aqueous extract (25 mg/mouse or nearly 1.25 g/kg body weight) was sufficient to bring about the hypoglycemic effect in diabetic mouse (Farzana, 2005). 10% crude extract was prepared by dissolving 10 gm powder of *P. marsupium* heartwood in 100 ml of distilled water. It was left overnight, boiled, and filtered. To prepare a distilled extract, the crude aqueous extract was distilled seven times (Ravana, 1962). Treatment with the *P. marsupium* crude extract and distillate was started from the 60th day after the administration of MSG.

17.2.3 MEASUREMENTS AND ESTIMATIONS

17.2.3.1 BODY WEIGHT AND LEE INDEX

The mouse was weighed using a top pan balance with an accuracy of 10 mg. The Lee Index (Lee, 1928) was calculated by dividing the cube root of the body weight by nose to anus length and multiplying by 1000. Bodyweight and Lee Index were calculated on the 0, 30th, 60th, 90th, and 120th days of the experiment.

17.2.3.2 BLOOD SUGAR LEVEL

The blood sugar was estimated using the glucose oxidase method (GOD) (Trinder, 1969). The serum was separated from the blood sample by centrifugation at 2000 rpm. 10 µl of the serum was mixed with 1 ml of GOD reagent and incubated for 10 min at room temperature. The optical density was measured at 505 nm, using a digital colorimeter.

17.2.3.3 SERUM INSULIN

The ADVIA Centaur Insulin assay is a two-site sandwich immunoassay using direct chemiluminescent technology (Dods, 1996) which uses constant amounts of two antibodies. The first antibody, in the Lite Reagent, was a monoclonal mouse anti-insulin antibody labeled with acridinium ester. The second antibody, in the solid phase, was a monoclonal mouse anti-insulin antibody, which was covalently coupled to paramagnetic particles. A direct relationship existed between the amount of insulin present in the sample and the amount of relative light units (RLUs) detected by the system.

17.2.3.4 HOMA INSULIN RESISTANCE

Insulin resistance was estimated by HOMA (homeostasis model assessment) developed in rodent models by Chen et al. (2008). HOMA index is an indicator of relative insulin resistance and was calculated by using the formula $HOMA_{IR}$ = fasting insulin (µU/ml) x fasting glucose (mmol/l)/22.5.

17.2.3.5 PHYTOCHEMICAL ANALYSIS AND IDENTIFICATION OF PHYTOCHEMICAL COMPONENTS

The GC-MS analysis was carried out on a GC Clarus 500 Perklin Elmer system and gas chromatograph interfaced to a mass spectrometer (GC-MS) instrument. The relative percentage amount of each component was calculated by comparing its average peak area (PA) to the total areas. The detection employed the NIST (National Institute of Standards and Technology) Ver.2.0-Year 2005 library. The compound prediction is based on Dr. Duke's Phytochemical and Ethnobotanical Databases by Dr. Jim Duke of the Agricultural Research Service/USDA. The interpretation of the GC-MS was conducted using the database of NIST having more than 62,000 patterns. The spectrum of the unknown component was compared with the spectrum of the known components stored in the NIST library. The name, molecular weight (MW), and structure of the components of the test materials were ascertained.

17.3 RESULTS AND DISCUSSION

This research work has been carried out to study the antidiabetic potential of the two types of extracts of *P. Marsupium* crude and distillate in the MSG induced diabetic obese mice. Serum insulin level in µU/ml in the control, MSG diabetic, *P. marsupium* crude extract fed diabetic and *P. marsupium* distillate fed diabetic albino mouse are shown in Figure 17.1. The results of this study reveal that the neonatal administration of MSG induces obesity and diabetes in the albino mouse, *M. musculus*. The MSG diabetic mice gained nearly 65% of its body weight during the 60 days of the experiment and the control mice gained only 29.5% of its body weight. The MSG injected animals also showed a higher Lee index.

The MSG diabetic mice are able to get over the hyperglycemic effect, when treated with the crude and distilled extracts of *P. marsupium* for a period of 60 days. The statistical analysis of the fasting blood sugar levels has shown significant decrease in the blood sugar level in the extract treated diabetic animals, from the 30th day onwards. The values of the fasting blood sugar in the distilled extract treated diabetic mouse on the 120th day are 115 ± 2.28 mg/dL which fall within the normal range of blood sugar. In summary, 60 days of the treatment with the *P. Marsupium* extracts has helped the diabetic mouse to regulate its blood glucose permanently. The anti-hyperglycemic potential of *P. marsupium* is well established. According to Joshi et al. (2004), *P. marsupium* decreased the blood glucose levels in non-insulin dependent diabetic rats. Vats et al. (2002) have also recorded similar effects in alloxan diabetic rats. The treatment of the MSG diabetic mice with the heartwood of the *P. Marsupium* crude extract has resulted in 66% of mortality in 120 days. As mentioned in the results, the analysis of blood sugar before mortality revealed severe hypoglycemic conditions. Such drastic decrease in the blood glucose level in the crude extract fed MSG diabetic mice could also be one of the causes for mortality. Severe hypoglycemic condition on the feeding of the *P. marsupium* crude extract is reported by Farzana (2005) in STZ mice.

The feeding of the *P. marsupium* distillate is found to reduce serum insulin resistance to the level of the control mice (Figure 17.2). Similar findings are reported by Grover et al. (2005) in type 2 diabetic patients, when administered *P. marsupium* bark extract. They have registered hyper-triglyceridaemia and hyperinsulinemia (insulin resistance) in type 2 diabetic patients. On the other hand, *P. marsupium* crude extract fed diabetic animals have shown a moderate decrease in insulin levels compared to the MSG animals.

Effects of *Pterocarpus marsupium* in the Management of Type 2 Diabetes 225

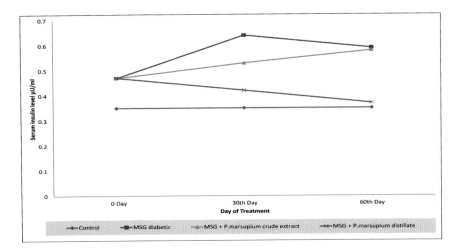

FIGURE 17.1 Serum insulin level in μU/ml in the control, MSG diabetic, *P. marsupium* crude extract fed diabetic and *P. marsupium* distillate fed diabetic albino mouse, *M. musculus* on 0, 30th and 60th day of the experiment.

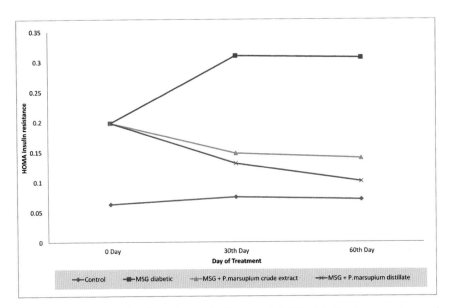

FIGURE 17.2 HOMA insulin resistance in the control, MSG diabetic, *P. marsupium* crude extract fed diabetic and *P. marsupium* distillate fed diabetic albino mouse, *M. musculus* on 0, 30th and 60th day of the experiment.

17.3.1 IDENTIFICATION OF PHYTOCHEMICAL COMPONENTS BY GAS CHROMATOGRAPHY-MASS SPECTROMETRY (GC-MS)

As there were distinct differences between the biological properties of the crude and the distilled extracts of *P. marsupium,* an attempt was made to find out the chemical nature of both of them. They have been subjected to GC-MS analysis. Twenty-three compounds were identified from the GC fractions of the crude extract of *P. marsupium* (Table 17.1). These compounds were identified by mass spectrometry attached with GC. The major compounds were 1,4-Benzenedicarboxylic acid, bis (2-methylpropyl) ester (25.21%), 1,2-Benzenedicarboxylic acid, butyl octyl ester (20.70%), 1,2-Benzenedicarboxylic acid, butyl 2-ethylhexyl ester(11.72%), 1,2-Benzenedicarboxylic acid, butyl 2-methylpropyl ester (6.12%), Phthalic acid, pentyl 2-pentyl ester (5.03%), 2-(3-Hydroxy-4-methoxyphenyl)-1,3-benzodioxane (2.37%) etc. Twenty compounds were identified from the GC fractions of the distilled extract. The major compounds were 1,2-Benzene dicarboxylic acid, butyl octyl ester (26.76%), Ethanol, 2-(4-phenoxyphenoxy)-benzoate (21.62%), 1,4-Benzenedicarboxylic acid, bis (2-methylpropyl) ester (9.82%), 1,2-Benzenedicarboxylic acid, butyl 8-methylnonyl ester (9.32%), Phthalic acid, pentyl 2-pentyl ester (6.23%), phthalic acid, neopentyl 2-pentyl ester (4.18%), Phthalic acid cyclo hexyl pentyl ester (2.28%), phthalic acid, monocyclohexyl ester (1.56%), etc.

A closer look at the result shows the distinct nature of the extracts. The distilled extract does not contain flavonoids, a fact which confirms with the phytochemical principle of distillation (Clarke and Haynes, 1975). Instead, their exclusive compounds are terpenoids, carbonic acid methyl phenyl ester, benzene sulphonic acid 4 methyl butyl ester and ethanol, 2-(4-phenoxyphenoxy) –benzoate.

Phytochemicals and their biological activities obtained through the GC-MS study in the distilled extract of the heartwood of *P. marsupium* are given in Table 17.2. The antidiabetic properties of flavonoids from *P. marsupium* are well established. The extracts of bark and heartwood of *P. marsupium* are reported to have an anti-diabetic effect (Gupta, 2010; Trivedi, 1971; Pandey and Sharma, 1975; Chakravarthy et al., 1980). The antiobese and antidiabetic properties of terpenoids are also well explained by Tsuyoshi Goto et al. (2010). Their findings highlighted that several bioactive terpenoids contained in herbal or dietary plants, which can modulate the activities of ligand-dependent transcription factors, peroxisome proliferator-activated receptors (PPARs). As PPARs are dietary lipid sensors that control energy homeostasis, daily intake of these terpenoids might be useful for the management for obesity-induced metabolic disorders, such as type 2 diabetes, hyperlipidemia, insulin resistance, and cardiovascular diseases. Ethanol, 2-(4-phenoxyphenoxy) –benzoate is a potent hypoglycemic, hypolipedemic, and anti-inflammatory agent (Link et al., 2005) and this is present in large proportion in the distilled extract. The other compound with antidiabetic potential is carbonic acid methyl phenyl ester. Carbonic acids with hypoglycaemic properties are identified by Geisen et al. (1978). The substances have an inhibitory effect on gluconeogenesis in the liver and lipolysis in adipose tissue. The above facts unravel an important conclusion that the distilled fraction of the aqueous extract contains specific chemical compounds with antidiabetic, antilipidemic, and anti-inflammatory properties. The most unique finding of this work is that the distilled extract is effective in MSG obese mice controlling diabetes directly through its anti-hyperglycemic properties and indirectly through managing obesity. Further, it is to be assumed that it will also have positive effects on the inflammated liver by fatty degeneration in obese condition. Though the crude extract also has compounds with these potential, it is exerting some kind of toxicity in the MSG diabetic condition. The toxic effect of a flavonoid, pterostilbene in the heartwood extract of *P. marsupium* is already reported (Haranath et al., 1958). As the distilled extract contains no flavonoids, the concern of toxicity is ruled out in distilled extract. As *P. marsupium* has different categories of chemical constituents, it may be further assumed that its antidiabetic property could be through more than one mechanism.

TABLE 17.1 Phytocomponents and Their Biological Activities Obtained Through the GC-MS Study in the Aqueous Crude Extract of the Heart Wood of *P. marsupium*

Name of the compound	Biological activity
Phenol	Antiviral, anti-allergic, antiplatelet, anticancerogenic, anti-inflammatory, antiproliferative, antioxidant
Benzofuran, 2,3-dihydro-	Antifungal, antimicrobial, antiaging
Phthalic anhydride	Antimicrobial, antihyperglycemic
Phenol, 2-propyl-	Antiviral, anti-allergic, antiplatelet, anticancerogenic, anti-inflammatory, antiproliferative, antiangiogenic, antioxidant
1,3-Benzenediol, 4-propyl-	Antioxidant, Neuroprotective effect

TABLE 17.1 *(Continued)*

Name of the compound	Biological activity
6-Isopropenyl-4,8a-dimethyl-1,2,3,5,6,7,8,8a-octahydro-naphthalen-2-ol	Antimicrobial, hypoglycemic
Phthalic acid, ethyl 2-pentyl ester	Antidiabetic, antiobesity, antilipidemic, antioxidant, antihyperglycemic
1,2-Benzenedicarboxylic acid, dipropyl ester	Antibacterial, antioxidant, hypoglycemic
1,4-Benzenedicarboxylic acid, bis(2-methylpropyl) ester	Anti-inflammatory, antifungal, anti-microbial
Phthalic acid, isobutyl 2-pentyl ester	Antidiabetic, antiobesity, antilipidemic, antioxidant, antihyperglycemic
1,2-Benzenedicarboxylic acid, butyl octyl ester	Antimicrobial, Antifouling, hepatoprotective
Phthalic acid, bis(2-pentyl) ester	Antidiabetic, antiobesity, antilipidemic, antioxidant, antihyperglycemic
1,2-Benzenedicarboxylic acid, butyl 2-methylpropyl ester	Anti-inflammatory, antifungal, anti-microbial
1,2-Benzenedicarboxylic acid, butyl 2-ethylhexyl ester	Antioxidant, antibacterial, anti-inflammatory and astringent
Benzenedicarboxylic acid, dipentyl ester	Anti-inflammatory, antifungal, anti-microbial
2-(3-Hydroxy-4-methoxyphenyl)-1,3-benzodioxane	Hypoglycemic, antihepatotoxicity

TABLE 17.2 Phytocompounds and Their Biological Activities Obtained Through the GC-MS Study in the Distilled Extract of the Heart Wood of *P. marsupium*

Name of the Compound	Biological Activity
Benzene,1-methyl-4-(1-methyl-2-propenyl)-	Antidiabetic, antiobesity, antifungal
Bicyclo[2.2.1]heptane, 1,7,7-trimethyl-	Anticonvulscent, hypoglycemic, anti-inflammatory
Carbonic acid, methyl phenyl ester	Antioxidant, antidiabetic, antibacterial
Benzenesulfonic acid, 4-methyl-, butyl ester	Hypolipidemic
Phthalic acid, cyclohexyl ethyl ester	Antidiabetic, antiobesity, antilipidemic, antioxidant, antihyperglycemic
Phthalic acid, ethyl 2-pentyl ester	Antidiabetic, antiobesity, antilipidemic, antioxidant, antihyperglycemic
1,2-Benzenedicarboxylic acid, dipropyl ester	Antibacterial, antioxidant, hypoglycemic
1,4-Benzenedicarboxylic acid, bis(2-methylpropyl) ester	Anti-inflammatory, antifungal, anti-microbial
Ethanol, 2-(4-phenoxyphenoxy)-, benzoate	Hypoglycemic, hypolipidemic, anti-inflammation, anti-analgesic

TABLE 17.1 *(Continued)*

Name of the Compound	Biological Activity
Phthalic acid, pentyl 2-pentyl ester	Antidiabetic, antiobesity, antilipidemic, antioxidant, antihyperglycemic
1,2-Benzenedicarboxylic acid, dipentyl ester	Anti-inflammatory, antifungal, anti-microbial
1,2-Benzenedicarboxylic acid, butyl 8-methylnonyl ester	Anti-inflammatory, antistress, antitumor, antioxidant.
1,2-Benzenedicarboxylic acid, butyl octyl ester	Antimicrobial, Anti-fouling, hepatoprotective
Phthalic acid, cyclohexylpentyl ester	Antidiabetic, antiobesity, antilipidemic, antioxidant, antihyperglycemic
Phthalic acid, monocyclohexyl ester	Antidiabetic, antiobesity, antilipidemic, antioxidant, antihyperglycemic

KEYWORDS

- **antilipidemic**
- **gas chromatography-mass spectrometry**
- **hyperinsulinemia**
- **monosodium glutamate**
- ***Pterocarpus marsupium***
- **type 2 diabetes**

REFERENCES

Ahmad, F., Khan, M. M., Rastogi, A. K., Chaubey, M., & Kidwai, J. R., (1991). Effect of (-) epicatechin on CAMP content, insulin release and conversion of proinsulin to insulin in immature and mature rat islets *in vitro*. *Ind. J. Ex. Bio., 29,* 516–520.

Broberger, C., Johansen, J., Johansson, C., Schalling, M., & Hokfelt, T., (1998). The neuropeptide Y/agouti gene-related protein (AGRP) brain circuitry in normal, anorectic, and monosodium glutamate-treated mice. *Proc. Natl. Acad. Sci., 95,* 15043–15048.

Calapai, G., Firenzuoli, F., Saitta, A., Squadrito, F., Arlotta, M. R., Costantino, G., & Inferrera, G., (1999). Antiobesity and cardiovascular toxic effects of *Citrus aurantium* extracts in the rat: A preliminary report. *Fitoterapia., 70,* 586–592.

Chakravarthy, B. K., Gupta, S., Gambhir, S. S., & Gode, K. D., (1980). Pancreatic β-cells regeneration. A novel anti-diabetic mechanism of *Pterocarpus marsupium* Rozb. *Indian J. Pharmacol., 12,* 123–127.

Chen, T., & Wong, Y. S., (2008). *In vitro* antioxidant and antiproliferative activities of selenium containing phycocyanin from selenium-enriched *Spirulin aplatensis*. *J. Agric Food Chem., 56,* 4352–4358.

Clarke, H. T., & Haynes, B., (1975). Separation of mixture of organic compounds. In: *A Handbook of Organic Analysis: Qualitative and Quantitative* (pp. 65–68). Edward Arnold Ltd., London.

Dawson, R., Pelleymounter, M. A., Millard, W. J., Liu, S., & Eppler, B., (1997). Attenuation of leptin-mediated effects by monosodium glutamate-induced arcuate nucleus damage. *Am. J. Physiol., 273,* E202–E206.

Djazayery, A., Miller, D. S., & Stock, M. A., (1979). Energy balances in obese mice. *Nutr. Metab., 23,* 357–367.

Dods, R. F., (1996). Diabetes mellitus. In: Kaplan, L. A., & Pesce, A. J., (eds.), *Clinical Chemistry: Theory, Analysis, and Correlation* (3rd edn., pp. 620–626) St. Louis: CV Mosby.

Farrell, G. C., & Larter, C. Z., (2006). Nonalcoholic fatty liver disease: From steatosis to cirrhosis. *Hepatology, 43*(1–2), S99-S112.

Farzana, K. F., (2005). Antidiabetic effect of *Pterocarpus marsupium* Roxb: A study in albino mouse, *Mus musculus*. PhD Thesis. Bharathidasan University, Tiruchirappalli.

Geisen, K., Hubner, M., Hitzel, V., Hrstka, V. E., Pfaff, W., Bosies, E., Regitz, G., Kuhnle, H. F., Schmidt, F. H., & Weyer, R., (1978). *Acylaminoalkyl Substituted Benzoic and Phenylalkane Acids with Hypoglycaemic Properties (Author's Transl.), 28*(7), 1081–1083.

Grover, J. K., Vats, V., & Yadav, S. S., (2005). *Pterocarpus marsupium* extract (Vijayasar) prevented the alteration in metabolic patterns induced in the normal rat by feeding an adequate diet containing fructose as sole carbohydrate. *Diabetes Obes. Metab., 7,* 414–420.

Grover, J. K., Yadav, S., & Vats, V., (2002). Medicinal plants of India with hypoglycemic potentials. *J. Ethnopharmacology, 81,* 81–100.

Gupta, S., Sharma, S. B., Singh, U. R., Bansal, S. K., & Prabhu, K. M., (2010). Elucidation of mechanism of action of *Cassia auriculata* leaf extract for its anti-diabetic activity in streptozotocin-induced diabetic rats. *J. Med. Food., 13,* 528–534.

Han, L. K., Sumiyoshi, M., Shang, J., Liu, M. X., Zhang, X. F., Zhang, Y. N., Okuda, H., & Kimura, Y., (2003). Anti-obesity action of *Salix matsudana* leaves (Part 1). Anti-obesity action by polyphenols of *Salix matsudana* in high fat-diet treated rodent animals. *Phytotherapy Research, 17,* 1188–1194.

Haranath, P. S. R. K., Ranganathrao, K., Anjaneyulu, C. R., & Ramnathan, J. D., (1958). Studies on the hypoglycaemic and pharmacological actions of some stilbenes. *Indian J. Medl. Sci., 12,* 85–89.

Harris, R. B. S., Mitchell, T. D., Yan, X., Simpson, J. S., & Redmann, J. R., (2001). Metabolic responses to leptin in obese db/db mice are strain dependent. *Am. J. Physiol., 281,* R115–132.

Hidaka, S., Okamoto, Y., & Arita, M., (2004). A hot water extract of *Chlorella pyrenoidosa* reduces body weight and serum lipids in ovariectomized rats. *Phytotherapy Research, 18,* 164–168.

Hirata, A. E., Andrae, I. S., Vaskevicius, P., & Dolnikoff, M. S., (1997). Monosodium glutamate (MSG) obese rats develop glucose intolerance and insulin resistance to peripheral glucose uptake. *Braz. J. Med. Biol. Res., 30,* 671–674.

Indian Council of Medical Research, (1998). Flexible dose open trial of Vijayasar in cases of newly-diagnosed non-insulin-dependent diabetes mellitus. Indian Council of Medical Research (ICMR), Collaborating Centers, New Delhi. *Indian J. Med. Res., 108,* 24–29.

Jezova, D., Kiss, A., Tokarev, D., & Skultetyova, I., (1998). Stress hormone release and proopiomelanocortin mRNA levels in neonatal rats treated with monosodium glutamate to induce neurotoxic lesions. *Stress Medicine, 14,* 255–260.

Joshi, M. C., Dorababu, M., Prabha, T., Kumar, M. M., & Goel, R. K., (2004). Effects of *Pterocarpus marsupium* on non-insulin-dependent diabetes mellitus induced rat gastric ulceration and mucosal offensive and defensive factors. *Indian Journal of Pharmacology, 36,* 296–302.

Kedar, P., & Chakrabarti, C. H., (1981). Blood sugar, blood urea, and Serum lipids as influenced by gurmar preparation, *Pterocarpus marsupium* and *Tamarindus indica* in diabetes mellitus. *Maharashtra Med. J., 28,* 165.

Lee, M. O., (1928). Determination of the surface area of the white rat with its application to the expression of metabolic results. *Am. J. Physiol., 89,* 24–33.

Link, J. T., Sorensen, B., Patel, J., Grynfarb, M., Goos-Nilsson, A., Wang, J., Fung, S., et al., (2005). Antidiabetic activity of passive nonsteroidal glucocorticoid receptor modulators. *J. Med. Chem., 48*(16), 5295–304.

Malecka-Tendera, E., & Mazur, A., (2006). Childhood obesity: A pandemic of the twenty-first century. *Int. J. Obes. (Lond), 30*(2), 1–3.

Maletinska, L., Shamas, T. R., Pirnik, Z., Kiss, A., Slaninova, J., Haluzik, M., & Zelezna, B., (2006). Effect of cholecystokinin on feeding is attenuated in monosodium glutamate obese mice. *Regul. Pept., 136,* 58–63.

Manickam, M., Ramanathan, M., Jahromi, M. A., Chansouria, J. P., & Ray, A. B., (1997). Antihyperglycemic activity of phenolics from *Pterocarpus marsupium*. *Journal of Natural Products, 60,* 609–610.

Maruthupandian, A., & Mohan, V. R., (2011). GC-MS analysis of some bioactive constituents of *Pterocarpus marsupium* Roxb. *International Journal of Chem. Tech. Research CODEN (USA): IJCRGG, 3,* 1652–1657. ISSN: 0974–4290

Matsuki, T., Horai, R., Sudok, R., & Iwakura, Y., (2003). Il-1 plays an important role in lipid metabolism-regulating insulin levels under physiological conditions. *Exp. Med., 198,* 877–888.

Mukherjee, P. K., Maiti, K., Mukherjee, K., & Houghton, P. J., (2006). Leads from Indian Medicinal plants with hypoglycaemic potentials. *J. Ethnopharmacology, 106,* 1–28.

Mukherjee, S. K., (1981). Indigenous drugs in diabetes mellitus. *J. Diab. Asso. Ind., 21,* 97–108.

Mukhtar, H. M., Ansari, S. H., Ali, M., Bhat, Z. A., & Naved, T., (2005). Effect of aqueous extract of *Pterocarpus marsupium* wood on alloxan-induced diabetic rats. *Pharmazie, 60,* 478–479.

Olney, J. W., (1969). Brain lesions, obesity, and other disturbances in mice treated with monosodium glutamate. *Science, 164,* 719–721.

Ona, Y., Hattori, E., Fukaya, Y., Imai, S., & Ohizumi, Y., (2006). Anti-obesity effect of *Nelumbonucifera* leaves extract in mice and rats. *Journal of Ethnopharmacology, 106,* 238–244.

Pandey, M. C., & Sharma, P. V., (1975). Hypoglycemic effect of bark of *Pterocarpu marsupium* Roxb. (Bijaka): A clinical study. *Med. Surg., 25,* 21–24.

Ravana, A. M., Govinthan, V. C., & Krishanan, A. A. M., (1962). *Raddiar Press and Book Depot.* Trivandrum.

Satyavati, G. V., Gupta, A. K., & Tandon, N., (1987). *Medicinal plants of India* (Vol. 2). Indian Council of Medical Research Publications, New Delhi.

Saxena, A., & Vikram, N. K., (2004). Role of selected Indian plants in management of type 2 diabetes: A review. *J. Alt. Comple. Med., 10,* 369–378.

Sheehan, E. W., Zemaitis, M. A., Slatkin, D. J., & Schiff, Jr. P. L., (2004). A constituent of *Pterocarpus marsupium* (-) epicatechin, as a potential hypoglycemic agent. *J. Natural Products, 45,* 232–234.

Trinder, P., (1969). Determination of blood glucose using 4-amino phenazone as oxygen acceptor. *J. Clin. Pathol., 22,* 246–248.

Tsuyoshi, G., Nobuyuki, T., Shizuka, H., & Teruo, K., (2010). *Various Terpenoids Derived from Herbal and Dietary Plants Function as PPAR Modulators and Regulate Carbohydrate and Lipid Metabolism PPAR Res.,* 483958. Published online 2010 June 3. doi: 10.1155/2010/483958. 2010.

Vats, V., Grover, J. K., & Rathi, S. S., (2002). Evaluation of antihyperglycemic and hypoglycemic effect of *Trigonella Foenum graecum* Linn., *Ocimum sanctum* Linn and *Pterocarpus marsupium* Linn in normal and alloxanized diabetic rats. *J. Ethnopharmacol., 79,* 95–100.

Youssef, M. H. M., (2002). Hematological effects of vitamin E in rats. *Zagazig University Med. J., 8,* 679–690.

CHAPTER 18

Antidiabetic and Antihyperlipidemic Activities of *Cyathea nilgiriensis* (Holttum) on STZ Induced Diabetic Rats

S. ELAVARASI,[1] G. REVATHI,[2] K. SARAVANAN,[2] and HORNE IONA AVERAL[1]

[1]P.G. and Research Department of Zoology,
Holy Cross College (Autonomous), Tiruchirappalli, Tamil Nadu, India,
E-mail: elavarasi888@gmail.com (S. Elavarasi)
ORCID ID: 0000-0001-9188-8188

[2]P.G. and Research Department of Zoology, Nehru Memorial College (Autonomous), Puthanampatti, Tiruchirappalli, Tamil Nadu, India

ABSTRACT

Herbal remedies are considered convenient for the management of diabetes due to their traditional acceptability, availability, and less side effects than oral hypoglycaemic agents. Tree fern, *Cyathea nilgiriensis* (CYATHEACEAE) is popular among the traditional healers of Kolli hill, Namakkal because of its medicinal value. The present research attempts to assess the antidiabetic and antihyperlipidemic effects of ethanolic extract of *Cyathea nilgiriensis* pith powder on STZ induced diabetic model. STZ induced albino rats were treated with (200 mg/kg of b.wt. dose) ethanol extract of *C. nilgiriensis* pith powder for 48 days. The treatment restored the elevated levels of body weight, blood glucose, and HbA_1C. Further, it significantly increased serum insulin level of STZ induced diabetic rats. Total cholesterol, triglycerides, LDL, and very low density lipoprotein (VLDL) were elevated and high density lipoproteins (HDL) level was decreased in diabetic rats. The elevation was markedly controlled near to normal level in extracts treated diabetic rats. Decreased HDL level was significantly increased by treatment with ethanol extract of *C. nilgiriensis*. From the above experimental responses, it is concluded that the *C. nilgiriensis* proved that it is one of the best promising and emerging drug against diabetes mellitus and its complications.

18.1 INTRODUCTION

Diabetes mellitus is a perplexing metabolic disorder which is distinguished by a disruption in the homeostasis amidst the restraint of glucose levels and insulin sensitivity (Hsu et al., 2013). It causes significant morbidity and mortality by different diabetic complications such as retinopathy, neuropathy, and nephropathy (Patel et al., 2011). The current treatment of diabetes with allopathic medicine provides good glycemic control. However, they cause a little complication and some undesirable side effects (Rang et al., 1991). Moreover, providing modern medical healthcare across the world (especially in developing countries) is still a far-reaching goal due to economic constraints. Consequently, there is a growing need to develop novel approaches towards the management and prevention of diabetes. In this context, it has been shown that despite access to conventional medical facilities, the use and dependence of complementary and alternative medicine continues to thrive (Johns et al., 1990; Berman et al., 1999). In the last few years, there has been an exponential growth in the field of herbal medicine and these drugs are gaining popularity both in developing and developed countries because of their natural origin and less side effects. Natural remedies from herbal plants are considered to be more effective and safe alternative treatment for diabetes mellitus (Gupta et al., 2005) due to their minimum side effects in clinical experiences and relatively low costs (Pari and Umamaheswari, 2000) as compared to oral hypoglycemic agents. Right from ancient times in India, indigenous herbal remedies such as Ayurveda and other Indian traditional medicines used many plants in the treatment of diabetes (Babu et al., 2006) and some of them were experimentally evaluated and active principles were isolated (Grover et al., 2002).

Ethnobotanical information of the tribal peoples of Kolli hills stated that pith of tree fern (*Cyathea nilgiriensis*) belonging to family CYATHEACEAE is given to the diabetic patients as powder form for the treatment of diabetes (Saravanan and Revathi, 2016). However, no experimental data are available. Thus, the widely used traditional medicinal plant *Cyathea nilgiriensis* was selected for the present study. The aim of the present study was to explore the antidiabetic potential of *C. nilgiriensis* pith powder in streptozotocin (STZ) induced diabetic albino rat models.

18.2 MATERIALS AND METHODS

18.2.1 COLLECTION OF PLANT MATERIALS

The *C. nilgiriensis* was collected from the places Mathikettan solai and Thembalam of Kolli hills (latitudes 11°55' 05" to 11°21' 10" N and longitudes 78°17' 05" to 78°27'45" E.), Tamil Nadu, India. Stem of this plant was cut into pieces and central part of the wood (pith) was collected and dried under room temperature and powdered with the help of mechanical grinder.

18.2.2 PREPARATION OF EXTRACTS

The pith powder of *C. nilgiriensis* was pulverized and extracted as a whole preparation in a Soxhlet apparatus using absolute ethyl alcohol. Extraction continued for 48 hours. The *C. nilgiriensis* extract was distillated to a dry mass using vacuum evaporator and the powder was stored in desiccator until use.

18.2.3 PHYTOCHEMICAL SCREENING

1. **Preliminary Phytochemical Screening:** The extracts of *C. nilgiriensis* pith powder was subjected to analyze the preliminary phytochemicals tests to find out the presence of secondary metabolites by standard procedures (Kokate, 1994; Harborne, 1973; Rajpal, 2002; Raaman, 2006).
2. **Identification of Bioactive Compounds by GC-MS Analysis:** Presence of individual phytocompounds in the study plant extract was analyzed using GC-MS (Thermo Fisher make, ITQ 900 model). One microliter of the sample was run in a DB-1 fused silica capillary column with helium (1 ml/min) as carrier gas, 250°C injector temperature, 280°C ion-source temperature and isothermal temperature 110°C (2 min), with an increase of 10°C/min to 200°C then 5°C/min to 280°C and 9 min to 280°C. The mass spectrum interpretation was performed using the library of National Institute of Standard and Technology (NIST) and the compounds were identified.

18.2.4 EVALUATION OF ANTIDIABETIC DIABETIC AND ANTIHYPERLIPIDEMIC ACTIVITY

1. **Experimental Setup:** Antidiabetic effect of ethanol extract of *C. nilgiriensis* pith powder was evaluated against STZ induced diabetic albino rats. Normal healthy male albino rats fasted for 12 hours were randomly divided into control and extract treated groups. All the rats were maintained as per the regulations of CPCSEA (Ethical Committee Approval No.790/03/ac/CPCSEA). They were divided into four groups of 4 rats each and caged in separate cages. The experimental set up was given below.
   - **Group-I:** Control (Non-diabetic rats).
   - **Group-II:** Diabetic rats (50 mg/kg b.wt of STZ).
   - **Group-III:** Diabetic rats treated with glibenclamide (5 mg/kg b.wt).
   - **Group-IV:** Diabetic rats treated with ethanol extract of *C. nilgiriensis* pith powder (200 mg/kg b.wt).
2. **Induction of Diabetes:** Diabetes to the rats belonging to group-II, group-III, and group-IV was made by single intraperitoneal injection of STZ (50 mg/kg of body weight). After 48 hours of injection of STZ, rats with blood glucose levels above 250 mg/dl were considered as diabetic rats and they were selected for the experimental studies.

    Antidiabetic potential of *C. nilgiriensis* was evaluated by analyzing abnormalities in serum blood glucose, glycosylated hemoglobin (HbA_1C), serum insulin level; experimental diabetic rats were compared with that of normal rats.
3. **Serum Blood Glucose:** Glucose reagent kit (Aspen Laboratories, New Delhi) was used to measure glucose level by the glucose oxidase-peroxidase (GOD-POD) method (Trinder, 1969) using semi autoanalyzer (Star 21 Plus).
4. **Glycosylated Hemoglobin (HbA_1C):** Quantitative determination of HbA_1C was made by cation exchange method (Gonen and Rubenstein, 1978) using commercially available GHB reagent kit (Medsource Ozone Biomedicals Pvt., Ltd., Haryana) using semi autoanalyzer (Star 21Plus).
5. **Serum Insulin:** It was estimated by chemiluminescence immune assay (CLIA) kit method (SIEMENS Medical Solutions Diagnostics Ltd., USA) using Centaur Immuno Assay Instrument, USA.

6. **Liver Glycogen:** It was quantitatively estimated in the liver tissue by the method of Kemp and Van Hejnigen (1954). Total cholesterol, triglycerides, HDL, and LDL were analyzed by Star 21plus biochemical auto-analyzer.

18.2.5 STATISTICAL ANALYSIS

The values were represented as Mean ± Standard deviation. Analysis of Variance (ANOVA) was used to compare the means of different experimental groups with normal groups. The post hoc test (Student-Newman Keuls (SNK) test) was performed to investigate the influence of the plant extract on various biochemical parameters in the extract-treated rats. All statistical analyses were performed by using windows based SPSS package (Statistical Package for Social Sciences/Statistical Product and Service Solutions).

18.3 RESULTS AND DISCUSSION

18.3.1 PHYTOCHEMICAL SCREENING

The phytochemical screening of ethanol extract of *C. nilgiriensis* pith powder revealed the presence of flavonoids, glycosides, tannin, protein, carbohydrate, saponin, phenols, triterpenoids, and steroids (Table 18.1). The results pertaining to GC-MS analysis led to the identification of number compounds from the GC fractionations of ethanol extract of *C. nilgiriensis* pith powder (Table 18.2 and Figure 18.1).

The four major compounds *viz.*, 2-furancarboxaldehyde, 5-(hydroxymethyl)-, n-Decanoic acid, Didodecyl phthalate, and squalene were identified with retention time (RT) of 6.44, 8.75, 12.70, 20.05 and 23.81. The fragmentations of the compounds are illustrated in GC-MS chromatogram (Figure 18.1).

The flavonoids, steroids, terpenoids, tannin, phenolic acids are recognized as antidiabetic compounds by many authors (Cherian and Augusti, 1995; Sayyah et al., 2004; Tanko et al., 2011). Rupasinghe (2003) reported that the saponins have antidiabetic principles. The terpenoids have also been shown to decrease blood sugar level in animal studies (Luo et al., 1999). Flavonoids, ellagic acids, phenolic acids, phytosterols, gallotanins, and other related polyphenols are reported to possess hypoglycemic, hypolipidemic, and antioxidant activities (Dhanabal et al., 2004). Presence of bitter principle in the plants has been implicated in the antidiabetic activities of many plants (Reher et al., 1991). Thus, the *C. nilgiriensis* pith powder may have antidiabetic activity due to the presence of flavonoids glycosides tannins and alkaloids.

TABLE 18.1 Preliminary Phytochemical Screening of Ethanol Extract of *Cyathea nilgiriensis*

S. No.	Name	Present/Absent
1.	Flavonoids	Present
2.	Alkaloids	Absent
3.	Tannin	Present
4.	Protein	Present
5.	Carbohydrate	Present
6.	Saponin	Present
7.	Glycosides	Present
8.	Phenols	Present
9.	Thiols	Present
10.	Sterols	Present
11.	Triterpenoids	Present

TABLE 18.2 Bioactive Components Identified from Ethanol Extract of *C. nilgiriensis* by GC-MS Analysis

S. No.	Name and Structure of the Compound	Molecular Formula	Molecular Weight	Retention time
1.	2-Furancarboxaldehyde,5-(hydroxymethyl)-	$C_6H_6O_3$	126	6.44
2.	n-Decanoic acid	$C_{10}H_{20}O_2$	172	12.70

TABLE 18.2 *(Continued)*

S. No.	Name and Structure of the Compound	Molecular Formula	Molecular Weight	Retention time
3.	Didodecyl phthalate	$C_{32}H_{54}O_4$	502	20.05
4.	Squalene	$C_{30}H_{50}$	410	23.81

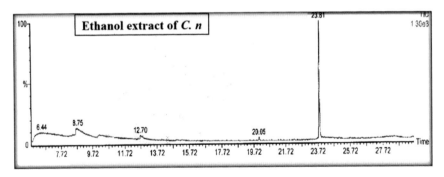

FIGURE 18.1 GC-MS chromatogram of ethanol extract of *C. nilgiriensis*.

18.3.2 ANTIDIABETIC ACTIVITY

Decreased level of body weight and increased level of blood glucose was observed in all diabetic rats. STZ was reported to cause a drastic reduction of insulin-producing β cells of islets of Langerhans thus inducing hyperglycemia (Elsner et al., 2000; Sharma and Garg, 2009). The increased level of blood glucose in the STZ induced diabetic rats might be due to the glycogenolysis or glyconeogenesis (Gupta et al., 2011). Chronic hyperglycemia, the primary clinical manifestation of diabetes mellitus is associated with the development of micro and macrovascular diabetic complications (Brownlee and Cermai, 1981). However, treatment with ethanol extract of *C. nilgiriensis* in the diabetic rat significantly ($p>0.005$) reduced the blood glucose level to near the normal level. It was very close to the blood glucose level of control rats and glibenclamide treated rats (Figure 18.2 and Table 18.3). The body weight was gradually increased in the extract-treated rats. HbA_1C is now considered as the most reliable marker of glycemic control in diabetes mellitus (Halim, 2003) and is used to identify the degree of oxidative stress in diabetic conditions (Gupta et al., 1997). The increased level of blood glucose stimulates non-enzymatic protein glycation which can lead to irreversible modification observed with the characterization of HbA_1C (Cohen and Wu, 1994). The HbA_1C level is very much higher in diabetic patients (Wolffenbuttel et al., 1996). The present result also exhibited an increased level of HbA_1C in diabetic rats. It might be due to the excessive production of glucose in the blood which further reacts with blood hemoglobin and prepared the glycated hemoglobin. Treatment with ethanol extract of *C. nilgiriensis* significantly decreased ($p>0.005$) the HbA_1C level close to control and standard drug-treated rats (Figure 18.2 and Table 18.3).

A partial or total deficiency of insulin affects carbohydrate metabolism which reduces the function of several key enzymes, including glucokinase (GK), phosphofructokinase, and pyruvate kinase resulting in impaired peripheral glucose utilization and augmented hepatic glucose production (Hikino et al., 1989). The insulin-dependent enzymes are also less active. The net effect is an inhibition of glycolysis and stimulation of gluconeogenesis leading to hyperglycemia (Vasudevan and Sreekumari, 1995). Administration of STZ to the rat exhibited insulin deficiency and produce hyperglycemia was observed in the present study. Ethanol extract of *C. nilgiriensis* administration in diabetic rats significantly increased ($p>0.005$) to the levels of glibenclamide treated and control rats. It may be due to increased pancreatic secretion from existing β cells. Flavonoids are known to regenerate the damaged beta cells in the STZ induced diabetic rats and act as insulin secretagogues (Chakravarthi, 1980; Geetha et al., 1994). Phytochemical analysis of this plant identified the presence of flavonoids. It has been isolated from the other plants and found to stimulate secretion or possess insulin-like effect (Marles and Farnsworth, 1995). Hence, treatment with this plant extract has an effect on protecting β cells and smoothing out fluctuation in glucose levels. The storage of liver glycogen

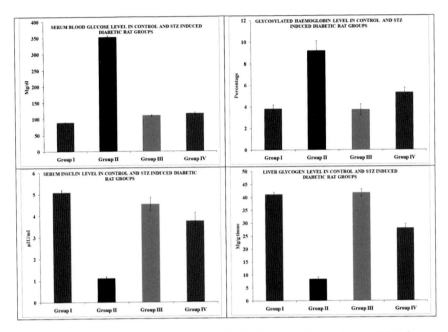

FIGURE 18.2 Effect of ethanol extract of *C. nilgiriensis* on blood glucose, HbA$_1$C, serum insulin, and liver glycogen levels in STZ induced diabetic rats.
Group I: Control.
Group II: STZ induced diabetic rats (50 mg/kg b.wt).
Group III: Glibenclamide (Standard drug) treated rats (5 mg/kg b.wt).
Group IV: Ethanol extract of *C. nilgiriensis* pith powder treated rats (200 mg/kg b.wt).

was markedly reduced in STZ induced diabetes rats, which directly affect the insulin and caused insulin deficiency (Gupta et al., 1997). Notable decrease of liver glycogen is found in diabetic rats which are proved by Bollen et al. (1998). Welihinda and Karuna (1986) also reported about the significant reduction in glycogen level in liver and skeletal muscle in diabetic rats. The glycogen level was restored after the treatment of ethanol extract of *C. nilgiriensis* which might be due to the increased secretion of insulin. Oral administration of ethanol extract of *C. nilgiriensis* pith powder recovered the glucose homeostasis during diabetic condition as evidenced by restoration of blood glucose, serum insulin and HbA$_1$C as well as liver glycogen content.

TABLE 18.3 Results of Student-Newman-Keuls (SNK) Post Hoc Test Show the Variations and Similarities in Glucose, HbA$_1$C, Serum Insulin, and Liver Glycogen Among the Different Group of Rats

Parameter	N	Subset for Alpha = 0.05			
Glucose (mg/dl)	4	One way ANOVA = $f_{3,12}$ = 6319.0 (p>0.005)			
		88.1 (I) ⟷	110.6 (III) ⟷	115.8 (IV) ⟷	353.3 (II) ⟷
HbA$_1$C (%)	4	One way ANOVA = $f_{3,12}$ = 71.76 (p>0.005)			
		3.7 (III)	3.8 (I)	5.3 (IV)	9.2 (II)
		⟵——————⟶		⟵—⟶	⟵⟶
Glycogen (mg/g/tissue)	4	One way ANOVA = $f_{3,12}$ = 667.14 (p>0.005)			
		8.4 (II) ⟷	28.0 (IV) ⟷	41.2 (I)	41.7 (III)
				⟵———————⟶	
Insulin (μIU/ml)	4	One way ANOVA = $f_{3,12}$ = 176.37 (p>0.005)			
		1.1 (II) ⟷	3.7 (IV) ⟷	4.6 (III) ⟷	5.1 (I) ⟷

Mean values are arranged in ascending order. Horizontal line connects similar mean.

18.3.3 ANTIHYPERLIPIDEMIC ACTIVITY

A continuous increase of blood glucose develops high odixative stress and free radicals react with lipids and cause lipid peroxidation. Further, the elevated level of oxidative stress can rise the hyperlipidemia (Manimegalai et al., 1993). Diabetes induced hyperlipidemia leads to excess mobilization of fat from the adipose due to the underutilization of glucose (Krishnakumar et al., 2000). In this study, after induction of diabetes, the serum cholesterol level was significantly raised. Diabetes lead with impaired carbohydrate metabolism and increased lipolysis causing accumulation of acetyl CoA

Antidiabetic and Antihyperlipidemic Activities of *Cyathea nilgiriensis*

which leads to cholesterol synthesis and ends with hyperlipidemia. Further, insulin deficiency results with hypercholesterolemia due to metabolic abnormalities (Murali et al., 2002). However, the treatment of ethanol extract of *C. nilgiriensis* to the diabetic rat significantly restored the lipid profile (Figure 18.3 and Table 18.4).

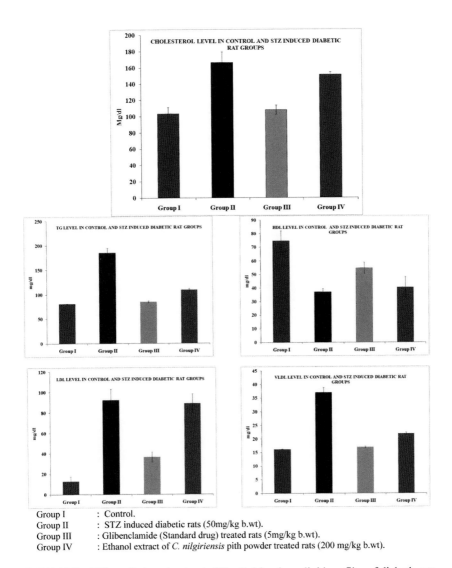

Group I   : Control.
Group II   : STZ induced diabetic rats (50mg/kg b.wt).
Group III : Glibenclamide (Standard drug) treated rats (5mg/kg b.wt).
Group IV   : Ethanol extract of *C. nilgiriensis* pith powder treated rats (200 mg/kg b.wt).

FIGURE 18.3 Effect of ethanol extract of C. nilgiriensis on lipid profiles of diabetic rats.

TABLE 18.4 Results of Student-Newman-Keuls (SNK) Post Hoc Test Show the Variations and Similarities in Lipid Profile Among the Different Group of Rats

Parameter	N	Subset for Alpha = 0.05			
Cholesterol (mg/dl)	4	One way ANOVA = $f_{3,12}$ = 56.0 (p>0.005)			
		103.3 (I)	108.0 (III)	151.0 (IV)	166.3 (II)
		←——————→		←——→	←——→
Triglyceride (mg/dl)	4	One way ANOVA = $f_{3,12}$ = 410.18 (p>0.005)			
		80.4 (I)	84.8 (III)	109.3 (IV)	185.0 (II)
		←——————→		←——→	←——→
HDL (mg/dl)	4	One way ANOVA = $f_{3,12}$ = 32.94 (p>0.005)			
		37.0 (II)	40.3 (IV)	54.5 (III)	74.5 (I)
		←——————→		←——→	←——→
LDL (mg/dl)	4	One way ANOVA = $f_{3,12}$ = 101.73 (p>0.005)			
		12.7 (I)	36.6 (III)	88.9 (IV)	92.3 (II)
		←——→	←——→	←————————→	
VLDL (mg/dl)	4	One way ANOVA = $f_{3,12}$ = 410.18 (p>0.005)			
		16.1 (I)	16.9 (III)	21.9 (IV)	37.0 (II)
		←——————→		←——→	←——→

Mean values are arranged in ascending order. Horizontal lines connect similar means.

Triglyceride is a chief standby energy for the body stored in adipose tissue. However, in diabetic condition increases the triglyceride and ketone bodies levels due to the raise of lipolysis and high synthesis of free fatty acid. The most common lipid abnormalities in diabetes are hypertriglyceridemia and hypercholesterolemia (Khan et al., 1995; Mitra et al., 1995). In the same way, triglycerides level was significantly increased in the STZ induced diabetic rats of present study. However, treatment of ethanol extract of *C. nilgiriensis* to the diabetic rat strongly reduced the triglyceride to level of normal rats and glibenclamide treated rats (Figure 18.3 and Table 18.4).

Low density lipid (LDL) is normally known as bad cholesterol because it can cause atherosclerosis and other related ailments. It is developed from VLDL-cholesterol. HDL cholesterol is commonly called as good cholesterol due its role in prevention of atherosclerosis by transporting the cholesterol from peripheral tissues to liver for excretion. Increase of total cholesterol and LDL levels decrease the HDL level (Mayne, 1996). High HDL cholesterol level is associated with low risk of coronary diseases (Santhosh et al., 2007). Significant increase in the cholesterol, triglycerides, LDL, and VLDL levels were noted in the STZ induced diabetic rats, while there was a significant ($p<0.005$) reduction in the HDL level. However, *C. nilgiriensis* extract treatment exhibited significant ($p<0.005$) decrease in total cholesterol, triglycerides, LDL, and VLDL levels, and notable raise in HDL level (Figure 18.3 and Table 18.4). There are many reports available on the hypolipidemic activity of phytochemicals such as saponins, flavonoids, phenolic compounds and triterpenoids (Leontowicz et al., 2000; Ogawa et al., 2005). Since the presence of saponins, flavonoids, and phenolic compounds in the pith of *C. nilgiriensis,* it has potent antidiabetic and hypolidemic efficiencies.

18.4 CONCLUSION

From the present findings, it is concluded that the ethanol extract of *C. nilgiriensis* has the potential antidiabetic activity and also able to counteract the hyperlipidemic state which occurs in diabetic conditions.

ACKNOWLEDGMENT

Authors greatly acknowledge the Management, the Principal, and Head of the Department of Zoology, Nehru Memorial College, Puthanampatti, for providing necessary facilities. The first author acknowledges the National Testing Service-India, Central Institute of Indian Languages, Manasagangotri, Mysore for financial support.

KEYWORDS

- **antidiabetic activity**
- **antihyperlipidemic activity**
- *Cyathea nilgiriensis*
- **diabetes mellitus**
- **phytochemical screening**
- **statistical analysis**

REFERENCES

Babu, P. A., Suneetha, G., Boddepalli, R., Lakshmi, V. V., Rani, T. S., Babu, Y. R., & Srinivas, K., (2006). A database of 389 medicinal plants for diabetes. *Bioinformation*, *1*(4), 130–131.

Berman, B. M., Swyers, J. P., & Kaczmarczyk, J., (1999). Complementary and alternative medicine: Herbal therapies for diabetes. *Journal of the Association for Academic Minority Physicians*, *10*, 10–14.

Bollen, M., Keppens, S., & Stalmans, W., (1998). Specific feature of glycogen metabolism in the liver. *Biochem. J.*, *336*, 19–31.

Brownlee, M., & Cerami, A., (1981). The biochemistry of the complications of diabetes mellitus. *Annu. Rev. Biochem.*, *50*, 385–432.

Chakravarthi, B. K., Gupta, S., Gambir, S. S., & Gode, K. D., (1980). Pancreatic beta-cell regeneration-A novel antidiabetic mechanism of *Pterocarpus marsupium* Roxb. *Indian Journal of Pharmacology*, *12*, 123–127.

Cherian, S., & Augusti, K. T., (1995). Insulin sparing action of leucopelargonidin derivative isolated from *Ficus bengalensin* Linn. *Indian J. Exp. Biol.*, *33*, 608–611.

Cohen, M. P., & Wu, V., (1994). Purification of glycated hemoglobin. *Methods Enzymol.*, *231*, 65–75.

Dhanabal, S. P., Koate, C. K., Ramanathan, M., Elango, K., & Suresh, B., (2004). The hypoglycemic activity of *Coccinia indica* and its influence on certain biochemical parameters. *Indian J. Pharmacol.*, *36*(4), 249–250.

Elsner, M., Guldbakke, B., Tiedge, M., Munday, R., & Lenzen, S., (2000). Relative importance of transport and alkylation for pancreatic beta-cell toxicity of streptozotocin. *Diabetologia.*, *43*, 1528–1533.

Geetha, B. S., Mathew, B. C., & Augusti, K. T., (1994). Hypoglycaemic effects of leucodelphinidin derivative isolated from *Ficus bengalensis* Linn. *Indian Journal of Physiological Pharmacology*, *38*, 220–222.

Gonen, B., & Rubenstein, A. H., (1978). *Diabetology*, *15*(1).

Groover, J. K., Yadav, S., & Vats, V., (2002). Medicinal plants of India with anti-diabetic potential. *J. Ethnopharmacol.*, *81*, 81–100.

Gupta, B. L., Nehal, M., & Baquer, N. Z., (1997). Effect of experimental diabetes on the activities of hexokinase, glucose-6-phosphate dehydrogenase, and catecholamines in rat erythrocytes of different ages. *Ind. J. Exp. Biol.*, *35*, 792–795.

Gupta, N., Agarwal, M., Bhatia, V., Sharma, R. K., & Narang, E., (2011). A comparative antidiabetic and hypoglycaemic activity of the crude alcoholic extracts of the plant *Leucas aspera* and seeds of *Pithecellobium bigeminum* in rats. *I.J.R.A.P.*, *2*, 275–280.

Gupta, R. K., Kesari, A. N., Watal, G., Murthy, P. S., Chandra, R., Maithal, K., & Tandan, V., (2005). Hyperglycaemic and antidiabetic effect of aqueous extract of leaves of. *Aannona squamosa* (L). *Current Science*, *88*, 1244–1254.

Halim, E. M., (2003). Effect of *Coccinia indica* (L) and *Abroma augusta* (L) on glycemia, lipid profile and on indicators of end organ damage in streptozotocin induced diabetic rats. *Indian J. Clin. Biochem.*, *18*, 54–63.

Harborne, J. B., (1973). *Phytochemical Methods*. Chapman and Hall Ltd., London.

Hikino, H., Kobayashi, M., Suzuki, Y., & Konno, C., (1989). Mechanisms of hypoglycemic activity of aconitan A, a glycan from *Aconitum carmic haeli* roots. *J. Ethnopharmacol.*, *25*, 295.

Hsu, H. C., Chiou, J. F., Wang, Y. H., Chen, C. H., Mau, S. Y., Ho, C. T., Cheng, P. J., Liu, T. Z., & Chen, C. H., (2013). Folate deficiency triggers on oxidative-nitrosative stress-mediated apoptotic cell death and impedes insulin biosynthesis in RINm5F pancreatic islet β-cells: Relevant to the pathogenesis of diabetes. *Plos One*, *8*(11), 1–13.

Johns, T., Kokwaro, J. O., & Kimanani, E. K., (1990). Herbal remedies of the Luo of Siaya District, Kenya establishing quantitative criteria for consensus. *Economic Botany*, *44*, 369–381.

Kemp, A., & Van Hejnigen, K. M., (1954). A colorimetric method for the determination of glycogen in tissues. *Biochem. J.*, *56*, 646–652.

Khan, B. A., Abraham, A., & Leelamma, S., (1995). Hypoglycaemic action of *Murraya koenigii* (curry leaf), *Brassical juncea* (mustard), mechanism of action. *Indian Journal of Biochemistry and Biophysics*, *32*, 106–108.

Kokate, C. K., (1994). *Pharmacognosy* (16th edn.). Nirali Prakasham Publication, Mumbai, India.

Krishnakumar, K., Augusti, K. T., & Vijayammal, P. L., (2000). Hypolipidaemic effect of *Salacia oblonga* Wall root bark in streptozotocin diabetic rats. *Med. Science*, *28*, 65–67.

Leontowicz, H., Gorinstein, S., Lojek, A., Leontowicz, M., Ciz, M., & Soliva-Fortuny, R., (2000). Comparative content of some bioactive compounds in apples, peaches and pears and their influence on lipids and antioxidant capacity in rats. *J. Nutr. Biochem.*, *13*, 603–610.

Luo, J., Cheng, J., & Yevich, E., (1999). Novel terpenoid-typenquinones isolated from *Pycanthu angolensis* of potential utility on the treatment of type-2 diabetes. *J. Pharmacol. Exptl. Therapy*, *288*, 529–534.

Manimegalai, A., Geetha, H., & Rajalakshmi, K., (1993). Effect of vitamin E on high fat diet induced hyperlipidemia in rats. *J. Experimental Biology*, *31*, 704.

Marles, J. R., & Fransworth, N. R., (1995). Antidiabetic plants and their active constituents. *Phytomedicine*, *1*, 32–36.

Mayne, P. D., (1996). *Clinical Chemistry in Diagnosis and Treatment*. Edward Arnold (A division of Hodder Headline Plc). London.

Mitra, S. K., Gopumadhavan, S., Muralidhar, T. S., Anturlikar, S. D., & Sujatha, M. B., (1995). Effect of D-400, a herbomineral preparation on lipid profile, glycated hemoglobin and glucose tolerance in STZ induced diabetes in rats. *Indian Journal of Experimental Biology*, *33*, 798–800.

Murali, B., Upadhyaya, U. M., & Goyal, R. K., (2002). Effect of chronic treatment with *Enicostemma littorale* in non-insulin dependent diabetic (NIDDM) rats. *J. Ethanopharmacol.*, *81*, 199–204.

Ogawa, H., Nakamura, R., & Baba, K., (2005). Beneficial effect of laserpitin, a coumarin compound from *Angelica keiskei* on lipid metabolism in stroke-prone spontaneously hypertensive rats. *Clin. Exp. Pharmacol. Physiol.*, *32*, 1104–1109.

Pari, L., & Umamaheswari, J., (2000). Antihyperglycaemic activity of *Musa sapientum* flowers: Effect on lipid peroxidation in alloxan diabetic rats. *Phytother. Res.*, *14*, 136–138.

Patel, D. K., Kumar, R., Prasad, S. K., Sairam, K., & Hemalatha, S., (2011). Antidiabetic and *in vitro* antioxidant potential of *Hybanthus enneaspermus* (Linn) F. *Muell in Streptozotocin-Induced Diabetic Rats*, *1*(4), 316–322.

Rajpal, V. (2002). *Standardization of Botanicals: Testing and Extraction Methods of Medicinal Herbs*. Eastern Publishers, New Delhi.

Raman, A., & Lau, C., (2006). Anti-diabetic properties and phytochemistry of *Momordica charantia* L. (Cucurbitaceae). *Phytomedicine*, *2*, 349–362.

Rang, H. P., & Dale, M. M., (1991). The endocrine system pharmocology. In: Nattrass, M., & Hale, P. T., (eds.), *Pharmacology* (pp. 504–508). Longman Group Ltd., UK.

Reher, G., Slijepcevic, M., & Krans, L., (1991). Hypoglycemic activity of triterpenes and tannins from *Sarcopoterium spinosum* and two *Sanguisorba* species. *Plant Med.*, *57*, A57–A58.

Rupasinghe, H. P., Jackson, C. J., Poysa, V., Di Berardo, C., Bewley, J. D., & Jenkinson, J., (2003). Soyasapogenol A and B distribution in soybean (*Glycine max* L. *Merr.*) in relation to seed physiology, genetic variability, and growing location. *J. Agric. Food Chem.*, *51*, 5888–5894.

Santhosh, K. S., Kesari, A. N., Gupta, R. K., Jaiswal, D., & Watal, G., (2007). Assessment of antidiabetic potential of *Cynodon dactylon* extract in streptozotocin diabetic rats. *J. Ethnopharmacol.*, *114*, 174–179.

Saravanan, K., & Revathi, G., (2016). *Isolation, Characterization, Evaluation of Antidiabetic Activity of Traditionally Used Herbal Plants on STZ Induced Diabetic Rats*. Project report submitted to UGC, New Delhi.

Sayyah, M., Hadidi, N., & Kamalinejad, M., (2004). Analgesic and anti-inflammatory activity of *Lactuca sativa* seed extract in rats. *J. Ethnopharmacol.*, *92*, 325–329.

Sharma, N., & Garg, V., (2009). Antidiabetic and antioxidant potential of ethanolic extract of *Butea monosperma* leaves in alloxan induced diabetic mice. *Indian Journal of Biochemistry and Biophysics*, *46*, 99–105.

Tanko, Y., Mabrauk, M. A., Adelaiye, A. B., Fatihu, M. Y., & Musa, K. Y., (2011). Anti-diabetic and some hematological effects of ethyl acetate and n-butanol fractions of *Indigofera pulchra* extract on alloxan-induced diabetic Wistar rats. *J. Diabetes Endocrinol.*, *2*, 1–7.

Trinder, P., (1969). Determination of glucose in blood using glucose oxidase with an alternative oxygen acceptor. *Ann. Clin. Biochem.*, *6*, 24–25.

Vasudevan, D. M., & Sreekumari, S., (1995). *Text Book of Biochemistry* (p. 282.). Jaypee Brothers Medical Publishers (P) Ltd., India.

Welihinda, J., & Karunanayake, E. H., (1986). Extra pancreatic effect of *M. charantia* in rats. *J. Ethnopharmacol.*, *17*, 247–255.

Wolffenbuttel, B. H., Giordino, D., Founds, H. W., & Bucala, R., (1996). Long term assessment of glucose control by hemoglobin-AGE measurement. *Lancet*, *347*, 513–515.

CHAPTER 19

Antidiabetic Activity of Drug Loaded Chitosan Nanoparticle

G. REVATHI,[1] S. ELAVARASI,[2] and K. SARAVANAN[1]

[1]*P.G. and Research Department of Zoology, Nehru Memorial College (Autonomous), Puthanampatti, Tiruchirappalli, Tamil Nadu, India, E-mail: revathiphdz@gmail.com (G. Revathi), ORCID: 0000-0002-6656-8219*

[2]*P.G. and Research Department of Zoology, Holy Cross College (Autonomous), Tiruchirappalli, Tamil Nadu, India*

ABSTRACT

Chitosan (CS) is a natural nontoxic biopolymer derived by the removal of an acetyl group (deacetylation) from chitin taken from the prawn shell. Chitosan nanoparticles (CNPs) are used as a drug carrier. It improves drug solubility, stability, enhances efficacy and reduces toxicity by releasing drug slowly. The present study was carried out to synthesis CS from prawn shell and preparing drug-loaded CNPs using polyherbal formulation (*Andrographis paniculata, Andrographis alata, Adhatoda zeylanica, Gymnema sylvestre, Syzygium cumini,* and *Justicia glabra*) and evaluated its antidiabetic efficiency. CNPs were synthesized by ionic gelation method. CS and drug-loaded CNPs were characterized by XRD pattern, Fourier transform infrared spectroscopy (FTIR) analysis, and SEM studies. Prepared CNPs showed spherical in shape, nano range particle size. The size of drug-loaded CNPs ranged from 37.6 nm to 39.5 nm. Nanoparticles (NPs) were found to be crystalline in nature confirmed by x-ray diffraction (XRD). The prepared drug-loaded CNPs exhibited 85% drug encapsulation efficiency. The present results suggested that drug-loaded CNPs could be used as an ideal carrier to deliver the antidiabetic drug to the specific target.

19.1 INTRODUCTION

Chitosan (CS) is the N-deacetylated product of chitin (Jaung et al., 2001), which is a natural polysaccharide mostly present in the shells of crustaceans and exoskeleton of insects (Lue et al., 2014). CS is biodegradable, biocompatible, bioactive, and polycationic substance, and has many significant biological and chemical properties (Denkbas et al., 2002). Thus, it is extensively used in industrial and medical fields (Selmer-Olsen et al., 1996). Even though CS has notable biological and medicinal properties, it reduces its applications due to its macromolecule nature. To overcome this drawback, nowadays chitosan nanoparticles (CNPs) is prepared from CS and effectively used in drug delivery system. Since CNP is prepared from biodegradable and biocompatible particles, it represents an effective option for controlled drug delivery. There is strong evidence of CNPs as effective drug carrier for many challenging drug delivery applications. Moreover, there is also indicative data regarding CS biocompatibility, which further supports the potential use in nanomedicine (Ajazuddin and Saraf, 2010). Using CNPs with plant extracts has revealed an advantageous strategy for herbal drugs (Bonifacio et al., 2014; No et al., 1989). Thus, the present research was conducted to formulate and develop strategies to improve the therapeutic efficacy of traditionally used polyherbal drug (includes *Andrographis paniculata, Andrographis alata, Adhatoda zeylanica, Gymnema sylvestre, Syzygium cumini,* and *Justicia glabra*) extracts employing CNPs as a carrier.

19.2 MATERIAL AND METHODS

Tripolyphosphate (TPP) was purchased from Loba chemical laboratory, Mumbai. Acetic acid and ethanol were purchased from Merck Specialties private limited, Mumbai. The double-distilled water was used for all the experiments.

19.2.1 PREPARATION OF CHITOSAN (CS)

CS was prepared from prawn (*Penaeus monodon*) shell by the method of No and Meyers (1989). The prawn shells were collected from local fish market (Tiruchirappalli). Shells were thoroughly washed with hot water to remove debris from the shells. Then cleaned shells were used for CS preparation. CS preparation was involved in three steps *viz.,* deproteinization, demineralization, and deacetylation (Vila et al., 2004):

1. **Deproteinization:** The prawn shells were deproteinized with addition of 3% NaOH solution [solid to solvent ratio of 1:10 (w/v)] and boiled for 1 hr with constant stirring. The boiled sample was allowed to cool at room temperature for 30 minutes. Then it was filtered and the filtrate was washed with tap water for 30 minutes. Then it was dried using an oven.

2. **Demineralization:** Deproteinized shell sample was demineralized with addition of 1N HCl [solid to solvent ratio of 1:25(w/v)] and allowed it for 60 minutes at room temperature. Then it was filtered under vacuum. The filtrate was washed for 30 min with tap water and oven-dried.

3. **Deacetylation:** It is the process to convert chitin to CS by removal of acetyl group. It is generally achieved by treatment with concentrated Sodium hydroxide solution (50%) at 90°C for 2 hours to remove acetyl groups from the polymer. Then the prepared CS was used for preparation of CNPs.

19.2.2 PREPARATION OF CHITOSAN NANOPARTICLES (CNPS)

The CS Nps were synthesized by ionic gelation method (Vimal et al., 2012; Sohini and Bhatt, 1996). CS was dissolved at a concentration of 3 mg/ml in 0.5% acetic solution. A solution of TPP at the concentration of 1.0 mg/ml was prepared with deionized water. Then 10 ml of sodium TPP solution was added drop by drop using a burette to 25 ml of the CS solution with constant stirring by magnetic stirrer. An opalescent suspension was formed spontaneously under the abovementioned conditions and they were by centrifuged at 16,000 rpm for a period of 30 min at 14°C. The supernatant was removed and the wet pellet of CS nanoparticles (NPs) was collected. The pellet was separately washed with 20%, 75%, and 99% ethanol. Then they were lyophilized with freeze-dried and the powder form CS NPs stored at 4°C for further studies.

19.2.3 ENCAPSULATION OF POLYHERBAL EXTRACT WITH CS NANOPARTICLES (NPS)

1. **Plant Collection and Extract Preparation:** *Andrographis paniculata* (Acanthaceae), *Andrographis alata* (Acanthaceae) *Adhatoda zeylanica* (Acanthaceae), *Gymnema sylvestre* (Ascelpidaceae), *Syzygium cumini* (Myrtaceae) *and Justicia glabra* (Acanthaceae) are included in the traditional polyherbal drug. Thus, they were selected and collected the respective plant parts from Kolli hills. The plants part were collected and dried under shade and powdered with the help of mechanical grinder. The plant powder was pulverized and extracted as a whole preparation by cold extraction method using ethanol (Nasti et al., 2009). Extract of polyherbal mixture was concentrated to a dry mass by vacuum evaporator and stored in desiccators until use.
2. **Encapsulation of Drug:** Polyherbal extract was incorporated into the TPP solution prior to CNPs formation. Ethanol extract of polyherbal mixture was added in the concentrations of 10% with respect to the total amount of CS used for particles preparation. Then the TPP solution with ethanol extract of polyherbal mixture was added to the CS solution drop by drop titration method with continuous stirring by magnet stirrer. After formation of opalescent suspension, drug-loaded CNPs were separated by centrifugation and further steps were followed as said above (synthesis of CNPs).
3. **Encapsulation Efficiency:** To check the encapsulation efficiency, the free extract that was present in the supernatant after centrifugation was measured in a UV spectrophotometer at 370 nm. The encapsulation efficiency was calculated using the following formula:

$$\text{Encapsulation efficiency} = \frac{\text{Total con. of drug in TPP} - \text{Con. of drug in supernatant}}{\text{Total drug}} \times 100$$

19.2.4 CHARACTERIZATION OF NANOPARTICLES (NPS)

1. **X-Ray Diffraction Study (XRD Study):** In order to confirm the crystalline or amorphous nature of drug in the CNPs, XRD analysis was carried out by studying the XRD pattern of CNPs and drug-loaded CNP. Samples were analyzed at NIT, Tiruchirappalli using Rigaku UltimaIII, XRD instrument.

2. **FTIR Analysis:** FTIR spectroscopy was measured for the determination of the types of bonds present in the CNPs and Polyherbal drug-loaded NPs. FTIR analysis was made at NIT, Tiruchirappalli.
3. **Morphological Study:** The morphological study of NPs was performed by field emission scanning electron microscope (FE-SEM). NPs were dispersed in water (5 ml) and sonicated for 3 minutes (E-Chom Tech, Taipei, Taiwan). Few drops of the prepared samples were put on double-sided adhesive tape fixed on the metal stub. After drying, the gold coating was performed for 80 seconds under vacuum. SEM analysis was made at Gandhigram Rural Institute (Deemed University), Gandhigram, Dindigul.

19.2.5 EVALUATION OF ANTIDIABETIC ACTIVITY

19.2.5.1 EXPERIMENTAL SET-UP

Normal healthy male albino rats fasted for 12 hours were randomly divided into four groups of four rats each and caged in separate cages. The experimental set up was given below:

- Group-I: Control (Non-diabetic rats).
- Group-II: Diabetic rats (50 mg/kg b.wt of streptozotocin (STZ)).
- Group-III: Diabetic rats treated with glibenclamide (5 mg/kg b.wt).
- Group-IV: Diabetic rats treated with polyherbal drug-loaded CNPs.

All the rats were maintained as per the regulations of CPCSEA (Ethical Committee Approval No.790/03/ac/CPCSEA). Antidiabetic effect of polyherbal drug-loaded CNPs was evaluated on the STZ induced diabetic albino rats. Induction of Diabetes to the group-II, group-III, and group-IV rats was made by single intraperitoneal injection of STZ (50 mg/kg of body weight). After 48 hours of injection of STZ, rats with blood glucose levels above 250 mg/dl were considered as diabetic rats and they were selected for the experimental studies. CNPs were administered to the rats for 15 days. Daily body weight, food intake, and water intake were measured every day at morning. After 15 days, the rats were sacrificed and blood was collected by cardiac punching method and serum was separated and used for biochemical analysis. The antidiabetic activity was evaluated by analyzing abnormalities in serum blood glucose level, glycosylated hemoglobin (HbA_1C) level and serum insulin level, and liver glycogen level in experimental diabetic rats.

1. **Serum Blood Glucose:** Glucose reagent kit (Aspen Laboratories, New Delhi) was used to measure glucose level by the glucose oxidase-peroxidase (GOD-POD) method (Trinder, 1969).
2. **Glycosylated Hemoglobin (HbA$_1$C):** Quantitative determination of HbA$_1$C was made by cation exchange method (Gonen and Rubenstein, 1978) using commercially available GHB reagent kit (Medsource Ozone Biomedicals Pvt., Ltd., Haryana).
3. **Serum Insulin:** It was estimated by the chemiluminescence immune assay (CLIA) kit method (SIEMENS Medical Solutions Diagnostics Ltd., USA) using the Centaur Immuno Assay Instrument, USA.
4. **Liver Glycogen:** It was quantitatively estimated in the liver tissue by the method of Kemp and Van Hejnigen (1954).

19.2.6 STATISTICAL ANALYSIS

To compare the means of different experimental groups with normal groups, analysis of variance (ANOVA) was performed. The post hoc test (Student-Newman Keuls (SNK) test) was performed to investigate the influence of the extract loaded CNPs on various biochemical parameters in the diabetic rats. All statistical analyses were performed by using windows based SPSS package (Statistical Package for Social Sciences/Statistical Product and Service Solutions).

19.3 RESULTS AND DISCUSSION

Zone of opalescent suspension was formed immediately upon mixing of TPP and CS solutions as molecular linkages were formed between TPP phosphates and CS amino groups. Zone of opalescent suspension was also formed after the addition of TPP in the polyherbal drug-containing CS. Formation of opalescent suspension after the addition of TPP in CS and polyherbal mixed TPP in CS solution indicates that the solution containing a suspension of colloidal particles (NPs). The cationic nature of CS made a notable achievement for the development of drug delivery systems. These cations contact with polyanions to form a gel through inter- and intramolecular cross-linkages. Ionic gelation is a simple and straightforward technique which involves the addition of alkaline phase and acidic phase. The adjustment of parameters such as concentration of CS, ratio of CS: TPP polymer and mixing condition influence the CNP size (Fàbregas et al., 2003; Qi et al., 2004).

19.3.1 X-RAY DIFFRACTION (XRD) ANALYSIS

XRD analysis was conducted for crystal phase identification and to compare the physical nature of polyherbal extract loaded CNPs and CNP. It is a non-destructive technique widely used for the characterization of crystalline materials. The XRD graph (Figure 19.1(A)) showed many peaks in 2θ range of 20° showing its crystalline structure of the drug-loaded CNPs.

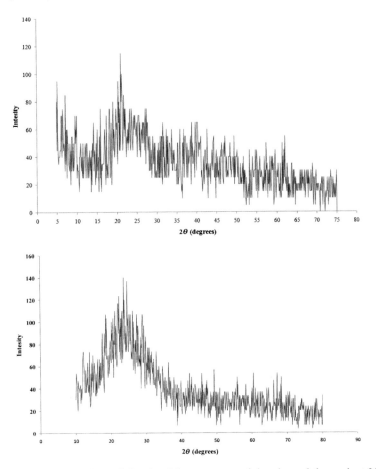

FIGURE 19.1 (A) XRD graph for the chitosan nanoparticles showed the peak at 20 theta degree. (B) XRD graph for the extract loaded chitosan nanoparticles showed the peak at 23 to 25 theta degrees.

XRD patterns of CNPs showed sharp peaks 2θ-scattered angles of 17, 20, and 21 (Figure 19.1(A)); these peaks were indicating the crystalline nature of drug. The polyherbal extract loaded CNP crystalline peak was increased in the drug-loaded nanoparticle formulation (Figure 19.1(B)). It indicates slight increase in crystalline nature of the extract loaded CNP form intensities of extract loaded CNPs peaks were also increased in the nanoparticle formulation. This increased intensity indicated the increased crystallinity of drug in nanoparticle form.

19.3.3 FTIR ANALYSIS

CS NPs were synthesized by mixing of CS and TPP. Polyherbal extract loaded CNPs were prepared by adding extract mixed TPP with CS solution. FTIR studies of CNPs and Polyherbal drug-loaded NPs were performed to characterize the chemical structure of NPs. FTIR spectra of CNPs and extract loaded NPs are shown in Figures 19.2(A) and 19.2(B). A band at 3895 cm^{-1} corresponds to the combined peaks of the (NH_2) and OH group stretching vibration in CS (Figure 19.2(A)). The band at 1801.67 cm^{-1}, is attributed to the $CONH_2$ group. The 1474.80 cm^{-1} peak of the C(NH_2) bending vibration was sharper than the peak at 1801.67 cm^{-1}, which shows the high degree of deacetylation of the CS. A shift from 3419 to 3427 cm^{-1} was observed, and the peak was sharper in the CNPs, which indicates that the hydrogen bonding is enhanced (Xu and Du, 2003). The intensities of ($CONH_2$) band at 1614.13 and 1538 cm^{-1} appeared which shows that the ammonium groups are cross-linked with TPP molecules (Knaul et al., 1999). Thus, it is suggested that polyphosphoric groups of sodium polyphosphate interact with the ammonium groups of CS, which serves to enhance both inter- and intramolecular interaction in CNPs. The IR spectrum of polyherbal extract loaded NPs (Figure 19.2(B)) showed the prominent peaks at 3943, 1793, 1474, and 1279 cm^{-1}. The broad peak at 3943 cm^{-1} corresponds to OH and NH stretching vibration, peak broadening indicates the presence of intermolecular hydrogen bonding.

19.3.3 SCANNING ELECTRON MICROSCOPY (SEM)

The morphological characteristics of CNPs and CS drug-loaded NPs are shown in Figure 19.3. The SEM of the drug-loaded NPs showed that the NPs have a spherical structure and aggregated. The spheres have mean diameters around 800 nm. The NPs dry powder consists of individual NPs, which touch each other, but retain their original size and shape.

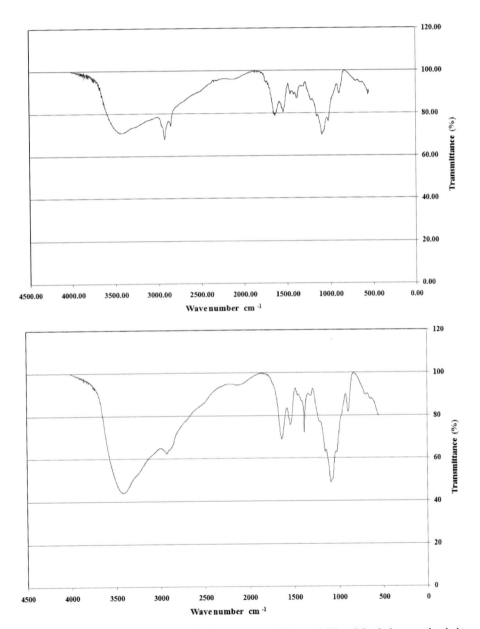

FIGURE 19.2 (A) FTIR spectra of chitosan nanoparticles and (B) polyherbal extract loaded chitosan nanoparticles.

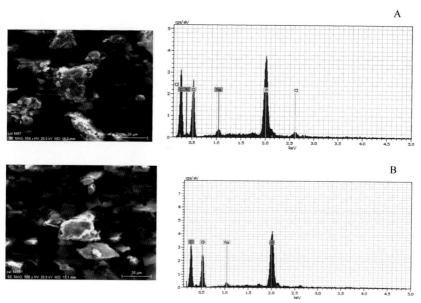

FIGURE 19.3 (A) SEM micrographs of chitosan nanoparticles (B) SEM micrographs of polyherbal extract loaded chitosan nanoparticles.

19.3.4 EVALUATION OF ANTIDIABETIC ACTIVITY OF DRUG LOADED CNPs

Induction of diabetes with STZ is associated with the characteristic loss of body weight which is due to increased muscle wasting (Swanston-Flat et al., 1990). The rate of protein synthesis in the liver and muscle was decreased in diabetic rats which account for the negative nitrogen balance and weight. In the present study, normal healthy animals were found to be stable in their body weight whereas diabetic rats showed heavy weight loss. However, drug-loaded CNPs treatment to the diabetic rats comparatively increased the initial weight loss. High consumption of water and food associated with high blood glucose levels and high frequency of urine output is a sign of the diabetic condition of STZ induced rats. In STZ induced diabetic rats, increased food consumption, and decreased body weight were also observed by Elavarasi (2015). It indicates polyphagic condition of diabetes (Hakim et al., 1997). It is interesting to note that drug-loaded CNPs treatment reduced the food intake and water intake of diabetic rats.

Administration of drug encapsulated CNPs to the diabetic rats for 11 days significantly ($p>0.005$) reduced the elevated level of blood glucose,

HbA$_1$C, and liver glycogen and significantly (p>0.005) increased the loss of insulin (Figure 19.4 and Table 19.1). Similarly, drug-loaded CNPs increased the loss of protein in the blood serum and tissues (Figure 19.4 and Table 19.1). This may be due to the fact that plant targets many signaling molecules, which diabetic cells highly rely on. A quick action of glipizide loaded CNPs was observed by Lokhande et al. (2013) in which a reduction in blood glucose levels within 1 h of treatment. Further, they stated that the sustained release CNPs of glipizide could be able to manage type II diabetes mellitus with reduced dose frequency, decreased side effects and improve patient compliance.

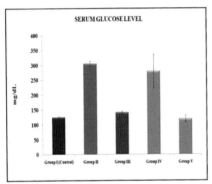
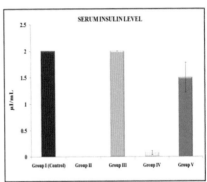
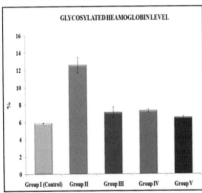
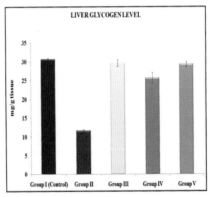

FIGURE 19.4 Effect of chitosan nanoparticles and drug-loaded CNPs on serum glucose, insulin, HbA$_1$C, and liver glycogen levels of control and STZ induced albino rats.
Group I: Control.
Group II: STZ induced diabetic rats (50 mg/kg b.wt).
Group III: Glibenclamide (Standard drug) treated rats (5 mg/kg b.wt).
Group IV: Chitosan nanoparticles treated rats (5 mg/kg b.wt).
Group V: Drug loaded Chitosan nanoparticles treated rats (5 mg/kg b.wt).

TABLE 19.1 Results of Student-Newman-Keuls (SNK) Post Hoc Test Show the Variations and Similarities in Glucose, HbA₁C, Liver Glycogen, and Serum Insulin Among Different Group of Rats with Response to the Treatment of Drug Loaded Chitosan Nanoparticles

Parameters	*Groups
Glucose (mg/dl)	One way ANOVA ($f_{4,10}$ = 186.9; $p>0.005$). Student-Newman-Keuls Post hoc test Subset or alpha = 0.05) 89.3 (I) — 122.0 (V) — 130.0 (III) — 188.3 (IV) — 278.3 (II)
HbA₁C (%)	One way ANOVA ($f_{4,10}$ = 27.99; $p>0.005$). Student-Newman-Keuls Post hoc test Subset or alpha = 0.05) 6.0 (I) — 6.5 (V) — 7.2 (III) — 7.4 (IV) — 12.7 (II)
Glycogen (mg/g tissue)	One way ANOVA ($f_{4,10}$ = 86.03; $p>0.005$). Student-Newman-Keuls Post hoc test Subset or alpha = 0.05) 11.5 (II) — 25.3 (IV) — 29.2 (V) — 29.6 (III) — 30.4 (I)
Serum insulin (µIU/ml)	One way ANOVA ($f_{4,10}$ = 102.87; $p>0.005$). Student-Newman-Keuls Post hoc test Subset or alpha = 0.05) 0.02 (II) — 0.2 (IV) — 0.5 (V) — 2.1 (I) — 2.1 (III)

Values are arranged in ascending order. Horizontal lines connect similar means.

*Groups:
I: Control rats.
II: Diabetic control rats.
III: Diabetic rats treated with glibenclamide.
IV: Diabetic rats treated with chitosan nanoparticles.
V: Diabetic rats treated with drug-loaded chitosan nanoparticles.

19.4 CONCLUSION

The prepared drug-loaded CNPs have a spherical shape with 85% drug encapsulation. Drug loaded CNPs could be used as an ideal carrier to deliver the polyherbal drug to the specific targets. It also showed good antidiabetic activity.

ACKNOWLEDGMENT

Authors thank the Management, Principal, and Head of the Department for providing necessary facilities and UGC, New Delhi for Financial Assistance.

KEYWORDS

- antidiabetic activity
- chitosan nanoparticles
- polyherbal drug
- scanning electron microscopy
- serum blood glucose
- x-ray diffraction analysis

REFERENCES

Ajazuddin, & Saraf, S., (2010). Applications of novel drug delivery system for herbal formulations. *Fitoterapia, 81,* 680–689.

Bonifácio, B. V., Silva, P. B., Ramos, M. A., Negri, K. M., Bauab, T. M., & Chorilli, M., (2014). Nanotechnology-based drug delivery systems and herbal medicines: A review. *Int. J. Nanomedicine, 9,* 1–15.

Denkbas, E. B., Kilicay, E., Birlikseven, C., & Ozturk, E., (2002). Magnetic chitosan microspheres: Preparation and characterization. *Reactive and Functional Polymers, 50,* 225–232.

Elavarasi, (2015). *Evaluation of Antidiabetic Potential of Traditionally Used Medicinal Plants, Cyathea Nilgiriensis (Holttum) and Pterocarpus Marsupium Roxb.* In STZ induced diabetic rat model. Ph. D thesis submitted to Bharatidasan University, Trichy.

Fàbregas, A., Miñarro, M., García-Montoya, E., Perez-Lozano, P., Carillo, C., Sarrate, R., Sanchez, N., Tico, J. R., & Sune-Negre, J. M., (2013). Impact of physical parameters on particle size and reaction yield when using the ionic gelation method to obtain cationic polymeric chitosan-tripolyphosphate nanoparticles. *Int. J. Pharm., 446,* 199–204.

Garcia-Fuentes, M., Torres, D., & Alonso, M. J., (2002). Design of lipid nanoparticles for the oral delivery of hydrophilic macromolecules. *Coll. Surf. B: Biointerf., 27,* 159–168.

Hakim, Z. S., Patel, B. K., & Goyal, R. K., (1997). Effects of chronic ramipril treatment in streptozotocin-induced diabetic rats. *Indian J. Physiol. Pharmacol., 41,* 353–360.

Juang, R. S., Wu, F. C., & Tseng, R. L., (2001). Solute adsorption and enzyme immobilization on chitosan beads prepared from shrimp shell wastes. *Bioresource Technology, 80,* 187–193.

Knaul, J. Z., Hudson, S. M., & Creber, K. A. M., (1999). *Journal of Polymer Science: Part B: Polymer Physics, 72,* 1079–1094.

Lokhande, A., Mishra, S., Kulkarni, R., & Naik, J., (2013). Formulation and evaluation of glipizide loaded nanoparticles. *Int. J. Pharm. Pharm. Sci., 5*(S4), 147–151.

Luo, Y., & Wang, Q., (2014). Recent development of chitosan-based polyelectrolyte complexes with natural polysaccharides for drug delivery. *Int. J. Biol. Macromol., 64,* 353–367.

Nasti, A., Zaki, N. M., De Leonardis, P., Ungphaiboon, S., Sansongsak, P., Rimoli, M. G., & Tirelli, N., (2009). Chitosan/TPP and chitosan/ TPP-hyaluronic acid nanoparticles: Systematic optimization of the preparative process and preliminary biological evaluation. *Pharm. Res., 26,* 1918–1930.

No, H. K., Meyers, S. P., & Lee, K. S., (1989). Isolation and characterization of chitin from crawfish shell waste. *Journal of Agricultural and Food Chemistry, 37,* 575–579.

Qi, L., Xu, Z., Jiang, X., Hu, C., & Zou, X., (2004). Preparation and antibacterial activity of chitosan nanoparticles. *Carbohydrate Research, 339,* 2693–2700.

Selmer-Olsen, E., Ratnaweera, H. C., & Pehrson, R., (1996). Novel treatment process for dairy wastewater with chitosan produced from shrimp-shell waste. *Water Science and Technology, 34,* 33–40.

Sohini, Y. R., & Bhatt, R. M., (1996). Activity of a crude extract formulation in experimental hepatic amoebiasis and in immunomodulation studies. *Journal of Ethnopharmacology, 54,* 119–124.

Vila, A., Sanchez, A., Janes, K., Behrens, I., Kissel, T., Vila Jato, J. L., & Alonso, M. J., (2004). Low molecular weight chitosan nanoparticles as new carriers for nasal vaccine delivery in mice. *European Journal of Pharmaceutics and Biopharmaceutics, 57,* 123–131.

Vimal, S., Taju, G., Nambi, K. S. N., Abdul, M. S., Sarath, B. V., Ravi, M., & Sahul, H. A. S., (2012). Synthesis and characterization of CS/TPP nanoparticles for oral delivery of gene in fish. *Aquaculture, 358/359,* 14–22.

Xu, Y., & Du, Y., (2003). Effect of molecular structure of chitosan on protein delivery. *International Journal of Pharmaceutics, 250,* 215–226.

CHAPTER 20

Antidiabetic Activity of Silver Nanoparticles Biosynthesized Using *Ventilago maderaspatana* Leaf Extract

P. KARUPPANNAN,[1] K. SARAVANAN,[2] and HORNE IONA AVERAL[1]

[1]*P.G. and Research Department of Zoology, Holy Cross College (Autonomous), Trichy, Tamil Nadu, India, Tel.: 9790298098, E-mail: vikramprabhu88@gmail.com (P. Karuppannan)*

[2]*P.G. and Research Department of Zoology, Nehru Memorial College (Autonomous), Puthanampatti, Trichy District, Pin – 621 007, Tamil Nadu, India*

ABSTRACT

This paper described the biosynthesis of silver nanoparticles (AgNPs) using *Ventilago maderaspatana* leaves, and explained their efficiency of antidiabetic activity. Biosynthesized Ag Nps were characterized by UV-Vis Spectroscope, FT-IR, transmission electron microscope (TEM), SEM, and EDAX. UV-Visible spectrum showed an absorption peak at nanoscale range of 431 nm which confirmed that the size of Ag particles are found at nanoscale range. FT-IR spectrum revealed the secondary metabolites found in the leaf extract responsibilities for the formation of nanoparticles (NPs). SEM study showed ball-shaped Ag Nps. TEM study revealed that size of biosynthesized AgNPs were found between the nanoscales of 10 to 50 nanometer. Results of the EDAX confirm the presence of silver in the biosynthesized NPs. Further, biosynthesized AgNPs effectively reduced the increased level of blood glucose, HbA$_1$C, total protein and albumin, and also restored level of serum insulin on STZ induced diabetic albino rats. Thus, *V. maderaspatana* leaf, extract may be useful for eco-friendly synthesis of Ag NPs and they can be used for antidiabetic drug preparation.

20.1 INTRODUCTION

Nanotechnology is emerging and fast growing science with application and preparation of nanomaterials (Albrecht et al., 2006). They usually ranging from 1 to 100 nm, and exhibit their remarkable potential in the field of biology and medicine based on the specific characters. Various nanoparticles (NPs) such as zinc, copper, gold, and silver nanoparticles (AgNPs) are used in various fields. Among the several metallic NPs, silver nanoparticle has been used in medical field because of its catalytic activity, good electrical conductivity and chemical stability (Sharma et al., 2009). Due to its demanding need, it has been prepared by many methods (Wang et al., 2005; Navaladian et al., 2007; Sreeram et al., 2008; Zamiri et al., 2011). Chemical synthesis of AgNps needs heavy pressure and high temperature and poisonous chemicals. They cause adverse effect when using in the biological system due to the toxic nature of capping agent present on the surface of the NPs. However, biological preparation of NPs is more safe, non-toxic to environment, and cheap. So, the present study was carried out to prepare AgNPs using leaves of *Ventilago maderaspatana* (*Rhamnaceae*) which is a medicinal plant used for the management of various ailments including diabetes.

20.2 MATERIALS AND METHODS

Analytical grade silver nitrate and ethyl acetate were used. *V. maderaspatana* leaves were collected from Kolli Hills a part of Eastern Ghats, Namakkal District, Tamil Nadu. The plant was identified as *Ventilago maderaspatana* (Gaertn.) (Family: Rhamnaceae) (Specimen access No. SJCBT2112) by the Rapinat Herbarium, St. Joseph College (Autonomous), Tiruchirappalli.

20.2.1 PREPARATION OF EXTRACT

V. maderaspatana leaves were collected, dried under shade and powdered. Then this powder was used to Soxhlet extraction using ethyl acetate solvent.

20.2.2 BIOSYNTHESIS OF SILVER NANOPARTICLES (AGNPS)

AgNPs synthesis was followed by the method of Vijaykumar et al. (2014). Freshly prepared ethyl acetate extract of *V. maderaspatana* leaves (10 ml) was added drop wise to 100 ml of silver nitrate (0.1 M) solution.

Simultaneously, stirred the mixture using the magnetic stirrer under 50°C for 30 minutes, continuously. Then the mixture was incubated in dark room at room temperature for 24 hrs. Changing in the color of solution was observed and recorded.

20.2.3 CHARACTERIZATION OF BIOSYNTHESIZED SILVER NANOPARTICLES (AGNPS)

The properties of AgNPs were analyzed using UV-VIS spectroscopy and FT-IR analysis. Morphological and chemical composition of AgNPs was studied using SEM, TEM, and EDAX.

20.2.4 EVALUATION OF ANTIDIABETIC ACTIVITY OF SILVER NANOPARTICLES (AGNPS) SYNTHESIZED USING V. MADERASPATANA LEAVES

20.2.4.1 ANIMAL SELECTION

Healthy adult male albino rats (*Rattus norvegicus*) with bodyweight 120 to 250 g were selected for this study. The rats were obtained from Tamil Nadu Veterinary and Animal Science University, Chennai, and maintained under a controlled environment. Before using the animal to experimental purpose, ethical approval from the Institutional Animal Ethical Committee (BDU/IAEC/2017/NE/43) was obtained.

20.2.4.2 INDUCTION OF DIABETES

Diabetes mellitus was induced by single intraperitoneal injection of streptozotocin (STZ) (50 mg/kg of body weight in double-distilled water) to overnight fasted albino rats (Al-Hariri, 2012). The rats with fasting blood glucose levels above 250 mg/dl were considered as a diabetic rat and they were selected for the experimental studies (Al-Hariri, 2012).

20.2.4.3 EXPERIMENTAL SETUP

The rats were divided into four groups and each group consists of three rats. The experimental sets up were:

- Group I: Normal control.
- Group II: Diabetic control.
- Group III: Treated with biosynthesized AgNPs (20 gm/kg body weight).
- Group VI: Treated with glibenclamide (5 mg/kg body weight).

After the 15 days of treatment, the experimental rats were sacrificed; the blood sample was collected and serum was separated by centrifugation method (5000 rpm for 10 minutes) for biochemical estimation.

20.2.4.4 BIOCHEMICAL ESTIMATION

Serum Insulin was calculated by chemiluminescence immune assay (CLIA) kit method (Science Medical solutions Diagnostics Ltd., USA) using Centaur Immuno Assay Instrument, USA. Serum Blood glucose level, glycosylated hemoglobin (HbA$_1$C), total protein and Albumin levels were tested using respective kits with Semi-Automatic Biochemistry Auto analyzer (Star 21Plus, Manufactured by Aspen Diagnostics Pvt., Ltd., New Delhi).

20.3 RESULTS AND DISCUSSION

20.3.1 SYNTHESIS OF AG NPS USING V. MADERASPATANA LEAVES

Among the various NPs, Ag Np is important and it has been widely studied. Ag Np possesses unique optical and electrical biological properties, and thus it is used in catalytic, biosensing, drug delivery, imaging, nano-device fabrication, and medicinal fields (Jain et al., 2008; Nair and Laurencin, 2007).

During the preparation of Ag Nps, dark brown color was formed after the addition of *V. maderaspatana* leaf extract in the silver nitrate solution due to the reduction of silver ions to AgNPs. Changing of color from colorless to dark brown shows the formation of AgNPs. It might be due to the reduction property of plant extract (Ahmad et al., 2003; Sengottaiyan et al., 2015). The Ag NPs exhibited yellowish-brown color in aqueous solution due to the excitation of surface plasmon vibrations. The reduction of AgNO$_3$ is due to the presence of phytochemicals such as alkaloids, phenolic compounds and flavonoids in the plant extracts (Mukherjee et al., 2001; Balaji et al., 2009; Iravani, and Zolfaghari, 2013). Earlier, Karuppannan and Saravanan (2016) reported the presence of these phytochemicals in the *V. maderaspatana* leaves. Thus, it is concluded that the plant, *V. maderaspatana* acted as capping agent to reduce the silver nitrate into Ag Np.

20.3.2 CHARACTERIZATION OF AG NPS USING *V. MADERASPATANA* LEAVES

20.3.2.1 UV-VISIBLE SPECTRUM

The UV-visible spectrum of AgNPs showed a plasmon resonance band at 431 nm (Figure 20.1). It authorized the development of AgNPs.

Due to the excitation of surface plasmon vibrations, brown color was developed after formation of AgNps. They are having γ max values between the range of 400 and 500 nm (Sastry et al., 1997). It is clearly indicated that Ag NPs exhibits a yellowish-brown color in aqueous solution due to excitation in UV-visible spectrum depending upon the particle size (Mock et al., 2002).

20.3.2.2 FT-IR ANALYSIS

FT-IR spectroscopy measurement was carried out to identity the possible involvement of the major functional groups for synthesis of AgNPs. Biological materials are known to interact with metal salts via functional groups and mediate their reduction to NPs. The spectrum of biosynthesized Ag NPs is represented in Figure 20.2. FT-IR spectrum showed several spectrums indicating the complex nature of the biological materials. The bands appear at 3914.60, 3766.23, 3462.82, 2361.18, 2078.40, 1636.22, 1379.17, and 672. 39 cm^{-1} were assigned to stretching vibration of -OH alcohols or -NH amines, phophric acid and ester stretch or combination of C-H stretch, alkynes, C = C stretch, C-H rock, C = O or aromatic CH, aromatic CH and CH alkene bends or C-Br or C-Cl, respectively. The present investigation was explained that the formation of AgNPs may be due to the reduction action of silver ions (Ag+) into nanoparticles (Ag0) by phytochemicals especially compounds with antioxidant properties found in *V. maderaspatana*. Presence of antioxidant activities in the *V. maderaspatana* was already established by Damayanth and Satyavati (2015).

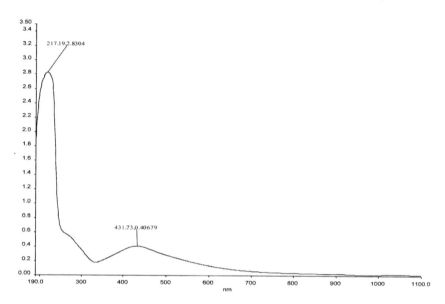

FIGURE 20.1 UV-Vis spectroscopy of biosynthesized silver nanoparticles using leaf ethyl acetate extract of *V. maderaspatana*.

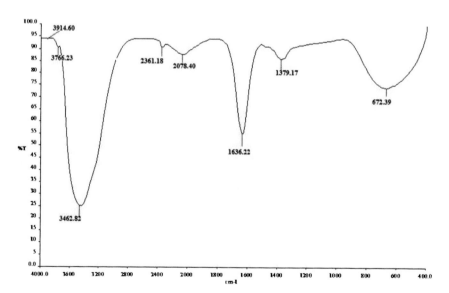

FIGURE 20.2 FT-IR results of biosynthesized silver nanoparticles using ethyl acetate extract of *V. maderaspatana*.

20.3.2.2 MORPHOLOGICAL CHARACTERS OF AG NPS

From the SEM images it can be observed that particles are spherical in shape and particles were uniformly distributed (Figure 20.3). Layer particles of Ag NPs are formed by aggregation of nanoparticles (NPs) which might be evaporation of solvent during sample preparation. For more evidence of silver in nanoscale level, the size and shape were analyzed using TEM (Figure 20.4). In the present study, the synthesized NPs were spherical in shape with varying sizes ranging from 20 to 50 nm (Figure 20.4). Thus, it is confirmed that the prepared Ag NPs are found to be minimum nanoscale. Earlier, Karuppannan and Saravanan (2017) prepared AgNPs using methanol extract of *V. maderaspatana* leaves with minimum nanoscale.

FIGURE 20.3 SEM images of biosynthesized silver nanoparticles.

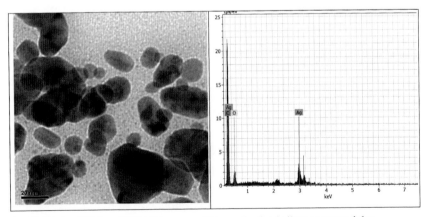

FIGURE 20.4 TEM and EDAX images of biosynthesized silver nanoparticles.

20.3.3 ANTIDIABETIC ACTIVITY OF SILVER NANOPARTICLES (AGNPS)

Diabetes mellitus is a major metabolic disorder characterized by high blood glucose levels. Improper management of diabetes may lead to the development of many chronic complications (Bailes, 2002). The pancreas is the primary sensing organ of dietary and energy states; it detects glucose concentrations in the blood and, in response to elevated blood glucose concentrations, insulin will be secreted. In diabetes conditions, this mechanism is impaired for several reasons (Alberti and Zimmet, 1998; Amos et al., 1997). Plant mediated treatment of diabetes is nowadays very popular due to its non-toxic, low cost and target-oriented treatment. The present study described the use of the medicinal plant, *V. maderaspatana* leaves in the preparation of silver nanoparticle and use of AgNPs in the treatment of diabetes. The present study has detected the antidiabetic activity of Ag Nps synthesized using *V. maderaspatana* in STZ induced type 2 diabetic rats. Analyzing the potential of antidiabetic agents with STZ induced hyperglycemia in rats is considered to be a good preliminary screening diabetic model and is used frequently (Ivorra et al., 1989). STZ brings the diabetic state by selectively destroying the pancreatic β-cells which make cells less in action and increase glucose level (Elsner et al., 2000). The present study also showed the high blood glucose level (454.0 ± 54.2 mg/dl) and low level of insulin after injection of STZ. However, treatment with Ag Nps for 15 days significantly reduced the blood glucose level (149.0 ± 7.77 mg/dl) near to the normal level. Treatment with standard drug (glibenclamide) also reduced the elevated level of glucose (167.3 ± 15.36 mg/dl). Treatment of Ag Nps also reduced HbA_1C level to normal level. On the other hand, the mean serum insulin level was increased to normal level after the treatment of *V.m* mediated AgNPs and standard drug glibenclamide. Insulin level in control, AgNps treated and standard drug-treated rats were more or less equal (Table 2001) but the insulin level of untreated diabetic rat was significantly differed from all other groups. Increased level of protein was (20.06 ± 0.85 mg/dl) found in STZ induced diabetic control group. Then, it was decreased to the normal level by the treatment of AgNPs and standard drug glibenclamide. Albumin level was increased in untreated diabetic rat (Table 20.1). In animal models of diabetes, biologically synthesized AgNPs and silver-coated gold nanoparticles (Ag@AuNPs) significantly reduced high blood glucose levels to normal (Subramani and Ahmed, 2012; Shaheen et al., 2016). Both gold nanoparticles (AuNPs) and Ag@AuNPs increased insulin levels, reduced insulin resistance, restored the lipid profile and reduced inflammatory markers compared with an untreated diabetic control group. The antidiabetic activity of Ag@AuNPs was greater than that of AuNPs (Shaheen et al., 2016). Thus, it is evident that AgNPs exhibit effective antidiabetic activity (Shaheen et al., 2016).

TABLE 20.1 Effect of *V. Maderaspatana* Mediated AgNPs Treatment on Levels of Glucose, HbA$_1$C, Serum Insulin, Total Protein, Albumin on STZ Induced Diabetic Albino Rats

Groups	Parameters				
	Glucose (mg/dl) Mean ± SE	HbA$_1$C (%) Mean ± SE	Insulin µIU/ml Mean ± SE	Total Protein (g/dl) Mean ± SE	Albumin (g/dl) Mean ± SE
Group I	137.5 ± 12.32 (114.0–172.0)	5.7 ± 0.50 (4.3–6.5)	8.3 ± 0.17 (7.9–8.7)	8.7 ± 0.32 (7.9–9.4)	7.0 ± 0.12 (6.6–7.2)
Group II	454.0 ± 54.2 (300.0–550.0)	12.8 ± 2.78 (4.9–17.9)	1.7 ± 0.42 (1.1–2.9)	20.6 ± 0.85 (19.1–23.0)	11.3 ± 0.63 (9.9–12.8)
Group III	149.0 ± 7.77 (132.0–168.0)	8.3 ± 0.44 (7.5–9.1)	6.6 ± 0.26 (5.9–7.1)	7.3 ± 0.45 (6.2–8.2)	6.4 ± 0.90 (3.8–7.8)
Group IV	167.3 ± 15.36 (136.0–197.0)	7.1 ± 0.21 (6.5–7.4)	7.7 ± 0.21 (7.2–8.2)	6.6 ± 0.29 (5.8–7.1)	5.8 ± 0.04 (5.7–5.9)

Values in the parenthesis are respective mean.
*Groups
Group I - Control
Group II - Diabetic control
Group III - Treated with 20 mg/kg/b.wt. of AgNPs
Group VI - Treated with 5 mg/kg/b.wt. of standard drug (glibenclamide).

20.4 CONCLUSION

From the study, it is concluded that the leaf ethyl acetate extract of *V. maderaspatana* possesses ability to synthesis the AgNPs without environmental hazard, and within the range of 10–50 nm in size. Moreover, the biosynthesized AgNPs effectively reduced the elevated levels of blood glucose, HbA$_1$C, Total protein and albumin, and also increased the level of serum insulin. Thus, the *V. maderaspatana* mediated AgNPs might be useful for the treatment of diabetes mellitus.

ACKNOWLEDGMENT

The authors thanks to the Management, Principal, and Head of the Department, Department of Zoology, Nehru Memorial College (Autonomous),

Puthanampatti for providing necessary facilities for conducting this work. The first author also thanks to UGC for providing fund do to this research work in the name UGC-RGNF.

KEYWORDS

- **antidiabetic activity**
- **biosynthesis**
- **nanomaterials**
- **silver nanoparticles**
- **standard drug glibenclamide**
- *Ventilago maderaspatana*

REFERENCES

Ahmad, A., Mukherjee, P., Senapati, S., Mandal, D., Khan, M. I., & Kumar, R., (2003). Extracellular biosynthesis of silver nanoparticles using the fungus *Fusarium oxysporum*. *Colloids and Surfaces, B: Biointerfaces, 28*, 313–318.

Alberti, K. G., & Zimmet, P. F., (1998). Definition, diagnosis, and classification of diabetes mellitus and its complications. Part 1: Diagnosis and classification of diabetes mellitus. Provisional report of a WHO consultation. *Diabet Med., 15*, 539–553.

Albrecht, M. A., Evans, C. W., & Raston, C. L., (2006). Green chemistry and the health implications of nanoparticles. *Green Chem., 8*, 417–432.

Al-Hariri, M. T., (2012). Comparison the rate of diabetes mellitus induction using STZ dissolved in different solvents in male rats. *J. Comp. Clin. Path. Re., 1*(3), 96–99.

Amos, A. F., McCarty, D. J., & Zimmet, P., (1997). The rising global burden of diabetes and its complications: Estimates and projections to the year 2010. *Diabet. Med., 14*, S7–85.

Bailes, B. K., (2002). Diabetes mellitus and its chronic complications. *AORN J., 76*, 265–82.

Balaji, D. S., Basavaraja, S., Bedre, M. D., Prabakar, B. K., & Venkataraman, A., (2009). Extracellular biosynthesis of functionalized silver nanoparticles by strains of cladosporium cladosporoides. *Colloids Surf B., 68*, 88–92.

Damayanth, D., & Satyavati, D., (2015). Antidiabetic, antihyperlipidemic and antioxidant properties of Roots of *Ventilago maderaspatana* Gaertn. On streptozotocin-induced diabetic rats." *IOSR-JPBS, 10*, 50–59.

Elsner, M., Guldbakke, B., Tiedge, M., Munday, R., & Lenzen, S., (2000). Relative importance of transport and alkylation for pancreatic beta-cell toxicity of streptozotocin. *Diabetologia, 43*, 1528–1533.

Iravani, S., & Zolfaghari, B., (2013). Green synthesis of silver nanoparticles using *Pinus eldarica* bark extrac. *Biomed. Res. Int.*, 639725.

Ivorra, M. D., Paya, M., & Villar, A., (1989). A review of natural products and plants as potential antidiabetic drugs. *J. Ethnopharmacol., 27*, 243–275.

Jain, P. K., Huang, X., El-Sayed, I. H., & El-Sayed, M. A., (2008). Noble metals on the nanoscale: Optical and photothermal properties and some applications in imaging, sensing, biology, and medicine. *Accounts of Chemical Research, 41,* 1578–1586.

Karuppannan, P., & Saravanan, K., (2017). Eco-friendly synthesis of silver nanoparticles using *Ventilago maderaspatna* (Gaertn.), their morphonlogical characterization. *International Journal of Chem. Tech. Research, 10*(9), 1–6.

Mock, J. J., Barbic, M., Smith, D. R., Schultz, D. A., & Schultz, S., (2002). Localized surface plasmon resonance effects by naturally occurring Chinese yam particles. *J. Chem. Phys., 119,* 6755.

Mukherjee, P., Ahmed, A., Mandal, D., Senapati, S., Sainkar, S. R., Khan, M. I., Parishcha, R., Ajaykumar, P. V., Alam, M., Kumar, R., & Sastry, M., (2001). Fungus-mediated synthesis of silver nanoparticle and their immobilization in the *Mycelia matrix*: A novel biological approach to nanoparticle synthesis. *Nano Lett., 1,* 515–529.

Nair, L. S., & Laurencin, C. T., (2007). Silver nanoparticle: Synthesis and therapeutic applications. *Journal of Biomedical Nanotechnology, 3,* 301–316.

Navaladian, S., Viswanathan, B., Viswanath, R. P., & Varadarajan, T. K., (2007). Thermal decomposition as route for silver nanoparticles. *Nanoscale Res. Lett., 2,* 44–48.

Sastry, M., Mayyaa, K. S., & Bandyopadhyay, K., (1997). pH dependent changes in the optical properties of carboxylic acid derivatized silver colloid particles. *Colloids Surf A., 127,* 221–228.

Sengottaiyan, A., Mythili, R., Selvankumar, T., Aravinthan, A., Kamala-Kannan, S., & Manoharan, K., (2015). Green synthesis of silver nanoparticles using *Solanum indicum* L. and their antibacterial, splenocyte cytotoxic potentials. *Research on Chemical Intermediates,* 10.1007/s11164-015-2199-7.

Shaheen, T. I., El-Naggar, M. E., & Hussein, J. S., (2016). Antidiabetic assessment: *In vivo* study of gold and core-shell silver-gold nanoparticles on streptozotocin-induced diabetic rats. *Biomed. Pharmacother., 83,* 865–75.

Sharma, V. K., Yngard, R. A., & Lin, Y., (2009). Silver nanoparticles: Green synthesis and their antimicrobial activities. *Adv. Colloid Interface Sci., 145,* 83–96.

Sreeram, K. J., Nidhin, M., & Nair, B. U., (2008). Microwave assisted template synthesis of silver nanoparticles. *Bull. Mater. Sci., 31,* 937–942.

Subramani, K., & Ahmed, W., (2012). Nanotechnology and the future of dentistry. In: *Emerging Nanotechnologies in Dentistry* (pp. 1–14). William Andrew Publishing, Boston.

VijayKumar, P. P. N., Pami, S. V. N., Kollu, P., Sayanaryana, K. V. V., & Shameem, U., (2014). Green synthesis and characterization of silver nanoparticles using *Boherhaavia diffusa* plant extract and their antibacterial activity. *Industrial Crops and Products, 52,* 562–566.

Wang, H., Qiao, X., Chen, J., & Ding, S., (2005). Preparation of silver nanoparticles by chemical reduction method. *Colloid Surf. A., 256,* 111–115.

Zamiri, R., Zakaria, A., Abbastabar, H., Darroudi, M., & Husin, M. S., (2011). Laser-fabricated castor oil-capped silver nanoparticle. *Int. J. Nanomed., 6,* 565–568.

CHAPTER 21

Hypoglycemic Activity of *Biophytum sensitivum* Whole Plant Extracts on Alloxan Induced Diabetic Rats

C. RENUKA,[1] K. SARAVANAN,[1] and P. KARUPPANNAN[2]

[1]*P.G. and Research Department of Zoology, Nehru Memorial College (Autonomous), Puthanampatti, Trichy District – 621 007, Tamil Nadu, India, E-mail: renukaphd@gamil.com (C. Renuka)*

[2]*P.G. and Research Department of Zoology, Holy Cross College (Autonomous), Tiruchirappalli, Tamil Nadu, India*

ABSTRACT

In the present study, medicinal plant *Biophytum sensitivum* was used to evaluate the hypoglycaemic activity against alloxan-induced diabetic albino rat model. Elevated levels of blood glucose, HbA$_1$C, and glycogen and decreased levels of insulin were observed in alloxan-induced diabetic rats. Treatment with different extracts of *B. sensitivum* and glibenclamide significantly increased the blood glucose, HbA$_1$C, and liver glycogen levels. Among the extracts, ethanol extracts of *B. sensitivum* recovered the diabetic condition. Thus, it is concluded that the herbal plant *B. sensitivum* used for the treatment of hyperglycemia and also considered as the promising drug candidate for the treatment of hyperglycaemic conditions.

21.1 INTRODUCTION

Diabetes mellitus is a non-communicable disease, which is caused by the abnormality of carbohydrate, protein, and lipid metabolism which is linked to low blood insulin level or insensitivity of target organs to insulin (Maiti et al., 2004). Diabetes mellitus can lead to several complications such as retinopathy, nephropathy, peripheral neuropathy, myopathy, encephalopathy, and the characteristic clinical signs and symptoms (American Diabetes Association (ADA), 2002). The current treatment of diabetes although provide good glycemic control, they do a little in preventing complications. Besides, these drugs are associated with unwanted side effects (Rang et al., 1991). There is a growing need to develop novel approaches towards the management and prevention of diabetes. The World Health Organization (WHO) has listed 21,000 plants, which are used for medicinal purposes around the world. Among these, 2500 species are in India, out of which 150 species are used commercially on a fairly large scale. The investigation on plant drugs will be a useful strategy in the discovery of new lead molecules eliciting improved activity by regulating the different mechanisms maintaining the carbohydrate metabolism and thus can be used in treating hyperglycemia of varied etiology (Robert and Thomas, 2006). *Biophytum sensitivum* is a small annual herb, medicinal plants belonging to family Oxalidaceae used in traditional oriental herbal medicines. It is well known for the hypoglycaemic properties of *B. sensitivum* and it is traditionally used for treatment of diabetes. But no scientific validation for antidiabetic activity of *B. sensitivum*. Hence, the present study was conducted to evaluate the antihyperglycaemic activity of *B. sensitivum* on alloxan-induced albino rat model.

21.2 MATERIALS AND METHODS

21.2.1 *EXPERIMENTAL ANIMALS*

Healthy male albino rats, *Rattus norvegicus* with the bodyweight ranged between 150 and 200 g were used for the present study. They were fed with standard pellet feed (Sai Durga Feeds & Foods, Bangalore) and water *ad libitum*. Ethical care principles were followed throughout the experimental period.

21.2.2 *COLLECTION OF PLANT MATERIALS*

In the present study, Sicker bud, *B. sensitivum* was selected to evaluate its effect on the treatment of diabetes. The *B. sensitivum* whole plant including aerial parts and roots were collected from fallow lands of nearby villages (Kottathur and Gandhi Nagar, Musiri, Taluk, Tiruchirappalli District, Tamil Nadu Lat. 11°06"; Long. 78°68"). The *B. sensitivum* whole plants were dried under shade and finely powdered. Then the extraction was prepared using different solvents like 70% ethanol, ethyl acetate, chloroform, and water.

21.2.3 INDUCTION OF DIABETES MELLITUS

Single intraperitoneal injection of alloxan (150 mg/kg of body weight in normal saline 0.9%; Rao et al., 1999) to the male albino rats was made to induce diabetes. After 48 hrs of alloxan injection, the rats with blood glucose levels above 250 mg/dl were considered as diabetic rats and they were chosen for the present investigation.

21.2.4 EVALUATION OF ANTIDIABETIC ACTIVITY

Antidiabetic activity of *Biophytum sensitivum* whole plant extracts was evaluated by analyzing abnormalities in serum blood glucose level (Sasaki et al., 1972), serum insulin level (Herbert et al., 1965), glycosylated hemoglobin (HbA$_1$C) level (Eross et al., 1984) and liver glycogen level (Kemp and Van Hejnigen, 1954) in the diabetic rats, using commercially available standard kits.

21.3 RESULTS

Alloxan injected rats showed diabetic condition with increased level of blood glucose (307.5 ± 4.70 mg/dl), HbA$_1$C level (6.3 ± 0.27%), and decreased level of liver glycogen (8.4 ± 0.47 mg/g) and serum insulin level (8.1 ± 0.20 µU/ml) (Table 21.1). Treatment with *B. sensitivum* extracts significantly reduced blood glucose level and glibenclamide, standard drug. However, the 70% ethanolic extract of *B. sensitivum* effectively reduced the blood glucose level near to the normal level (84.5 ± 1.37 mg/dl). Similarly, the administration of 70% ethanolic extract of *B. sensitivum* and glibenclamide remarkably reduced the HbA$_1$C level very close to control rats. The decreased level of liver glycogen level in alloxan-induced diabetic rats markedly increased after the treatment of *B. sensitivum* extracts. Aqueous ethanolic extract of *B. sensitivum* decreased in a greater level of liver glycogen when compared to other extracts (10.4 ± 0.19 mg/g). In alloxan-induced diabetic rats the serum insulin level was found to be very low (8.1 ± 0.20 µU/ml) when compared to control rats (37.0 ± 0.39 µU/ml). Conversely, the administration of 70%

ethanolic extract, ethyl acetate extract and ethanolic extract of *B. sensitivum* restored the serum insulin level towards close to normal level. Among the extracts, 70% ethanolic extract of *B. sensitivum* was greater in increasing the serum insulin level of diabetic rats (31.8 ± 0.63 µU). From the results, it is understood the 70% ethanolic extract of *B. sensitivum* effectively reduced the blood glucose level and HbA$_1$C level, and increased the levels of insulin and liver glycogen in alloxan-induced diabetic albino rat model (Table 21.1 and Figure 21.1).

TABLE 21.1 Effect of *B. senitivum* Extracts on Blood Glucose, Insulin, HbA$_1$C, and Liver Glycogen Levels (Mean ± SE) in Alloxan Induced Diabetic Rats

Groups	Parameters			
	Glucose (mg/dl)	HbA$_1$C (%)	Serum Insulin (µU/ml)	Liver Glycogen (mg/g)
I	88.1 ± 1.44	2.8 ± 0.09	37.0 ± 0.39	14.5 ± 0.23
II	307.5 ± 4.70	6.3 ± 0.27	8.1 ± 0.20	8.4 ± 0.47
III	103.9 ± 1.94	4.8 ± 0.23	23.7 ± 0.71	11.0 ± 0.27
IV	84.5 ± 1.37	2.9 ± 0.08	31.8 ± 0.63	10.4 ± 0.19
V	104.4 ± 2.28	4.6 ± 0.34	29.6 ± 0.37	9.6 ± 0.33
VI	83.0 ± 1.32	3.1 ± 0.21	37.4 ± 0.33	12.1 ± 0.27
VII	303.7 ± 3.80	6.9 ± 0.21	8.0 ± 0.28	8.3 ± 0.28

*Groups: I = Control (non-diabetic rats); II = Diabetic control; III = Diabetes + ethanol extract; IV = Diabetes + 70% ethanol extract; V = Diabetes + ethyl acetate extract; VI = Diabetes + glibenclamide; VII = Diabetes + DMSO.

21.4 DISCUSSION

Herbal preparations are "natural" and thus they are basically safe. However, they can be more powerful and more toxic if they used wrongly (Bateman et al., 1998). Several authors studied the antidiabetic and hypoglycaemic activity of many plants. Antidiabetic activity of *Leucas aspera* and seeds of *Pithecellobium bigeminum* in alloxan induced and streptozotocin (STZ) induced diabetic rats studied by Gupta et al. (2011). Ethanolic extracts of the plant *L. aspera* and *P. bigeminum* were capable of exhibiting significant

Hypoglycemic Activity of *Biophytum sensitivum* Whole Plant Extracts

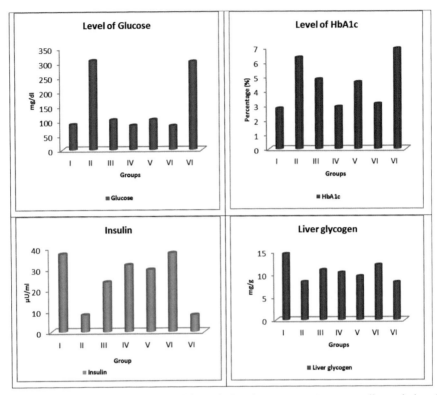

FIGURE 21.1 Hypoglycaemic activity of *Biophytum sensitivum* on alloxan-induced diabetic albino rats.

reduction in blood glucose level. Tatiya et al. (2011) evaluated hypoglycaemic potential of *Bridelia retusa* bark in albino rats. Their findings indicated that methanolic extract of bark of *B. retusa* (200 and 400 mg/kg body weight), petroleum ether extract of bark of *B. retusa* (200 and 400 mg/kg body weight) and n-butanol extract of *B. retusa* (200 and 400 mg/kg body weight) significantly lowered the blood sugar level. In the present study, *B. sensitivum* whole plant extracts was used to evaluate its effects in the treatment of diabetes. Alloxan induction to the rats highly reduced the insulin level due to the preventing the secretion of insulin for beta cell and results increased glucose level in blood. The HbA_1C level was found to be higher in diabetic condition and it was observed by many authors (Cohen and Wu, 1994; Wolffenbuttel et al., 1996). Diabetic rats showed increased level of blood glucose accompanied by elevated liver glycogen in diabetic rats that suggest poor glycemic control mechanism. Glycogen is the primary intracellular storable form of glucose. The synthesis of glycogen by the liver is impaired in diabetic condition. Notable decrease of liver glycogen is found in diabetic rats which are proved by (Bollen et al., 1998). Welihinda and Karuna (1986) also reported about the significant reduction in glycogen level in liver and skeletal muscle in diabetic rats. Treatment of 70% ethanol extract of *B. sensitivum* extracts decreased blood glucose level in the alloxan-induced diabetic rats because of restoration of serum insulin level. Further, elevated levels of HbA_1C and decreased level of liver glycogen were recovered after the administration of *B. sensitivum* extracts. These mechanisms of this plant are due to the presence of alkaloid, flavonoid, terpenoids, tannin, etc., in this plant. The presence of these phytochemicals in the whole plant of *B. sensitivum* was reported by Renuka (2011).

21.5 CONCLUSION

From the present findings, it is understood that the extracts of *B. sensitivum* have the potential to the hypoglycaemic activity which occurs in alloxan-induced diabetic rats. Further aqueous ethanolic extracts possessed the good hypoglycaemic activity against alloxan-induced albino rats.

ACKNOWLEDGMENT

The authors thank the Management, Principal, and Head of the Department of Zoology, Nehru Memorial College (Autonomous), Puthanampatti, Trichy for providing necessary facilities. The first author acknowledges the UGC for providing financial support in the name of RGNF.

KEYWORDS

- **alloxan**
- **antidiabetic activity**
- ***Biophytum sensitivum***
- **diabetes mellitus**
- **hypoglycaemic**
- **medicinal plants**
- ***Rattus norvegicus***

REFERENCES

American Diabetes Association, (2002). Diabetic Nephropathy. *Diabetes Care, 25*, S85–S89.

Bateman, J., Chapman, R. D., & Simpson, D., (1998). Possible toxicity of herbal remedies. *Scoff. Med. J., 43*, 7–15.

Bollen, M., Keppens, S., & Stalmans, W., (1998). Specific feature of glycogen metabolism in the liver. *Biochem. J., 336*, 19–31.

Cohen, M. P., & Wu, V., (1994). Purification of glycated hemoglobin. *Methods Enzymol., 231*, 65–75.

Eross, J., Kreutzmann, D., Jimenez, M., Keen, R., Rogers, S., Colwell, C., Vines, R., & Sinik, M., (1984). Colorimetric measurement of glycosylated protein in whole blood cells, plasma and dried blood. *Ann. Clin. Biochem., 21*, 519–522.

Gupta, N., Agarwal, M., Bhatia, V., Sharma, R. K., & Narang, E., (2011). A comparative antidiabetic and hypoglycaemic activity of the crude alcoholic extracts of the plant *Leucas aspera* and seeds of *Pithecellobium bigeminum* in rats. *I.J.R.A.P., 2*, 275–280.

Herbert, V., Lau, K. S., Gottieb, C. W., & Bleicher, J. J., (1965). Coated charcoal immunoassay of insulin. *J. Clin. Endocrinol. Metab., 25*, 1375–1384.

Kemp, A., & Van Hejnigen, K. M., (1954). A colorimetric method for the determination of glycogen in tissues. *Biochem. J., 56*, 646–652.

Maiti, R., Jana, D., Das, U. K., & Ghosh, D., (2004). Antidiabetic effect of aqueous extract of seed of *Tamarindus indica* in streptozocin-induced diabetic rats. *J. Ethnopharmacology, 92*, 85–91.

Rang, H. P., & Dale, M. M., (1991). The endocrine system pharmocology. In: Nattrass, M., & Hale, P. T., (eds.), *Pharmacology* (pp. 504–508). Longman Group Ltd., UK.

Rao, B. K., Kesavulu, M. M., Giri, R., & Appa Rao, C., (1999). Antidiabetic and hypolipidemic effects of *Momordica cymblaria* Hook fruit powder in alloxan-diabetic rats. *J. Ethnopharmacol., 67*, 103–119.

Renuka, C., (2011). Effect of *Biophytum sensitivum* extracts in the treatment of experimental diabetic albino rats. *PhD Thesis*. Submitted to Bharathidasan University, Tiruchirappalli, Tamil Nadu, India.

Robert, W. M., & Thomas, V., (2006). Antihyperlipidemic agent. In: Brunton, L. L., & Lazo, S. J., (eds.), *Goodman and Gillman's the Pharmacological Basis of Therapeutics* (11th edn., p. 933). New York: Mc Graw-Hill.

Sasaki, T., Matsy, S., & Sonae, A., (1972). Effect of acetic acid concentration on the color reaction in the O-toluidine boric acid method for blood glucose estimation. *Rinshbo Kagaku, 1*, 346–353.

Tatiya, A. U., & Saluja, A. K., (2011). Evaluation of phytochemical standards and *in vitro* antioxidant activity of tannins rich fraction of stem bark of Bridelia retusa (Li). *International Journal of Pharm. Tech. Research, 2*(1), 649–655.

Welihinda, J., & Karunanayake, E. H., (1986). Extra pancreatic effect of *M. charantia* in rats. *J. Ethnopharmacol., 17*, 247–255.

Wolffenbuttel, B. H., Giordino, D., Founds, H. W., & Bucala, R., (1996). Long term assessment of glucose control by hemoglobin-AGE measurement. *Lancet, 347*, 513–515.

CHAPTER 22

Antidiabetic Activity of *Ventilago maderaspatana* Leaf Extracts on STZ Induced Diabetic Rats

P. KARUPPANNAN,[1] K. SARAVANAN,[2] and P. PREMALATHA[2]

[1] *P.G. and Research Department of Zoology, Holy Cross College (Autonomous), Trichy, Tamil Nadu, India, Tel.: 9790298098, E-mail: vikramprabhu88@gmail.com*

[2] *P.G. and Research Department of Zoology, Nehru Memorial College (Autonomous), Puthanampatti, Trichy District, Pin – 621 007, Tamil Nadu, India*

ABSTRACT

Diabetes mellitus is a disorder of carbohydrate metabolism characterized by a high level of blood glucose. It is due to the deficiency or inefficiency of insulin secretion, action or both. Many synthetic drugs are for the treatment of diabetes. But they cause unwanted side effects. Thus, there is an urgent need for alternative medicine for the management of diabetes without any side effects. *Ventilago maderaspatana* is traditionally used by tribal people for the treatment of various diseases including diabetes. However, no scientific studies are available. Hence, the present study was carried out to study the antidiabetic activity of leaf extract of *V. maderaspatana* on STZ induced diabetic albino rats. Treatment with of *V. maderaspatana* leaf extracts drastically decreased the blood glucose and HbA_1C levels, and remarkably raised the level of serum insulin in STZ induced diabetic rats. It also lowered the total protein and albumin levels. Since the leaf extracts of *V. maderaspatana* possessed good antidiabetic activity against STZ induced albino rat model, it can be used as a drug candidate for diabetes.

22.1 INTRODUCTION

Diabetes mellitus is a chronic disease worldwide with a prevalence varying from 1 to 8%. The occurrence of diabetes is increased globally, is likely to attain 300 million patients in the year 2025 with India projected to have the largest number of diabetic cases (Gupta et al., 2003). Diabetes mellitus is defined as a heterogeneous metabolic disorder which alters the carbohydrates, fat, and protein metabolisms. Diabetes mellitus is a condition when the amount of blood glucose in the blood is too high due to the insufficiency or inefficiency of insulin. Insulin is a polypeptide hormone secreted by beta cells of Islets of Langerhans. Insulin allows the glucose enter into the body cells and helps to burn and produce the energy (Gulliford, 1994). Diabetes mellitus affects almost all tissues in the body and is related with significant complications of multiple organs including the eyes, nerves, kidneys, and blood vessels. Several oral hypoglycaemic synthetic drugs (sulfonylureas and biguanides) are available for the treatment of diabetics. However, they cause side effects and no permanent cure (Noor et al., 2008). Thus, it is very important to find new drugs for the treatment of diabetes. Many scientists are focused to explore the plant natural products which have secondary metabolites with anti-diabetic effect as the alternative drug for the treatment of diabetes (Noor et al., 2008). *Ventilago maderaspatana* is a medicinal plant belongings to the family Rhamnaceae. It is used for the treatment of many ailments. The seeds of this plant are given to diabetic patients to lower the blood glucose (Nandkarni, 2000; Pavani et al., 2012). However, no literatures are available on antidiabetic activity of leaves of *V. maderaspatana.* Hence, the present study was conducted to evaluate the antidiabetic activity leaf extracts of *V. maderaspatana* on STZ induced albino rat model.

22.2 METHODOLOGY

The analytical grade chemicals and reagents were used for the biochemical analysis. Medicinal plant *Ventilago maderaspatana* was collected from Kolli Hills a part of Eastern Ghats, Namakkal District, Tamil Nadu, South India and authenticated (Specimen access No. SJCBT2112), by the Rapinat Herbarium, Dept. of Botany, St. Joseph College, Trichy.

22.2.1 PREPARATION OF EXTRACT

Shadow dried *V. maderaspata* leaves were powdered for Soxhlet extraction. Ethanol, ethyl acetate, and chloroform solvents were used for extraction.

22.2.2 ANIMAL SELECTION

Healthy adult male albino rats (*Rattus norvegicus*) with bodyweight ranged from 120 to 250 g were selected for this study. The rats were obtained from Tamil Nadu Veterinary and Animal Science University, Chennai, and maintained under a controlled environment. Animal handling and experiments were followed by the Institutional Animal Ethical Committee (BDU/IAEC/2017/NE/43).

22.2.3 INDUCTION OF DIABETES

A single intraperitoneal injection of streptozotocin (STZ) (50 mg/kg of body weight in double-distilled water) was made for induction of Diabetes mellitus to albino rats (Al-Hariri, 2012). The rats with fasting blood glucose levels above 250 mg/dl were considered as a diabetic rat and they were selected for the experimental studies (Al-Hariri, 2012).

22.2.4 EXPERIMENTAL SETUP

The rats were divided into 5 groups and each group consists of 3 rats. The experimental set up were
- **Group I:** Normal control;
- **Group II:** Diabetic control;
- **Group III:** Treated with leaf ethanol extract of *V. maderaspatana* (400 gm/kg body weight);
- **Group IV:** Treated with leaf ethyl acetate extract of *V. maderaspatana* (400 gm/kg body weight);
- **Group V:** Treated with leaf chloroform extract of *V. maderaspatana* (400 gm/kg body weight);
- **Group VI:** Treated with glibenclamide (5 mg/kg body weight).

Extracts were orally administered through an oral gavage needle.

22.2.5 ESTIMATION OF ANTIDIABETIC ACTIVITY OF V. MADERASPATANA

After the 30 days of extract treatment, the experimental rats were sacrificed; the blood sample was collected through cardiac punch. Then the serum was separated by centrifugation method (5000 rpm for 10 minutes).

22.2.6 BIOCHEMICAL ESTIMATION

Serum Insulin was calculated by chemiluminescence immune assay (CLIA) kit method (Science Medical solutions Diagnostics Ltd., USA) using Centaur Immuno Assay Instrument, USA. Serum Blood glucose level, glycosylated hemoglobin (HbA$_1$C), total protein and Albumin levels were tested using the Semi-Auto analyzer (Star 21Plus, Manufactured by Aspen Diagnostics Pvt., Ltd., New Delhi).

22.2.7 HISTOLOGICAL STUDY

The pancreas tissue was taken to the histological study.

22.3 RESULTS

The mean level of blood glucose, HbA$_1$C, serum insulin, serum protein, and albumin are given in Table 22.1. Injection of STZ to the albino rats highly increased blood glucose level and HbA$_1$C, and decreased insulin level. After administration of *V. maderaspatana* extracts for 30 days remarkably reduced the blood glucose. Similarly, the glibenclamide treated group rats exhibited significant decrease in glucose level. The mean level of HbA$_1$C level was significantly decreased in standard drug, glibenclamide treated group and ethanol, ethyl acetate, and chloroform extracts of leaves of *V. maderaspatana* treated groups. The mean level of insulin was found to be low in STZ induced diabetic groups. It was recovered to normal level after the treatment of standard drug glibenclamide and *V. maderaspatana*. An increased level of protein was (20.06 ± 0.85 g/dl) observed in STZ induced diabetic control group. It was decreased to the normal level after the treatment of standard drug glibenclamide. Similarly, administration of *V. maderaspatana* extracts significantly decreased to normal level. An increase (12.8 ± 1.03 g/

dl) in the albumin level was noticed in the STZ induced diabetic control rats when compared to control rats (7.2 ± 0.31 g/dl). The glibenclamide treated rats exhibited a high decrease (6.2 ± 0.31 g/dl) in their albumin level when compared to diabetic control rats. Treatment of *V. maderaspatana* extracts to the diabetic rats drastically decreased elevated albumin level. Among the extracts treatment, 400 mg/kg bwt ethyl acetate extract of *V. maderaspatana* treatment for 30 days to the STZ induced diabetic albino rats exhibited good recovery to all parameter studied *viz.* blood glucose level, HbA_1C level, insulin level, serum protein, and serum albumin.

TABLE 22.1 Effect of Different Extracts of *V. Maderaspatana* on Levels of Glucose, HbA_1C, Serum Insulin, Total Protein, Albumin on STZ Induced Diabetic Albino Rats

Groups	Parameters				
	Glucose (mg/dl)	HbA_1C (%)	Insulin (µIU/ml)	Protein (g/dl)	Albumin (g/dl)
Group I	137.5 ± 12.32	5.7 ± 0.50	8.3 ± 0.17	8.7 ± 0.32	7.2 ± 0.31
	(114.0–172.0)	(4.3–6.5)	(7.9–8.7)	(7.9–9.4)	(6.6–8.1)
Group II	454.0 ± 54.15	14.6 ± 0.24	1.7 ± 0.10	20.6 ± 0.85	12.8 ± 1.03
	(300.0–550.0)	(13.9–14.9)	(1.6–2.0)	(19.1–23.0)	(10.8–15.6)
Group III	275.0 ± 70.53	5.1 ± 0.68	6.4 ± 0.29	9.6 ± 0.4	5.5 ± 0.51
	(135.0–360.0)	(4.0–6.3)	(5.9–6.9)	(8.9–10.2)	(4.5–6.2)
Group IV	145.0 ± 5.20	6.1 ± 0.26	7.7 ± 0.21	6.8 ± 0.18	5.6 ± 0.75
	(136.0–154.0)	(5.6–6.5)	(7.4–8.1)	(6.5–7.1)	(4.8–7.1)
Group V	178.0 ± 22.86	6.2 ± 0.08	6.2 ± 0.35	9.8 ± 0.61	6.4 ± 1.00
	(140.0–219.0)	(6.2–6.4)	(5.7–6.9)	(8.7–10.8)	(5.3–8.4)
Group VI	167.23 ± 15.36	7.1 ± 0.21	7.7 ± 0.21	6.9 ± 0.90	6.2 ± 0.31
	(136.0–197.0)	(6.5–7.4)	(7.2–8.2)	(5.2–9.4)	(5.7–7.2)

Values in the parenthesis are range of respective mean.
*Groups: **Group I:** Control; **Group II:** Diabetic control; **Group III:** Treated with 400 mg/kg/b.wt. of leaf ethanol extract of *V. maderaspatana*; **Group VI:** Treated with 400 mg/kg/b. wt. of leaf ethyl acetate extract of *V. maderaspatana*; **Group V:** Treated with 200 mg/kg/b. wt. of leaf chloroform extract of *V. maderaspatana*; **Group VI:** Treated with standard drug glibenclamide 5 mg/kg of body weight.

22.3.1 EFFECT OF TEST HERBAL DRUGS ON PANCREAS

The pancreas of control rats showed normal structure of islet cells, large duct, smaller ducts, and acini. The islets are composed of normal acini are lined by round to oval cells with moderate cytoplasm and small round to oval nuclei. The diabetic rats showed severe damage on pancreas, including depletion of pancreatic islet cells. Further, it showed atrophy of β cells and vascular degenerative changes in the islets. The area of islets was reduced markedly in diabetic rats which were well restored in glibenclamide treated diabetic rats. This showed close similarity to that of control rats (Figure 22.1). The ruptured tissues of pancreas were slightly restored. They were also preserved the islet cells and acini. Pancreas sections showed no changes in size of islets and hyperchromic nucleus. This showed no similarity to that of control rats. The damaged tissues of pancreas were partially restored. They were also preserved the islet cells with increased β cells. The acinar cells were seen to be normal. There was also a relative increase of granulated and normal beta cells in the diabetic group treated with this extract. This showed close resemblance with that of control rats.

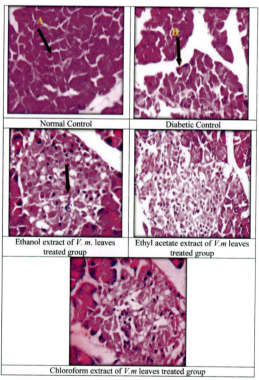

(A) Pancreatic capillaries (B) Necrosis cell (C) Normal acini

FIGURE 22.1 Effect of *V. maderaspatana* extracts treatment on histoarchitecture of pancreas of STZ induced diabetic rats.

22.4 DISCUSSION

Diabetes is a metabolic disorder with severe insulin deficiency or insulin resistance resulting in the impairment of hyperglycemia, the hallmark of diabetes mellitus is largely as a result of hepatic glucose over production (Gupta et al., 1999). It is characterized by elevated blood glucose and metabolic abnormalities associated with structural and functional alteration in various organs (Kopelman and Hatman, 1998). A significant increase in blood glucose and HbA_1C levels and decreased in serum insulin and total hemoglobin levels were observed in STZ induced diabetic rats when compared with normal. STZ was reported to cause a drastic reduction of insulin-producing β cells of islets of Langerhans thus inducing hyperglycemia (Elsner et al., 2000; Sharma and Garg, 2009). The increased level of blood glucose in the STZ induced diabetic rats might be due to the glycogenolysis or glyconeogenesis (Gupta et al., 2011). Zhang et al. (2008) also reported increased levels of glucose in urine of STZ-induced diabetic rat due to gluconeogenesis and glycogenolysis. After the treatment of different extracts of leaf extracts of *V. maderaspatana* the blood glucose level was reduced. Among the extracts, ethyl acetate extract of leaf of *V. maderaspatana* significantly reduced to more or less normal level. It might be due to the presence of bioactive compounds *viz.,* flavonoids, polyphenols, tannins (Sanders et al., 2001; Ghosh et al., 2007). HbA_1C is now considered as the most reliable marker of glycemic control in diabetes mellitus (Halim, 2003) and is used to identify the degree of oxidative stress in diabetic conditions (Gupta et al., 1997). Diabetic rats showed a higher level of HbA_1C indicating their poor glycemic control. The STZ induced diabetic rats shown an increased level of blood glucose accompanied by elevated HbA_1C and decreased insulin level and liver glycogen in diabetic rats suggest poor glycemic control mechanism. STZ induced diabetes rat enhanced the level of glycated hemoglobin (A1c) due to excessive production of glucose in the blood which further reacts with blood hemoglobin and prepared the glycated hemoglobin. In the present study, *V. maderaspatana* extracts treatment significantly reduced the HbA_1C level in STZ induced diabetic rats. This result was coincided with previous studies that reduction in HbA_1C levels during the diabetes treatment considerably reduced microvascular complications (Calisti and Tognetti, 2005). Insulin plays a crucial role in lowering blood glucose level by enhancing glycogenesis in liver and muscles. Glycogen content of skeletal muscles and liver markedly decreases in diabetes. Insulin deficiency decreases the uptake of glucose by cells. A partial or total deficiency of insulin causes a derangement in carbohydrate metabolism that decreases the activity of several key enzymes, including glucokinase (GK), phosphofructokinase, and pyruvate kinase resulting in impaired peripheral glucose utilization and augmented hepatic glucose production (Hikino et al., 1989). In the present study, the administration of STZ to the rat exhibited insulin deficiency and produced hyperglycemia. Insulin level was increased in *V. maderaspatana* leaf and bark extracts treated diabetic rats which were more or less similar to the levels of glibenclamide treated and control rats. It is due to the regenerative properties of different extracts of *V. maderaspatana*. Daisy et al. (2009) reported that the insulin levels in diabetic rats increased (50%) after treatment with dihydroxy glycemic triacetate and that might be due to the stimulation of insulin secretion from regenerated or residual B-cells. Injection of STZ to the rat significantly altered the total protein and albumin level. Alderson et al. (2004) explained the reduction in total protein in diabetic condition due to significant increase in protein excretion. It is mainly due to the formation of a Schiff base between amino group of lysine (and sometimes arginine) residues and excess glucose molecules in blood to form glycoalbumin. Hypoalbuminemia is one of the factors responsible for the onset of ascites related to liver fibrosis. Insulin deficiency leads to decreased protein content (Manchester, 1972). In the present study, there was an alteration in total protein content and serum albumin in STZ induced diabetic rats. Treatment of ethyl acetate extract of *V. maderaspatana* to the diabetic rats restored the protein and albumin levels to the normal level. The increased level of serum protein and albumin in herbal extracts treated diabetic rats are presumed to be due to increased protein catabolism and gluconeogenesis during diabetes (Palanivel et al., 2001).

22.5 CONCLUSION

From the study, it is concluded that the leaf extracts *V. maderaspatana* recovered the glucose homeostasis during diabetic conditions as evidenced by the restoration of blood glucose, HbA_1C and serum insulin. Thus, the plant may be useful for the development of the antidiabetic drug.

ACKNOWLEDGMENT

The authors thanks to the Management, Principal, Nehru Memorial College (Autonomous), Puthanampatti for their valuable support. The first author thanks the University Grant commissions (UGC) for providing funds in the name of UGC-RGNF.

KEYWORD

- **antidiabetic activity**
- **diabetes mellitus**
- **glycoalbumin**
- **herbal plant**
- **streptozotocin**
- ***Ventilago maderaspatana***

REFERENCES

Alderson, N. L., Chachich, M. E., Frizzell, N., Canning, P., & Metz, T. O., (2004). Effect of antioxidants and ACE inhibition on chemical modification of proteins and progression of nephropathy in streptozotocin-diabetic rats. *Diabetologic, 47*, 1385–1395.

Al-Hariri, M. T., (2012). Comparison the rate of diabetes mellitus induction using STZ dissolved in different solvents in male rats. *J. Comp. Clin. Path. Res., 1*(3), 96–99.

Calisti, L., & Tognetti, S., (2005). Measure of glycosylated hemoglobin. *Acta Bio. Med. Atenei. Parm., 76,* 59–62.

Daisy, P., Eliza, J., & Farook, K. A. M., (2009). A novel dihydroxy gymnemic triacetate isolated from Gymnema sylvestre possessing normoglycemic and hypolipidemic activity on STZ-induced diabetic rats. *J. Ethnopharmacol., 126*, 339–344.

Elsner, M., Guldbakke, B., Tiedge, M., Munday, R., & Lenzen, S., (2000). Relative importance of transport and alkylation for pancreatic beta-cell toxicity of streptozotocin. *Diabetologia., 43,* 1528–1533.

Ghosh, D., & Konishi, T., (2007). Anthocyanins and anthocyanin-rich extracts: Role in diabetes and eye function. *Asia Pac. J. Clin. Nutr., 16,* 200–208.

Gulliford, M. C., (1994). Health and health care in the English-speaking Caribbean. *J. Pub. Health Med., 16,* 263–269.

Gupta, B. L., Nehal, M., & Baquer, N. Z., (1997). Effect of experimental diabetes on the activities of hexokinase, glucose-6-phosphate dehydrogenase, and catecholamines in rat erythrocytes of different ages. *Ind. J. Exp. Biol., 35,* 792–795.

Gupta, D., Raju, J., & Baquer, N. Z., (1999). Modulation of some gluconeogenic enzyme activities in diabetic rat liver and kidney. Effect of antidiabetic compounds. *Indian Journal of Experimental Biology, 37,* 196–199.

Gupta, N., Agarwal, M., Bhatia, V., Sharma, R. K., & Narang, E., (2011). A comparative antidiabetic and hypoglycaemic activity of the crude alcoholic extracts of the plant *Leucas aspera* and seeds of *Pithecellobium bigeminum* in rats. *I.J.R.A.P., 2,* 275–280.

Gupta, O. P., & Phatak, S., (2003). Pandamic trends in prevalence of diabetes mellitus and associated coronary heart disease in India- their causes and prevention. *Int. J. Diabetes Dev. Countries*, *23*, 37–50.

Halim, E. M., (2003). Effect of *Coccinia indica* (L) and *Abroma augusta* (L) on lycaemia, lipid profile and on indicators of end-organ damage in streptozotocin-induced diabetic rats. *Indian J. Clin. Biochem.*, *18*, 54–63.

Hikino, H., Kobayashi, M., Suzuki, Y., & Konno, C., (1989). Mechanisms of hypoglycaemic activity of aconitan-A, a glycan from *Aconitum carmichaeli* roots. *J. Ethnopharmacol.*, *25*, 295.

Kopelman, P. G., & Hatman, G. A., (1998). Diabetes: Exploding type II. *Lancet*, *1*(5), 352.

Manchester, K. L., (1972). Effect of insulin on protein synthesis. *Diabetes*, *21*(2), 447.

Nandkarni, A. K., (2000). *Materia Medica* (2nd edn., p.1). Tarun Enterprises: New Delhi.

Noor, A., Gunasekaran, S., Manickam, A. S., & Vijayalakshmi, M. A., (2008). Antidiabetic activity of *Aloe vera* and histology of organs in streptozotocin-induced diabetic rats. *Current Science*, *94*(8), 1070–1076.

Palanivel, R., Thangavel, M., Selvendran, K., & Sakthisekaran, D., (2001). Insulinomimetric effect of ammonium para tungstate on protein metabolism in streptozotocin-induced diabetic rats. *Biomed.*, *21*, 23–30.

Pavani, M., Sankara Rao, M., Mahendra, N. M., & Appa, R. C., (2012). Ethnobotanical explorations on anti-diabetic plants used by tribal inhabitants of seshachalam forest of Andhra Pradesh, India. *Indian J. Fundamental Appl. Life Sci.*, *2*, 100–105.

Sanders, R. A., Rauscher, F. M., & Watkins, J. B., (2001). Effects of quercetin on antioxidant defense in streptozotocin-induced diabetic rats. *J. Biochem. Mol. Toxicol.*, *15*, 143–149.

Sharma, N., & Garg, V., (2009). Antidiabetic and antioxidant potential of ethanolic extract of *Butea monosperma* leaves in alloxan-induced diabetic mice. *Indian Journal of Biochemistry and Biophysics*, *46*, 99–105.

Zhang, H., Wang, J., Liu, Y., & Sun, B., (2015). Peptides derived from oats improve insulin sensitivity and lower blood glucose in streptozotocin-induced diabetic induced mice. *Journal of Biomedical Sciences*, *4*, 1–7.

CHAPTER 23

Isolation, Identification, and Molecular Docking of Antidiabetic Compounds of *Cyathea nilgiriensis* (Holttum)

S. ELAVARASI,[1] G. REVATHI,[2] and K. SARAVANAN[2]

[1]*P.G. and Research Department of Zoology, Holy Cross College (Autonomous), Trichy, Tamil Nadu, India,*
E-mail: elavarasi888@gmail.com

[2]*P.G. and Research Department of Zoology, Nehru Memorial College (Autonomous), Puthanampatti, Trichy District, Tamil Nadu, India*

ABSTRACT

In the present study, phytochemical, and bioactive compounds were identified in the ethylacetate extract of *Cyathea nilgiriensis* pith powder. Phytochemical studies showed the presence of flavonoids, alkaloids, glycosides, carbohydrate, phenols, and steroids. Totally four bioactive compounds were identified by GC-MS analysis. Among the 4 compounds, 1,2-Benzenedicarboxylic acid, butyl 2-ethylhexyl ester was selected as ligand. Molecular docking of ligand with diabetic target (protein tyrosine phosphatase 1 beta) was performed by autodock ver. 4.2.1. The ligand, 1,2-Benzenedicarboxylic acid butyl 2-ethyl hexyl ester was strongly bound with target protein tyrosine phosphatase 1 beta. From the result, it is concluded that the bioactive compound 1,2-benzene dicarboxylic 2 ethyl hexyl ester may be used as the drug candidate for the treatment of diabetes mellitus after clinical validation.

23.1 INTRODUCTION

Right from ancient times in India, indigenous herbal remedies such as Ayurveda and other Indian traditional medicines used many plants in the treatment of diabetes (Babu et al., 2006). In the indigenous Indian system of medicine, good numbers of plants were mentioned for the cure of diabetes and some of them have been experimentally evaluated and active principles were isolated (Grover et al., 2002). WHO (1980) has also recommended the evaluation of the effectiveness of plants in conditions where there are no safe modern drugs (Upadhaya and Pandey, 1984). A report of World Ethnobotanical information stated that about 880 plants may possess hypoglycemic activity (Aguilar et al., 1998) and nearly 343 plants have been reported in scientific papers (Lamba et al., 2000).

Phytochemical compounds were isolated and their antidiabetic effects were evaluated by many scientists (Sachedewa and Kremani, 2003; Gupta et al., 2005; Noor et al., 2008). According to the previous investigations, there are some medicinal plants and herbs which exert hypoglycemic effects such as *Balanites aegyptiaca* fruits (Kamel et al., 1991), *Trigonella foenum-graecum* seeds (Ali et al., 1995; Renuka, 2009), *Acacia nilotica* and *Acacia farnesiana* (Wassel et al., 1992). Additionally, there are reports indicating that worldwide more than 1200 species of plants have been recorded as traditional medicine for diabetes (Marles and Farnsworth, 1995). Although most of these species have not undergone rigorous scientific evaluation, more than 80% of those have been tested and showed antidiabetic activity (Marles and Farnsworth, 1995). In kollihills of Tamil Nadu, India, the tribal people used the tree fern (*Cyathea nilgiriensis*) pith powder for the treatment of diabetes but it was not scientifically validated. Thus, the present study was carried out to identify the diabetic compounds and they were docking with diabetic targets.

Molecular docking is the best computational tools used for the prediction of interaction between two molecules such as target and ligand. It is extensively used for the discovery of new drug candidate for many ailments (Morris et al., 1998). Diabetes mellitus is a condition where the amount of glucose in the blood is too high due to insufficiency of insulin or inefficient of insulin. Insulin is a polypeptide hormone secreted by beta cells of the pancreas that allows the glucose to enter into body cell and helps to burn glucose and provides energy. The current treatment of diabetes with allopathic medicine provides good glycemic control. However, they cause a little complication and some undesirable side effects (Rang et al., 1991). Natural remedies from herbal plants are considered to be more effective and safe alternative treatment for diabetes mellitus (Gupta et al., 2005) due to their minimum side effects in clinical experiences and relatively low costs (Pari and Umamaheswari, 2000) as compared to oral hypoglycemic agents.

Thus, the present study was focused to in silico evaluation of antidiabetic activity of bioactive compound isolated from *C. nilgiriensis* against diabetic target.

23.2 MATERIALS AND METHODS

23.2.1 COLLECTION OF PLANT MATERIALS

The *C. nilgiriensis* was collected from the places Mathikettan solai and Thembalam of Kolli hills ((latitudes 11°55' 05" to 11°21' 10" N and longitudes 78°17' 05" to 78°27'45" E.), Tamil Nadu, India. Stem of this plant was cut into pieces and central part of the wood (pith) was collected and dried under room temperature and powdered with the help of mechanical grinder. The extraction was prepared by Soxhlet apparatus using ethyl acetate as a solvent. Then the solvent was concentrated using distillation unit.

23.2.2 PHYTOCHEMICAL SCREENING

Phytochemical screening of *C. nilgiriensis* was done by standard method (Kokate, 1994; Harborne, 1973; Rajpal, 2002; Raaman, 2006). Bioactive compounds from *C. nilgiriensis* were identified by GC-MS analysis.

23.2.3 DOCKING ANALYSIS

AutoDock 4.2.1 was used for the docking simulations in this study (Morris et al., 1998). The 3D structure of target protein was retrieved from protein data bank (PDB). The structure of ligands was sketched using Chemsketch software and optimized using "Prepare Ligands" in the AutoDock 4.0 for docking studies.

The docking was performed using AutoDock Ver. 4.2.1.

AutoDock 4.0 was used to predict the ligands bound structurally with target and evaluate the biding energy of ligand and target by scoring function. The search algorithm was based on the Lamarckian genetic algorithm and the results were analyzed using binding energy. Binding energy was calculated by Van der Waals interactions, H bond and electrostatic interactions. The docking result was visualized and analyzed using the software UCSF CHIMERA.

23.3 RESULTS AND DISCUSSION

23.3.1 PHYTOCHEMICAL SCREENING

The phytochemical analysis under study exhibited the presence of flavonoids, alkaloids, glycosides, carbohydrate, phenols, and steroids in ethylacetate extract of *C. nilgiriensis* pith powder (Table 23.1). Different phytochemicals have been found to possess a wide range of activities, which may help in protection against chronic diseases. Phytochemicals such as saponins, terpenoids, flavonoids, tannins, steroids, and alkaloids have anti-inflammatory effects (Manach et al., 1996; Latha et al., 1998; Liu, 2003; Akindale and Adeyemi, 2007). Glycosides, flavonoids, tannins, and alkaloids have hypoglycaemic activities (Oliver, 1980; Cherian and Augusti, 1995). Rupasinghe et al. (2003) have reported that saponins possess hypocholesterolemic and antidiabetic properties. The spectrum profile of GC-MS confirmed the presence of seven major compounds with retention time (RT) of 10.80, 18.65, 19.89, 20.71, and 34.05 (Table 23.2). The fractions of the compounds are illustrated in Figure 23.1. Among these 1,2-Benzenedicarboxylic acids, butyl cyclohexyl ester was selected for docking studies.

TABLE 23.1 Preliminary Phytochemical Screening of Ethyl Acetate Extracts of *Cyathea Nilgiriensis*

S. No.	Phytochemicals	Status
1.	Flavonoids	(+)
2.	Alkaloids	(+)
3.	Tannin	(−)
4.	Protein	(−)
5.	Carbohydrate	(+)
6.	Saponin	(−)
7.	Glycosides	(+)
8.	Phenols	(+)
9.	Thiols	(−)
10.	Sterols	(+)
11.	Triterpenoids	(−)

(+) = Present;
(−) = Absent.

Isolation, Identification, and Molecular Docking of Antidiabetic Compounds

TABLE 23.2 Bioactive Components Identified from Ethylacetate Extract of *C. Nilgiriensis* by GC-MS

S. No	Name and Structure of the Compound	Molecular Formula	Molecular Weight	Retention Time
1.	5-hydroxymethylfurfural	$C_6H_6O_3$	126	10.80
2.	1,2-Benzenedicarboxylic acid, butyl cyclohexyl ester	$C_{18}H_{24}O_4$	304	18.65
3.	Dibutyl phthalate	$C_{16}H_{22}O_4$	278	19.89
4.	1,2-Benzenedicarboxylic acid, butyl 2-ethylhexyl ester	$C_{20}H_{30}O_4$	334	20.71

TABLE 23.2 *(Continued)*

S. No	Name and Structure of the Compound	Molecular Formula	Molecular Weight	Retention Time
5.	Olean-12-ene-3,15,16,21,22,28-hexol, (3á, 15á, 16á, 21á, 22á)-	$C_{30}H_{50}O_6$	506	34.05

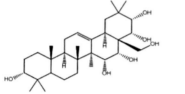

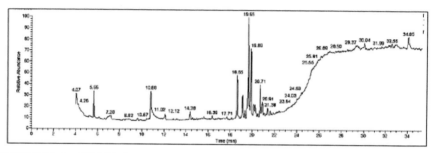

FIGURE 23.1 GC-MS chromatogram of ethylacetate extract of *C. nilgiriensis*.

23.3.2 DOCKING STUDIES

In silico models have potential use in the discovery and optimization of novel molecules, with affinity to the target, clarification of absorption, distribution, metabolism, excretion, and toxicity properties as well as physiochemical characterization (Ekins et al., 2007). Docking is an important in the study of protein-ligand interaction properties such as binding energy, geometry complementarity, electron distribution, hydrogen bond donor-acceptor, hydrophobicity, and polarizability thus molecular docking contribute a major role in the drug discovery in the identification of innovative small molecular scaffold, exhibiting the important properties with selectivity for the target (Krvoat et al., 2005). The ligand butyl 2 ethyl hexyl ester1 is identified in the ethyl acetate extract of *C. nilgiriensis*. Figure 23.2 shows the 3D structure of Ligand and Figure 23.3 shows the 3D structure of target protein (5K9w). Result of docking analysis of protein tyrosine phosphatase 1 beta (5K9W) with ligand 1,2-benzene dicarboxylic 2 ethyl hexyl ester is shown in Figure 23.4. Ligand 1,2-Benzenedicarboxylic 2 ethyl hexyl ester was effectively bound with target protein tyrosine phosphatase 1 beta by H bond with minimum binding energy (-8.96 kcal/mol) (Figure 23.5). Considering docking results of the protein tyrosine phosphatase 1 beta compounds, the tested inhibitor was stably posed in similar pocket domains of PTP1B residues including GLY 45. Detailed binding mode analysis with docking simulation showed that the inhibitors can be stabilized by the simultaneous establishment of multiple hydrogen bonds and van der Waals contacts in the pocket site. Protein tyrosine phosphatase 1 beta is a negative regulator of insulin signaling. It is a new target for drug design against type II diabetes for its management of insulin resistance. It plays an important regulator of the insulin signaling pathway. The receptor binds IGF-I have tyrosine kinase activity which binds with high affinity to activate many cellular proliferations.

FIGURE 23.2 3D structure of ligand 1,2-benzene dicarboxylic 2 ethyl hexyl ester.

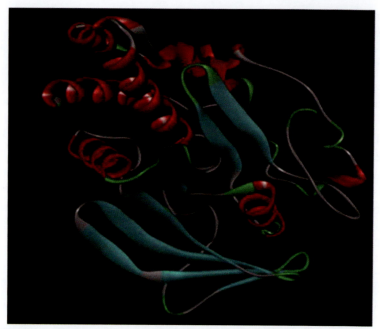

FIGURE 23.3 3D structure of target protein tyrosine phosphatase 1 beta.

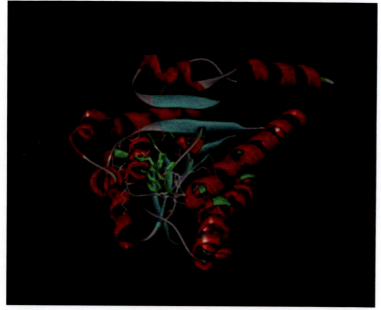

FIGURE 23.4 Docking of 1,2-benzene dicarboxylic 2 ethyl hexyl ester with tyrosine phosphatase 1 beta.

Isolation, Identification, and Molecular Docking of Antidiabetic Compounds

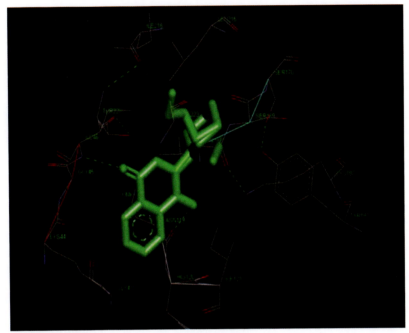

FIGURE 23.5 Bonding of ligand with target protein.

23.4 CONCLUSION

In conclusion, the ligand butyl 1,2-Benzenedicarboxylic 2 ethyl hexyl ester identified from the ethyl acetate extract of *C. nilgiriensis* was effectively bound with target protein tyrosine phophatase 1 beta with minimum binding energy. Thus, the bioactive candidate 1,2-Benzenedicarboxylic 2 ethyl hexyl ester may be used for the treatment of diabetes mellitus and related diseases.

ACKNOWLEDGMENTS

The authors thank the Management, the Principal and Head of the Department of Zoology, Nehru Memorial College, Puthanampatti, for providing necessary facilities to do this research work successfully. The first author acknowledges the National Testing Service-India, Central Institute of Indian Languages, Manasagangotri, Mysore for financial support.

KEYWORDS

- **antidiabetic properties**
- *C. nilgiriensis*
- **docking**
- *In silico* models
- **phytochemical screening**
- **saponins**

REFERENCES

Aguilar, F. J. A., Ramos, R. R., Guitterez, S. P., Contreras, A. A., Weber, C. C. C., & Saenz, J. L. F., (1998). Study of the antihyper glycemic activity effect of plants used as antidiabetics. *J. Ethnopharmacol., 61,* 101–110.

Akindele, A. J., & Adeyemi, O. O., (2007). Antiinflammatory activity of the aqueous leaf extract of *Byrsocarpus coccineus. Fitoterapia, 78,* 25–28.

Ali, L., Azad khan, A. K., & Hassan, Z., (1995). Characterization of the hypoglycaemic effects of Trigonella foenum graecum seed. *Planta Med., 61,* 358–360.

Babu, P. A., Suneetha, G., Boddepalli, R., Lakshmi, V. V., Rani, T. S., Babu, Y. R., & Srinivas, K. A., (2006). Database of 389 medicinal plants for diabetes. *Bioinformation, 1*(4), 130–131.

Cherian, S., & Augusti, K. T., (1995). Insulin sparing action of leucopelargonidin derivative isolated from *Ficus bengalensin* Linn. *Indian J. Exp. Biol., 33,* 608–611.

Ekins, S., Mestres, J., & Testa, B., (2007). *In silico* pharmacology for drug discovery: Applications to targets and beyond. *Br. J. Pharmacol., 152*(1), 21–37.

Groover, J. K., Yadav, S., & Vats, V., (2002). Medicinal plants of India with anti-diabetic potential. *J. Ethnopharmacol., 81,* 81–100.

Gupta, R. K., Kesari, A. N., Watal, G., Murthy, P. S., Chandra, R., & Tandon, V., (2005). Nutritional and hypoglycemic effect of fruit pulp of *Annona squamosa* in normal healthy and alloxan-induced diabetic rabbits. *Ann. Nutr. Metab., 49,* 407–413.

Harborne, J. B., (1973). *Phytochemical Methods* (pp. 49–188). Chapman and Hall, Ltd., London.

Hsu, H. C., Chiou, J. F., Wang, Y. H., Chen, C. H., Mau, S. Y., Ho, C. T., Cheng, P. J., Liu, T. Z., & Chen, C. H., (2013). Folate deficiency triggers on oxidative-nitrosative stress-mediated apoptotic cell death and impedes insulin biosynthesis in RINm5F pancreatic islet β-cells: Relevant to the pathogenesis of diabetes. *Plos One, 8*(11), 1–13.

Kamel, M. S., Ohtant, K., Kurokawa, T., Assaf, M., El-Shanawany, M. A., Ali, A. A., Kasai, R., Ishibashi., S. S., & Tanaka, O., (1991). Studies on *Balanitesa egyptiaca* fruits, an antidiabetic Egyptian folk medicine. *Chemical and Pharmaceutical Bulletin, 9,* 1229–1233.

Kokate, C. K., (1994). *Pharmacohnosy* (16th edn.). Nirali Prakasham Publication, Mumbai, India.

Krovat, E. M., Steindl, T., & Lange, T., (2005). Recent advances in docking and scoring. *Curr. Computer-aided Drug Des., 1,* 93–102.

Lamba, S. S., Buch, K. Y., Lewis, H., & Lamba, H. J., (2000). Photochemical as potential hypoglycemic agents. *Studies in Natural Products Chemistry, 21,* 457–495.

Latha, R. M., Geetha, T., & Varalakshmi, P., (1998). Effect of *Vernonia cinerea* less flower extract in adjuvant-induced arthritis. *General Pharmacol., 31,* 601–606.

Liu, R. H., (2003). Health benefits of fruit and vegetables are from additive and synergistic combinations of phytochemicals. *Am. J. Clin. Nutr., 78,* 517S–520S.

Manach, C., & Regerat, F. O., (1996). Texier Bioavailability, metabolism and physiological impact of 4-oxo-flavonoids. *Nutr. Res., 16,* 517–544.

Marles, J. R., & Fransworth, N. R., (1995). Antidiabetic plants and their active constituents. *Phytomedicine, 1,* 32–36.

Morris, G. M., Goodsell, D. S., Halliday, R. S., Huey, R., Hart, W. E., Belew, R. K., & Olson, A. J., (1998). Automated docking using a Lamarckian genetic algorithm and empirical binding free energy function. *J. Comput. Chem., 19,* 1639–1662.

Noor, A., Gunasekaran, S., Manickam, S. A., & Vijayalakshmi, M. A., (2008). Antidiabetic activity of Aloe Vera and histology of organs in streptozotocin-induced diabetic rats. *Curr. Sci., 94,* 1070–1076.

Oliver, B., (1980). Oral hypoglycaemic plants in West Africa. *J. Ethnopharmacol., 2,* 119–127.

Pari, L., & Umamaheswari, J., (2000). Antihyperglycaemic activity of *Musa sapientum* flowers: Effect on lipid peroxidation in alloxan diabetic rats. *Phytother. Res., 14,* 136–138.

Patel, D. K., Kumar, R., Prasad, S. K., Sairam, K., & Hemalatha, S., (2011). Antidiabetic and *in vitro* antioxidant potential of *Hybanthus enneaspermus* (Linn) F. Muell in streptozotocin-induced diabetic rats. *Asian Pac. J. Trop Biomed., 1*(4), 316–322.

Raaman, N., (2006). *Phytochemical Techniques* (p. 318). New India Publishing. Medical.

Rajpal, V., (2005). *Standardization of Botanicals* (2nd edn., p. 284) Eastern Publisher.

Rang, H. P., & Dale, M. M., (1991). The endocrine system pharmacology. In: Nattrass, M., & Hale, P. T., (eds.), *Pharmacology* (pp 504–508). Longman Group Ltd., UK.

Renuka, C., Ramesh, N., & Saravanan, K., (2009). Evaluation of the antidiabetic effect of Trigonella foenum-graecum seed powder on alloxan-induced diabetic albino rats. *International Journal of Pharm. Tech. Research, 1,* 1580–1584.

Rupasinghe, H. P., Jackson, C. J., Poysa, V., Di Berado, C., Bewley, J. D., & Jenkinson, J., (2003). Soyasapogenol A and B distribution in Soybean (*Glycine Max* L. Merr.) in relation to seed physiology, genetic variability, and growing location. *J. Agric. Food Chem., 51,* 5888–5894.

Sachdewa, A., & Khemani, L. D., (2003). Effect of hibiscus rosa-sinensis Linn. ethanol flower extract on blood glucose and lipid profile in streptozotocin-induced diabetes in rats. *J. Ethnopharmacol., 89,* 61–66.

Upadhayay, V. P., & Pandey, K., (1984). Ayurvedic approach to diabetes mellitus and its management by indigenous resources. In: Bajaj, J. S., (ed.), *Diabetes Mellitus in Developing Countries* (pp. 375–377). Inter Print, New Delhi.

Wassel, G. M., Abd El-Wahab, S. M., Aboutabl, E. A., Ammar, N. M., & Afifi, M. S., (1992). Phytochemical examination and biological studies of *Acacia nilotica* L. wild and *Acacia fanesiana* L. wild growing in Egypt. *Egyptian Journal of Pharmaceutical Sciences, 33,* 327–340.

WHO, (1980). Expert committee on diabetes mellitus. *World Health Organization Tech. Rep. Series.* (p. 649).

CHAPTER 24

Antidiabetic Potentials of Major Indian Spices: A Review

PRATIBHA, HEMA LOHANI, S. ZAFAR HAIDER, and NIRPENDRA K. CHAUHAN

Center for Aromatic Plants (CAP), Industrial Estate Selaqui – 248011, Dehradun, Uttarakhand, India, Tel.: +91-9450743795, E-mails: zafarhrdi@gmail.com, zafarhaider.1@rediffmail.com, ORCID: https://orcid.org/0000-0002-3061-9264 (Z. Haider)

ABSTRACT

In the recent few years, there is an appreciable increase towards the use of plant-derived products worldwide as they are natural in origin and with few side effects. About 10% of the population all over the world is affected by diabetes mellitus and this percentage is increasing day by day. A review was conducted to pile up the information about Indian spices bearing, bioactive constituent such as polyphenols, tannins, saponins, and flavonoids, responsible for its anti-diabetic property and the method analyzed for the treatment of antidiabetic activity such as α-Amylase and α-Glucosidase inhibition assay, O-toluidine method, glucose oxidase-peroxidase method (GOD/POD) and ELISA method (enzyme-linked immunosorbent assay). The present review enumerates the number of spices described in this clearly demonstrated the importance of herbal plants parts in the treatment of diabetes mellitus, which may help the researchers for further researches in the medicinal properties of the Indian Spices and provide them information in approaching the efficiency and potency of the spices used normally in the kitchen.

24.1 INTRODUCTION

The modern world is facing an epidemic problem related to health issues, diabetes is one of them. According to the Global Report on Diabetes, an estimated 422 million peoples in the world are suffering from diabetes and the figure predicted to be increased (WHO, 2016). Diabetes, a serious metabolic disease is characterized by increased blood glucose levels due to decreased insulin secretion, or when the body cannot effectively use the insulin (American Diabetes Association (ADA), 2014). More than 80% of death from Diabetes mellitus occurs in low- and middle-income countries (Mathers and Loncar, 2006). Diabetics can go through other complications such as nephropathy, neuropathy, retinopathy, cardiovascular disorders, and hypertension causing long-term damage, dysfunction, and eventually the failure of organs, especially the eyes, kidneys, nerves, heart, and blood vessels due to free radical formation during the condition of hyperglycemia (Huang et al., 2005). Oral hypoglycemic agents/drugs may be effective for glycemic control, but they come with their numerable side effects such as liver disorders, flatulence, abdominal pain, renal tumors, hepatic injury, acute hepatitis, abdominal fullness, and diarrhea. Furthermore, they are not even safe during pregnancy (Olatunde et al., 2014). Therefore, there is an increasing need for the development of a natural and safe product without side effects.

Various bio-active components present in the spices are responsible for potential anti-diabetic properties. Spices that exhibit hypoglycaemic, hypolipidaemic, and antioxidant activities may have potential roles in the treatment of diabetes (Otunola & Afolayan, 2015). Spices are mainly different plant parts such as bark, leaves, flower, bud, and seed which are being used from ancient times in cuisines, desserts, and traditional foods. In addition, spices are used as a traditional medicine to decrease the health complications. Many spices such as Cinnamon, Cumin, Coriander, Fennel, Ajwain, Cardamom, Clove, Curry Leaves, Nutmeg, Pepper, Caraway, and Bay Leaves, etc. are having numerous health beneficial and medicinal properties, for instance, antidiabetic, antioxidant, anti-inflammatory, wound healing, anti-melanogenic, anti-tumor, anti-cancer, anti-repellent, antitussive, antiplatelet activity, and anti-nephrotoxic activity. The massive amount of herbs and spices has been described for the treatment of diabetes throughout the world (Kesari et al., 2005).

Spices are found to possess many medicinal, health beneficial and favorable properties but the present chapter mainly focuses on the potential anti-diabetic properties of the traditionally used major Indian spices considering the detail about the method used (*in-vitro* and *in-vivo*) and their conclusion specified by the researcher to conduct the antidiabetic study.

24.2 ANTIDIABETIC POTENTIALS IN MAJOR INDIAN SPICES

A number of investigations on the antidiabetic potential of major spices were carried out by workers and their findings pointed out that some spices and their bioactive constituents have a very high potential in countering life-threatening problems. A review of scientific literatures was carried out to explore the previous works on the antidiabetic potential of major spices by various *in-vitro* and *in-vivo* assays methods, such as α-Amylase and α-Glucosidase inhibition assay, O-toluidine reagent, GOD/POD method, Radioimmunoassay kit, enzymatic colorimetric (GOD-PAP) method, enzyme-linked immunosorbent assay (ELISA), Accu-chek sensor glucometer, and spectrophotometrically, etc. described in Table 24.1. Studies revealed that most of the spices exhibit effective antidiabetic activity against the studied in vivo methods on various used diabetic models such as streptozotocin (STZ) and alloxan-induced diabetic rats or mouse as diabetic models. Some authors had also used hereditary diabetic mice, for example, KK-Ay mice. The fruits are commonly used plant parts as a spice and other parts (leaf, stem, bark, roots, and seeds) were also used for therapeutics.

Some other studies also were conducted for furthermore description of the antidiabetic potential of spices. Abdulrazaq et al. (2012) demonstrated that after 15 and 30 days, the aqueous extract of ginger in STZ-diabetic rats caused 38% and 68% reduction in plasma glucose level at a dose of 500 mg/kg. The ethanolic extract of Garlic (*Allium sativum* L.) had also shown better antidiabetic activity when compared with the known antidiabetic drug glibenclamide (600 mg/kg), for 14 days at different oral doses (0.1, 0.25, and 0.5 g/kg body wt.) reported by Eidi et al. (2006). The (80%) ethanolic extract of garlic reduces the serum glucose concentration to 98.1% in comparison to the diabetic control 168.6% (Bokaeian et al., 2010). In a similar way, the aqueous extract of *Allium sativum* L. roots had significantly decreased the glucose diffusion inducing its hypoglycemic activity at a various concentrations (10, 20, 40 g/L) (Younas and Hussain, 2014). Ethanolic extract of Black pepper leaves was reported to have antidiabetic property; the study revealed that for 21 days there was a significant decrease in the serum blood Glucose in male albino rats (Onyesife et al., 2014). The aqueous extract also shows an appreciable decrease in the blood glucose level at a dose of 50 mg/kg for 8 weeks in alloxan-induced diabetic Wister albino rats reported by Sarfraz et al. (2017). The aqueous extract of *Piper longum* at a dose of 200 mg/kg for 30 days had considerably decreased the fasting glucose level in the STZ induced diabetic rats and the decrease in the level of blood glucose is mainly due to increased secretion of insulin by the remnant beta cells of islets of Langerhans indicating its potential as antidiabetic agent (Nabi et al., 2012). The aqueous extract of *Origanum vulgar* at a dose of 20 mg/kg, after single or 15 daily doses had significantly reduced the blood glucose level in STZ-induced diabetic rats (Lemhadri, 2004). *Murraya koenigii* possesses significant hypoglycemic potential in STZ-induced diabetic rats and its extract were found to be more effective when compared to glibenclamide, an antidiabetic drug (Arulselvan et al., 2006). A similar study disclosed that there was a significant decrease in the blood glucose level at a dose of 400 mg/kg in the alloxan treated diabetic rats (Abubakar et al., 2014). The gene expression by DNA microarray analysis revealed that clove acts like insulin by reducing phosphoenolpyruvate carboxykinase (PEPCK) and glucose 6-phosphatase (G6Pase) gene expression (Parle and Khanna, 2011). Clove bud powder also reduced the blood glucose level in the STZ-induced diabetic rats when compared to the control at a dose of 35 mg/kg for 30 days (Adefegha et al., 2014). A study resulted by Dhandapani et al. (2002) for 6 weeks at an oral dose of 0.25 g/kg shows deduction in the level of blood glucose and increment in the level of total hemoglobin. Another study revealed that the Cuminaldehyde and Cuminol were two main bioactive components responsible for the elevation of the insulin secretion level to 3.34 and 3.58 fold for 25 µg/ml (Patil et al., 2013). Fennel (saunf) essential oil also exhibit antidiabetic property which was evidenced by the study on STZ-induced diabetic rats at a dose of 30 mg/kg had made improved pathological changes in pancreas and kidney (El-Soud et al., 2011). Nutmeg at a dose of 450 mg/kg had reduced both elevated triglyceride (TG) and cholesterol level to 47% and 66.7% (Arulmozhi et al., 2007). In a single dose, there was a significant reduction in the blood glucose level (8.7%) of STZ-induced diabetic rats. by the fruits of nutmeg (Ahmad et al., 2008). There was a subsequent decrease in the blood glucose level by the aqueous extract of rosemary on STZ-induced diabetic rats at a dose of 200 mg/kg for

21 days (Emam, 2012). Similar study depict that 0.01% aqueous solution of the distilled extract inhibited α-glucosidase (maltase and sucrase) in the small intestine of STZ-induced diabetic mice with IC_{50} values of 290 µg/mL (maltase inhibitory activity) and 150 µg/mL (sucrase inhibitory activity) and an active compound was isolated and identified to be 5,7,3',4'-tetrahydroxy flavone (luteolin) to cause the inhibition by Rosemary (Koga et al., 2006). The rhizome of turmeric consist of bio-active component Ar-Turmerone also showed potent α-glucosidase (IC50 = 0.28 µg) and α-amylase (IC50 = 24.5 µg) inhibition (Sreenivasan et al., 2015). Curcumin present in turmeric also acts as an antidiabetic agent due to its anti-inflammatory and antioxidative properties, by affecting the blood enzymes (Seo et al., 2008). In another study, curcumin was found to decreases the blood glucose and plasma glycoprotein level in diabetic rats (Murugan and Pari, 2007a). The rhizomes of *Curcuma longa* (turmeric) with *Emblica Officinalis* (Indian gooseberry) in equal proportion at a dose of 0.9 g/kg after 30 days show a considerable decrease in the fasting glucose plasma level showing anti-diabetic property (Rao et al., 2013). The ethanolic extract of turmeric causes an elevation in the blood glucose level of type 2 diabetic KK-A(y) mice at a dose of 0.2 or 1.0 g/100 g diet (El-Masry, 2012). Curcuminoid and sesquiterpenoid extracted by the ethanolic and hexane solvent from the rhizome of turmeric also had shown antidiabetic activity against type 2 diabetic KK-A(y) mice for 6 weeks by reducing the blood glucose level significantly (Nishiyama et al., 2005). The ethanolic extract of cinnamon was found to decrease the blood sugar level at different doses 100, 200 and 300 mg/kg after 21 days during wound healing when the diabetic rats were treated (Soni et al., 2013). Cassia and true cinnamon are also found to have direct antidiabetic potency. The extract of cassia causes a reduction in the fasting glucose level and elevation in the insulin secretion which effectively than true cinnamon (Verspohl et al., 2005). Ferulic acid, main constituent responsible for Asafoetida (Heeng or hing) antioxidant and antidiabetic activity reduces the blood glucose level significantly at a dose of 0.05% in KK-Ay mice (Ohnishi et al., 2004). Similar results were also obtained from another study by Jung et al. (2007) where ferulic acid suppresses the blood glucose level in the db/db mice and appreciably increased the plasma insulin. The antidiabetic activity of coriander leaf and stem against alloxan-induced diabetic rats revealed its potent activity to reduce blood glucose level when compared with the standard clinically available drug glibenclamide to a significant level. The bioactive component such as phenols, saponins, tannins, and flavonoids are chiefly responsible for its activity against diabetes (Sreelatha

and Inbavalli, 2012). Several research studies shown that the decrease in the blood glucose level was due to increase in the insulin secretion considerably from the beta-cells of the pancreas in the diabetic rats by the celery seeds (*Apium graveolens*) as well as by *T. indica* (Maiti, 2005; Veermanic et al., 2008). Another study revealed that the n-butanol extract of celery seeds reduced the decreased the level of glucose and increased the lowered glutathione (GSH) concentration, and prevented lipid-peroxidation-induced liver damage in STZ-induced diabetic rats after 21 days (Jabbar et al., 2012). Dill tablet had also demonstrated the antidiabetic activity for 100 mg/kg and 300 mg/kg the reduction in the blood glucose level was found to be 210.26 ± 6.38 mg/ml and 150.26 ± 6.38 mg/ml when compared with diabetic control (Oshaghi et al., 2015). Parsley (*Petroselinum crispum*), had also reduced the blood glucose level to a considerable level in comparison to control diabetic rats at a dose of 2 g/kg of body weight for 42 days (Ozsoy-Sacan et al., 2006). Spices are not only used alone but also in the form of mixture or paste of many spices. The study conducted by Otunola and Afolayan (2015), revealed that the combined mixture of spices such as garlic, ginger, and pepper in equal proportions also contributes there potential against diabetes. The aqueous extract of these spices at different doses (200 mg/kg and 500 mg/kg) in alloxan-induced diabetic rats for 7 days reduced the elevated fasting blood glucose level significantly. Spices (Cinnamon, clove, and ginger) essential oil reduced the blood glucose level and increased the insulin level significantly after 8 weeks in diabetic rats (Shukri et al., 2010).

24.3 BIO-CHEMICAL CONSTITUENT OF SPICES RESPONSIBLE FOR ANTIDIABETIC PROPERTIES

Various bio-active components present in the spices such as piperine, curcumin, and its derivatives, gingerol, polyphenols such as flavonoids are responsible for their antidiabetic properties, which were confirmed by the literature studies. Their chemical structures are shown in Figure 24.1. Phenols and tannins are the main components responsible for the potent antioxidant activity and which in turn can decrease the oxidative stress in diabetic-induced rats (Soni et al., 2013). The polyphenolic components such as flavonoids are responsible for the inhibition of alpha-glucosidase activities. The structure of various bio-active components possessed by the spices is presented in Table 24.1. Curcumin (1,7-bis(4-hydroxy-3-methoxyphenyl)-1,6-heptadiene-3,5-dione) is a bio-active component of Turmeric

TABLE 24.1 Antidiabetic Properties in Major Indian Spices

S. No.	Common Name	Botanical Names and Family	Part Used as Spice	Extract/Active Components	Method	Inferences	References
1.	Small Cardamom	*Elettaria cardamomum* Maton (Zingiberaceae)	Fruit, Seed	Aqueous and Methanol	α-Amylase inhibition assay and α-Glucosidase inhibition assay	α-Amylase inhibition shown by aqueous and methanol was 82.99 and 39.93%, respectively. α-Glucosidase inhibition shown by aqueous and methanol was 10.41% and 13.73%.	Ahmed et al. (2017)
2.	Large Cardamom, Bari Ilaichi	*Amomum subulatum* Roxb. (Zingiberaceae)	Fruit, Seed	Acetone and Methanol/ protocatechuic acid.	Mercodia Insulin ELISA	Both the extract had shown increment in insulin level (acetone: 18.55 ± 0.37 ng/dl and methanol: 18.66 ± 1.59 ng/dl and a decrease in the blood glucose level (Acetone: 127.87 ± 0.56 μu/ml and methanol: 85.59 ± 1.89 μu/ml).	Vavaiya et al. (2012)
3.	Black Pepper, Kali Mirch	*Piper nigrum* L. (Piperaceae)	Fruit	Piperine	Accu-Chek sensor glucometer	At a dose of 20 mg/kg for 14 days, had shown a significant decrease in the blood glucose level (204.00 ± 7.95 mg/dl) as compared to control.	Atal et al. (2012)

TABLE 24.1 (Continued)

S. No.	Common Name	Botanical Names and Family	Part Used as Spice	Extract/Active Components	Method	Inferences	References
			Seeds	Aqueous	Glucose oxidase	After 10 days, at a dose of 400 mg/kg, the decrease in the fasting blood glucose level was found to be 118.50 ± 3.37 mg/dl as compared to control. At a dose of 0.5 ml per rat/day the reduction in the blood glucose level was found to be 129 ± 8.0.	Shukla and Rangari (2015) Kaleem et al. (2005)
4.	Pepper Long	*Piper longum* L. (Piperaceae)	Fruit	n-Hexane (HS), Ethyl acetate (EAS), Residual Ethanolic Extract (RES)	–	After 28 days at a dose of 200 mg/kg and 400 mg/kg (bw p.o once a day), RES had shown a significant reduction in the blood glucose level 206.14 ± 5.01 mg/dl and 183.30 ± 4.10 mg/dl. RES also had shown an appreciable increase in the plasma insulin level 8.99 ± 1.07 µU/ml (200 mg/kg BW p.o once a day) and 11.16 ± 0.88 µU/ml (400 mg/kg BW p.o once a day).	Kumar et al. (2011)

TABLE 24.1 (Continued)

S. No.	Common Name	Botanical Names and Family	Part Used as Spice	Extract/Active Components	Method	Inferences	References
				Ethanol	O-toluidine reagent; Insulin: ELISA method using Boehringer Manheim GmbH Kit	The decrement in the level of blood glucose was 91.99 ± 7.60 mg/dl and the increment in the plasma insulin level was 16.69 ± 0.30 µU/ml at a dose of 300 mg/kg of body weight for 45 days.	Manoharan et al. (2007)
			Roots	Hexane, Ethylacetate, Methanol, Aqueous	glucose oxidase-peroxidase method	The maximum reduction in the fasting blood glucose was shown by aqueous extract with 75% (85.5 ± 13.5 mg/dl) decrease when compared to control at a dose of 200 mg/kg for 6 hrs.	Nabi et al. (2013)
5.	Cumin, Jeera	*Cuminum cyminum* L. (Apiaceae)	Fruit, seed	-	Enzymatic oxidation	For 14 days, 5% cumin resulted in the deduction of elevated blood glucose level to 160 ± 2.0 mg/dl compared to diabetic control.	Ahmed et al. (2013)

TABLE 24.1 (Continued)

S. No.	Common Name	Botanical Names and Family	Part Used as Spice	Extract/Active Components	Method	Inferences	References
				Cuminaldehyde	Aldose reductase and alpha-glucosidase inhibition assay	The IC50 value for aldose reductase was found to be 0.00085 mg/ml and for 0.5 mg/ml for alpha-glucosidase on Sprague Dawley rats.	Lee (2005)
				Aqueous	Blood Glucose: Glucose oxidase-peroxidase (GOD-POD)	For 8 hrs at a dose of 300 mg/kg the mean percent reduction of the serum glucose level was found to be 15.38 ± 0.58.	Sushruta et al. (2006)
6.	Celery	Apium graveolens L. (Apiaceae)	Leaf, Fruit	Ethanol	Ames One Touch Glucometer	After 43 days the deduction in plasma glucose was found to be 139.6 ± 1.65 mg/dl and the increase in the plasma insulin level was 8.913 ± 4.0 µl/ml.	Niaz et al., (2013)
				n-Butanol	Glucose MR® kit	The SD diabetic rats' shows decrease in the blood glucose after 6 days and reached to normal after 12 days at a dose of 60 mg/kg.	Jabbar and Basim (2013)

TABLE 24.1 (Continued)

S. No.	Common Name	Botanical Names and Family	Part Used as Spice	Extract/Active Components	Method	Inferences	References
						The subsequent decrease in the elevation of blood glucose (125.03 mg/dl) was found when compared with the diabetic control (320.21 mg/dl).	Jabbar et al. (2012)
7.	Fennel, Saunf	*Foeniculum vulgare* Mill. (Apiaceae)	Fruit, seeds	Aqueous	Blood Glucose: glucose oxidase-peroxidase (GOD-POD); Insulin: Ubi-Magiwel Kit	After 35 days, at a dose of 150 mg/kg and 300 mg/kg, the decrease in the elevation of glucose level was 110.4 ± 0.64 mg/dl and 101.4 ± 0.34 mg/dl. The insulin secretion was also found to increased for 150 mg/kg (17.7 ± 0.23 µU/ml) and 300 mg/kg (18.5 ± 0.28 µU/ml).	Anitha et al. (2014)
						For 8 hrs at a dose of 300 mg/kg the mean percent reduction of the serum glucose level was found to be 12.15 ± 2.88.	Sushruta et al. (2006)

TABLE 24.1 (Continued)

S. No.	Common Name	Botanical Names and Family	Part Used as Spice	Extract/Active Components	Method	Inferences	References
8.	Aniseed	*Pimpinella anisum* L. (Apiaceae)	Fruit	Hexane, Ethyl acetate, Methnol, Aqueous, Benzene, n-butanol	α-Amylase and α-Glucosidase inhibition assay	The IC50 value for α-amylase and α-glucosidase was 0.12 mg/ml and 0.15 mg/ml respectively for ethyl acetate showing highest antidiabetic activity. The sequence for the antidiabetic activity was as follows for 500 µg/ml: Ethyl acetate (94% and 87%) > hexane (93% and 86%) > benzene (91% and 85%) > methanol (84% and 83%) > aqueous (81% and 79%) > n-butanol (75% and 77%).	Shobha et al. (2013)
						At an oral dose of 2 g per day for 60 days resulted in the reduction in the fasting blood glucose level 122.16+0.9 mg/dl in two patients.	Shobha and Andulla, (2018)

TABLE 24.1 (Continued)

S. No.	Common Name	Botanical Names and Family	Part Used as Spice	Extract/Active Components	Method	Inferences	References
9.	Caraway	*Carum carvi* L. (Apiaceae)	Fruit	80% Ethanol	Glucose: Glucose oxidase-peroxidase (GOD-POD); Insulin: radioimmunoassay kit	There was a subsequent decrease in the serum glucose level and increase in the plasma insulin level for 1, 3, 5 hrs at a dose of 0.2 g/kg, 0.4 g/kg and 0.6 g/kg of body weight.	Eidi et al. (2010)
				Aqueous	Blood Glucose: Glucose oxidase-peroxidase (GOD-POD)	For 8 hrs at a dose of 300 mg/kg the mean percent reduction of the serum glucose level was found to be 23.73 ± 0.94.	Sushruta et al. (2006)
10.	Dill	*Anethum graveolens* L. (Apiaceae)	Fruit, seeds	Aqueous, Ethanol	Glucometer (Dr. Morepen Gluco Monitor)	At a dose 3.04 g/kg for 15 days, there was a subsequent decrease in the fasting Blood Glucose by the aqueous extract than the ethanol extract.	Mishra (2013)
				Aqueous	Blood Glucose: Glucose oxidase-peroxidase (GOD-POD)	For 8 hrs at a dose of 300 mg/kg the mean percent reduction of the serum glucose level was found to be 14.35 ± 3.79.	Sushruta et al. (2006)

TABLE 24.1 (Continued)

S. No.	Common Name	Botanical Names and Family	Part Used as Spice	Extract/Active Components	Method	Inferences	References
11.	Parsley	*Petroselinum crispum* Mill. (Apiaceae)	Leaf	Aqueous	O-toluidine method	After 42 days, the decline in the blood glucose level was found to be 110.9 ± 45.1 mg/dl.	Yanardag et al. (2003)
12.	Clove, Laung	*Syzygium aromaticum* L. (Myrtaceae)	Flower buds	Aqueous (raw and gamma irradiated extract)	Radioimmunoassay kit	Decline on the glucose level (129.60 ± 5.3 mg/dl; 127.80 ± 6.2 mg/dl) and elevation on the insulin level (26.63 ± 3.75 µU/ml; 26.80 ± 4.34 µU/ml) of the raw and gamma-irradiated albino rats.	Hamza and Elshahat (2011)
			Leaves	80% Acetone (free and bound phenolic extract)	Alpha-amylase inhibition assay, alpha-glucosidase inhibition assay, and lipid peroxidation assay	The free phenolic extract has reported higher activity with IC50 value 497.27 (α-amylase inhibition) and the bound phenolic extract has a higher activity for α-glucosidase inhibition with IC50 127.51 (200–800 µg/ml).	Adefegha and Oboh (2012)

Antidiabetic Potentials of Major Indian Spices: A Review

TABLE 24.1 (Continued)

S. No.	Common Name	Botanical Names and Family	Part Used as Spice	Extract/Active Components	Method	Inferences	References
13.	Curry Leaf, Kari patta	*Murraya koenigii* L. (Rutaceae)	Leaves	Aqueous	GOD/POD method (Glucose oxidase and Peroxidase method) (Blood glucose label in Alloxan-induced diabetic rats)	Significant reduction in the blood glucose level to 224 after treatment (7 days) which was 240.1 before treatment.	Bhopal et al. (2015)
					Glucometer	The % of blood glucose reduction was found to be 85% for 200 mg/kg and 83% for 400 mg/kg.	Al-Ani et al. (2017)
14.	Asafoetida, Heeng	*Ferula assafoetida* L. (Apiaceae)	Oleogum resin from rhizome	Aqueous (boiling water)	Analysis of Glucose: Nelson Somogyi's method, Analysis of Insulin: Radioimmunoassay kit	It had decreased the Blood Glucose level 6.75 ± 0.31 mmol/l and increased insulin level (0.48 ± 0.05 ng/ml) significantly, in albino rats at a dose of 0.2 g/kg for 14 days.	Abu-Zaiton (2010)
					Spectrophotometrically	50 mg/kg dose had reduced the Blood glucose level significantly after 4 weeks than the 100 mg/kg and 300 mg/kg.	Akhlaghi et al. (2012)

TABLE 24.1 (Continued)

S. No.	Common Name	Botanical Names and Family	Part Used as Spice	Extract/Active Components	Method	Inferences	References
				Methanol, Ethanol, Aqueous	α-Glucosidase inhibition assay	The α-Glucosidase inhibition was found to be 28% of the aqueous extract.	Yarizade et al. (2017)
15.	Ginger, Adrak	*Zingiber officinale* Rosc. (Zingiberaceae)	Rhizome	Aqueous (Hot)	Enzymatic colorimetric (GOD-PAP) method and ultrasensitive insulin assay	At the doses, 500 mg/kg and 1000 mg/kg the reduction in the blood glucose level was found to be 133.20 ± 5.07 mg/dl and 127.40 ± 6.39 mg/dl and increase in the insulin level was 0.214 ± 0.37 μU/ml and 0.056 ± 0.047 μU/ml.	Al-Qudah et al. (2016)
				50% Ethanol	GOD/POD method (Glucose oxidase method) and ELISA	In 6 weeks, the blood glucose had decreased to a considerable level from the control ($6.9 + 1.0$ vs. $8.5 + 0.7$ mmol/l).	Shadli et al. (2014)
				Aqueous	–	Dose 500 mg/kg for 42 days revealed that the extract had lowered the serum glucose level to 118.2 ± 1.6 mg/dL.	Jafri et al. (2011)

TABLE 24.1 (Continued)

S. No.	Common Name	Botanical Names and Family	Part Used as Spice	Extract/Active Components	Method	Inferences	References
				Oleo-resin	Accu-Chek sensor glucometer	At a dose of 400 mg/kg, the decrease in the fasting blood glucose level was found to be 131.83 ± 3.14 mg/dl when compared to the control.	Shukla and Rangari (2015)
				Juice	Glucometer	After a 12th day, there was a subsequent decrease in the blood glucose level to 5.47 ± 0.59 mmol/L.	Sultana et al. (2014)
					Glucometer (Accu-Chek sensor from Roche diagnostic corporation)	The reduction in the percentage of blood glucose on day 1 was 10.6% and which steadily increased to 77.1% on day 42.	Asha et al. (2011)
16.	Turmeric, Haldi	*Curcuma longa* L. (Zingiberaceae)	Rhizome	Ethyl acetate, methanol, and Aqueous4	Alpha-amylase inhibition and alpha-glucosidase inhibition	The ethyl acetate exhibit highest alpha-amylase inhibition (IC50 0.4 µg/ml) and alpha-glucosidase inhibition (IC50 71.6 µg/ml).	Lekshmi et al. (2014)

TABLE 24.1 (Continued)

S. No.	Common Name	Botanical Names and Family	Part Used as Spice	Extract/Active Components	Method	Inferences	References
				Aqueous	Haemoglucotest (glucose strips)	The blood glucose concentration had decreased from 355 ± 1.7 to 200 ± 1.2 mg/dl within 6 weeks.	El-Bahr and Al-Azraqi (2014)
					Spectronic 21 spectrophotometer	For 28 days (dose: 200 mg/kg) it had decreased the level of Blood glucose to appropriate level (7.83 ± 0.01 mmol/L) when compared with the diabetic control.	Olatunde et al. (2014)
				99.99% Ethanol	-	Decline in the blood glucose level to 141.33 ± 10.49 (300 mg/kg) and 114.12 ± 11.29 (500 mg/kg) in diabetic rats for 28 days.	Kumar (2013)
				Curcumin	Fasting Glucose: Glucometer; Insulin: Enzyme-Linked Immunosorbent Assay (ELISA)	The fasting blood Glucose level was found to be 12.53 ± 0.42 mmol/L and the Insulin level 2.08 ± 0.17 ng/ml.	Maha (2013)

TABLE 24.1 *(Continued)*

S. No.	Common Name	Botanical Names and Family	Part Used as Spice	Extract/Active Components	Method	Inferences	References
				Spent turmeric oleoresin	α-Amylase inhibition and α-Glucosidase inhibition assay	The IC50 obtained for α-Amylase 0.16 ± 0.05 µg/ml was and for α-Glucosidase was 0.71 ± 0.1 µg/ml.	Nampoothiri (2012)
				-	Blood analysis: Cyanmethemo-globin method and spectrophotometri-cally	The level of glucose in diabetic rats was reduced to 90.3 ± 6 mg/dl.	Mohamed et al. (2009)
				Acetone (spent turmeric)	GOD/POD kit (Glucose oxidase method)	The reduction in the fasting blood glucose level was 247.7 ± 17.1 mg/dl.	Kumar et al. (2005)
17.	Cinnamon	*Cinnamomum zeylanicum* Breyn (Lauraceae)	Bark	Methanol	α-Glucosidase inhibition assay and α-Amylase inhibition assay	The %inhibition of wheat α-Amylase at varying concentration (20–100) was found to be ranging from 11.53–38.46% and the IC50 value–130.55 ± 10.5. The% inhibition of yeast α-Glucosidase at varying concentration (20–100) was found to be ranging 10.25–35.40% and with IC50 value 140.01 ± 10.08.	Nair et al. (2013)

TABLE 24.1 (Continued)

S. No.	Common Name	Botanical Names and Family	Part Used as Spice	Extract/Active Components	Method	Inferences	References
				n-hexane, ethyl acetate, methanol, chloroform, and dichloromethane	α-Glucosidase inhibition assay and α-Amylase inhibition assay.	The IC50 value ranged from 0.5 to 8.7 µg/ml and 37.1 to 52.5 µg/ml for inhibition of α-Glucosidase and α-Amylase	Salehi et al. (2013)
			Leaves	Ethanol	Glucose oxidase method	For 5 hrs at different doses (100 mg/kg, 150 mg/kg and 200 mg/kg) had resulted in the reduction of blood level 202.14 ± 0.42 mg/dl, 195.31 ± 1.58 mg/dl and 190.22 ± 1.91 mg/dl respectively.	Tailang et al. (2008)
18.	Cassia	*Cinnamomum cassia* Blume (Lauraceae)	Bark	Aqueous	Glucose oxidase-peroxidase strip and Glucometer	The decrease in mean fasting blood glucose level was found to be 203.5 ± 13.47 on 10th and 191.5 ± 12.72 on the 15th day as compared to diabetic control at an oral dose of 60 mg/kg.	Kamble and Rambhimaiah, (2013)

TABLE 24.1 (Continued)

S. No.	Common Name	Botanical Names and Family	Part Used as Spice	Extract/Active Components	Method	Inferences	References
19.	Tejpat, Bay leaf, Cinnamon,	*Cinnamomum tamala* Nees & Eberm (Lauraceae)	Bark	Methanol and Aqueous	α-Amylase assay	The% inhibition of methanol and aqueous extract was found to be 97.49% and 93.78% respectively, and IC50 value for both was 1.80 and 5.53.	Kumanan et al. (2010)
			Leaves	Ethanol	Glucose oxidase-peroxidase strip (Accue-check® diagnostic kit), glucometer	The decrease in the fasting blood glucose level was found to be 188.5 ± 2.42 mg/dl compared diabetic control.	Bisht and Sisodia, (2011)
					Alpha amylase inhibition activity (*In-vitro*) and GOD-POD (glucose diagonastic kit; *In-vivo*)	After 21 days the reduction in the blood glucose level was found to be 184.5 ± 16.73 mg/dl for 200 mg/kg and 159.83 ± 13.22 mg/dl for 400 mg/kg.	Palanisamy et al. (2011)
					Glucose oxidase (Accue-check glucometer)	The decline in the blood glucose level was 140.7 ± 0.166 mg/dl when compared with the diabetic control.	Soni et al. (2013)

TABLE 24.1 *(Continued)*

S. No.	Common Name	Botanical Names and Family	Part Used as Spice	Extract/Active Components	Method	Inferences	References
				Aqueous	Nelson method	For 3 weeks (20 days), at a dose of 250 mg/kg the decrease in the fasting blood glucose level 73.20 ± 1.86.	Chakraborty and Das, (2010)
20.	Coriander, Dhaniya	*Coriandrum sativum* L. (Apiaceae)	Leaf & Fruit	Aqueous	Glucometer	At a dose 250 mg/dl and 500 mg/dl after 21 days the reduction in blood glucose level 136.83 ± 1.99 mg/dl and 128.67 ± 2.40 mg/dl.	Naquvi et al. (2012)
			Leaves	Methanol	Glucometer	At a dose of 200 mg/kg and 400 mg/kg for 21 days the blood glucose reduced to 172.92 ± 5.43 mg/dl and 142.55 ± 2.29 mg/dl when compared to diabetic control.	Mazhar and Mazumder (2013)
			Seeds	–	Glucometer-elite commercial test (Bayer); based on the glucose oxidase method	Decrease in the blood glucose level was found to be 308+15.12 mg/dl (0.5% CSE) and 288.5+44.22 mg/dl (1% CSE) for the 7 days.	Naveen and Khanum (2012)

TABLE 24.1 (Continued)

S. No.	Common Name	Botanical Names and Family	Part Used as Spice	Extract/Active Components	Method	Inferences	References
				Aqueous	Glucose oxidase method	The decreased in the concentration of glucose level was found to be 27.9 ± 3.40 mmol/l.	Gallagher et al. (2003)
						For 8 hrs at a dose of 300 mg/kg the mean percent reduction of the serum glucose level was found to be 12.84 ± 3.64.	Sushruta et al. (2006)
21.	Garlic, Lassan	*Allium sativum* L. (Alliaceae)	Bulb	Aqueous	Spectro-photometrically	For 2, 5 and 7 weeks at the dose of 500 mg/kg elevation in the serum glucose level were found to be 29%, 68%, and 57%.	Thomson et al. (2007)
				Ethanol	Glucometer	After 4 weeks at a dose of 200 mg/kg, the reduction in blood glucose level was found to be 9.33 ± 0.80 mmol/L.	Mamun et al. (2017)

TABLE 24.1 *(Continued)*

S. No.	Common Name	Botanical Names and Family	Part Used as Spice	Extract/Active Components	Method	Inferences	References
22.	Tamarind, Imli, Ambli	*Tamarindus indica* L. (Caesalpiniaceae)	Leaf	Aqueous methanolic extract	Digital Glucometer	After 6 hrs, at a dose of 50 mg/kg, 100 mg/kg and 200 mg/kg, the deduction in the blood glucose level was found to be 124 ± 1.08, 115 ± 6.93 and 107 ± 4.09.	Ramchander et al. (2012)
			Fruit	Methanol	Spectro-photometrically/Dunnett's test	After 5 days at a dose of 200 mg/kg/day and 400 mg/kg/day, the decrease in the elevation of blood glucose was found to be 6.12 ± 0.21 and 5.62 ± 0.27 when compared with the control (diabetic).	Nahar et al. (2014)
				Hydro-methanolic extract	Glucometer; Insulin: Immunosorbent assay (ELISA kit)	After 14 days the decrease in the elevated fasting blood glucose level was found to be 88.4 ± 7.6 mg/dl.	Maiti et al. (2014)

TABLE 24.1 *(Continued)*

S. No.	Common Name	Botanical Names and Family	Part Used as Spice	Extract/Active Components	Method	Inferences	References
				Aqueous	Blood glucose: O-toluidine method; Insulin: ELISA (Enzyme-Linked ImmunoSorbent Assay) method using the Boehringer Mannheim kit.	The reduction in the level of fasting blood glucose and elevation in the plasma insulin was 106.9 ± 14.3 mg/dl and 11.98 ± 0.78 µU/ml in Albino Wistar male rats for 45 days at a dose of 500 mg/kg of body weight.	Manoharan et al. (2009)
23.	Nutmeg	*Myristica fragrans* Houtt. (Myristicaceae)	Seed	Aqueous	Glucometer	There is a subsequent decrease in the level of fasting blood glucose at a dose of 35 mg/kg of body weight.	Oseni and Idowu, (2014)
						The decrease in the glucose level of alloxan-induced diabetic rats for 30% aqueous extract, meal compound with 5% nutmeg and 10% nutmeg were found to be 98 ± 0.03 mg/dl, 87 ± 0.023 mg/dl, and 80 ± 0.021 mg/dl.	Oyindamola et al. (2017)

TABLE 24.1 (Continued)

S. No.	Common Name	Botanical Names and Family	Part Used as Spice	Extract/Active Components	Method	Inferences	References
				Petroleum ether	Touch Glucometer	At a dose, 200 mg/kg after 4 h the decrease in the mean blood glucose level was from 56.00 ± 2.38 to 49.75 ± 2.05 mg%.	Somani and Singhai (2008)
24.	Allspice	*Pimenta dioica* L. (Myrtaceae)	Leaf	Methanol	Blood Glucose: O-toluidine; Plasma Insulin: ELISA method (Enzyme-Linked Immunosorbent Assay) using Boehringer Manheim Kit (Boehringer analyzer, ES300)	At a dose of 75 mg/kg and 150 mg/kg of body wt. the reduction in the blood glucose level and elevation in the plasma insulin level was 135.24 ± 5.84 mg/dl, 105.23 ± 8.81 mg/dl, 14.21 ± 0.70 µU/ml and 15.31 ± 1.20 µU/ml.	Yogalakshmi and Vaidehi (2015)
25.	Rosemary	*Rosmarinus officinalis* L. (Lamiaceae)	Leaf	Aqueous	Glucometer	After 21 days, at a dose of 200 mg/kg, the decrease in the blood glucose level was found to be 86.4 ± 3.6 mg/dl.	Khallil et al. (2012)

TABLE 24.1 *(Continued)*

S. No.	Common Name	Botanical Names and Family	Part Used as Spice	Extract/Active Components	Method	Inferences	References
				Powder	Glucose Enzokit	After 8 weeks at different doses 2, 5, 10 g/kg the decrease in the fasting glucose were found to be 111 ± 13 mg/dl (11.23%), 107 ± 11 mg/dl (15.74%) and 103 ± 11 mg/dl (18.25%).	Labban et al. (2014)
26.	Oregano	*Origanum vulgare* L. (Lamiaceae)	Leaf	Aqueous	Winn-Deen et al. (1988)	Decreased blood glucose level was found to be 81.65 ± 1.049 and there is significant increase in the insulin level was also observed at a dose of 20 mg/kg after 6 weeks.	Nema and Omimah (2013)

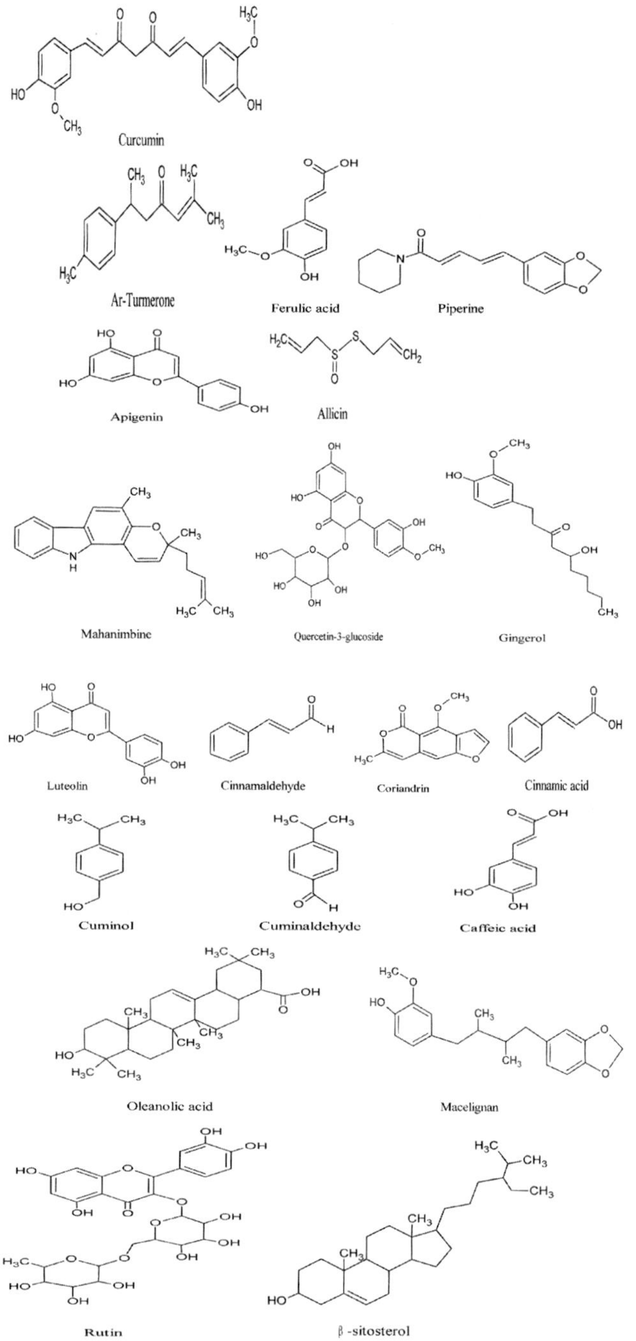

FIGURE 24.1 Structure of the bioactive and phytochemical compounds of the spices.

possessing anti-diabetic properties (Chattopadhyay et al., 2004; Murugan and Pari, 2007a). Other analogous constituents such as curcuminoids, sesquiterpenoids, and Ar-Turmerone (2-methyl-6-(4-methylphenyl) hept-2-en-4-one) are also responsible for suppressing the blood glucose level in diabetic-induced rats (Sreenivasan et al., 2015; Nishiyama et al., 2005). Gingerol (1-[4′-hydroxy-3′-methoxyphenyl]-5-hydroxy-3-decanone), and zingerone are the chemical component isolated from the *Zingiber officinale* responsible for its remarkable antidiabetic activity (Otunola & Afolayan, 2015; Arul et al., 2016). Piperine (5-(1,3-benzodioxol-5-yl)-1-piperidin-1-ylpenta-2,4,dien-1-one), a bioactive constituent of black pepper is mainly responsible for suppressing the diabetic in rats (Shukla and Rangari, 2015). Mahanimbine (3,5-dimethyl-3-(4-methylpent-3-enyl)-11H-pyrano[3,2-a]carbazole), is a phytochemical alkaloid carbazole component possessed by curry leaves which are found to be mainly responsible for its anti-diabetic properties (Chandra et al., 2015). Allicin, S-allyl cysteine sulphoxide (3-prop-2-enylsulfinylsulfanylprop-1-ene), a sulfur-containing amino acid in garlic (*Allium sativum* L.) have potential to reduce the blood sugar level in diabetic-induced rats (Augusti and Sheela, 1996). Characterization of clove revealed the presence Oleanolic acid (10-hydroxy-2,2,6a,6b,9,9,12a-heptamethyl-1,3,4,5,6,6a,7,8,8a,10,11,12,13,14b-tetradecahydropicene-4a-carboxylic acid) present in the essential oil possess anti-diabetic, anticancer, and antimicrobial activity (Mittal et al., 2014). Due to the presence of phenols and tannins in the extract of Asafoetida, suppress the blood glucose level in the STZ alloxan induced rats. Ferulic acid (4-Hydroxy-3-methoxycinnamic acid) is a component responsible for the antioxidant activity and diabetic suppressing activity shown by Asafoetida (Akhlaghi et al., 2012). The bioactive components such as Cinnamaldehyde (3-phenylprop-2-enal), Eugenol (2-methoxy-4-prop-2-enylphenol), Cinnamic acid (3-phenylprop-2-enoic acid), coumarin, and procyanidins were found to be mainly responsible for the antioxidant and antidiabetic activity of cinnamon (Vanschoonbeek, 2006; Qin, 2010). *Coriandrum sativum* consists of various biochemical active compounds, mainly isocoumarin (1H-2-Benzopyran-1-one) and most significantly coriandrin (4-Methoxy-7-methyl-5H-furo(2,3-g)benzopyran-5-one) possessing antidiabetic activity (Lal et al., 2004). Celery seeds have the highest content of flavonoids such as apigenin (5,7-Dihydroxy-2-(4-hydroxyphenyl)-4H-chromen-4-one) and luteolin (2-(3,4-Dihydroxyphenyl)-5,7-dihydroxy-4H-chromen-4-one), the main active principles which were found to exhibit antidiabetic activity in them (Naiz et al., 2013). Due to presence of some crucial biocomponents (polyphenols) such as n-Hexacosane, eicosanoic acid, β-sitosterol (17-[(2R,5R)-5-ethyl-6-methylheptan-2-yl]-10,13-dimethyl-2,3,4,7,8,9,11,12,14,15,16,17-dodecahydro-1H-cyclopenta[a]phenanthren-3-ol), octacosanyl ferulate (3-(4-hydroxy-3-methoxyphenyl) prop-2-enoate), 21-oxobehenic acid, and pinitol (6-methylcyclohexane-1,2,3,4,5-pentol) *Tamarindus indica* (tamarind) plays a potent role in medical sciences as an anti-microbial, anti-inflammatory, anti-lipid peroxidative, antioxidant, and anti-diabetic activity (Manoharan et al., 2009; Zohrameena et al., 2017). Phytochemical screening of *Piper longum* L. revealed that the main flavonoids which are responsible for the increment in the secretion of insulin and glycogen content (Kumar et al., 2011). *The Cuminum cyminum* phytochemical analysis shows Cuminaldehyde and cuminol ((4-propan-2-ylphenyl) methanol) are the chief components reliable for its antidiabetic property (Lee, 2005; Patil, 2013; Singh et al., 2017). Presence of phenols, tannins, and saponins provide antidiabetic activity in Aniseeds (Shobha and Andulla, 2018) The major phenolic compounds which are responsible for the antidiabetic activity of aniseeds are mainly rutin (2-(3,4-dihydroxyphenyl)-5,7-dihydroxy-3-[3,4,5-trihydroxy-6-[[3,4,5-trihydroxy-6-methyloxan-2-yl]oxymethyl]oxan-2-yl]oxychromen-4-one), caffeic acid (3-(3,4-dihydroxyphenyl)prop-2-enoic acid) and luteolin (Shobha et al., 2013). The active component which is responsible for the antidiabetic activity in nutmeg and in Rosemary to cause the inhibition was found to be Macelignan (7-(3,4-methylenedioxyphenyl)-7′-(4-hydroxy-3-methoxyphenyl)-8,8′-dimethylbutane) and luteolin (5,7,3′,4′-tetrahydroxy flavones) (Han et al., 2008; Koga et al., 2006).

24.4 CONCLUSION

It is clear that antidiabetic properties reported in the historical records are also supported by the latest researches and studies on the medicinal properties of spices. The *in vitro* and *in vivo* analysis confirmed the antidiabetic activity of different spices found in the Indian kitchen. In addition, to their antidiabetic properties spices also establish to possess many additional medicinal, curative properties. Further advanced studies and deeper insights are needed to explore and determine efficient bio-components held in the spices for the discovery of their medicinal properties. The abundance of information that is currently available in respect of the medicinal properties of the Indian spices makes them important in the field of drug science, and in pharmaceutical areas.

KEYWORDS

- *Allium sativum*
- antidiabetic
- antimicrobial activity
- bioactive constituents
- bio-chemical constituent
- eicosanoic acid
- sesquiterpenoids

REFERENCES

Abdulrazaq, N. B., Cho, M. M., Win, N. N., Zaman, R., & Rahman, M. T., (2012). Beneficial effects of ginger (*Zingiber officinale*) on carbohydrate metabolism in streptozotocin-induced diabetic rats. *Br. J. Nutr., 108*, 1194–1201.

Abubakar, N. A., Oise, A. E., & Saidu, A. N., (2014). Phytochemical constituents and hypoglycemic effect of aqueous and ethanolic extracts of *Murraya koenigii* in alloxan-induced diabetic rats. *Journal of Dental and Medical Sciences, 13*(9), 08–12.

Abu-Zaiton, A. S., (2010). Anti-diabetic activity of Ferula assafoetida extract in normal and alloxan-induced diabetic rats. *Pak J. Biol Sci., 13*, 97–100.

Adefegha, S. A., & Oboh, G., (2012). *In vitro* inhibition activity of polyphenol-rich extracts from *Syzygium aromaticum* (L.) Merr. & Perry (Clove) buds against carbohydrate hydrolyzing enzymes linked to type 2 diabetes and Fe^{2+}-induced lipidperoxidation in rat pancreas. *Asian Pac. J. Trop. Biomed., 2*(10), 774–781.

Adefegha, S. A., Oboh, G., Adefegha, O. M., Boligon, A. A., & Athayde, M. L., (2014). Antihyperglycemic, hypolipidemic, hepatoprotective, and antioxidative effects of dietary clove (*Syzygium aromaticum*) bud powder in a high-fat diet/streptozotocin-induced diabetes rat model. *J. Sci. Food Agric., 94*, 2726–2737.

Ahmad, R., Srivastava, S. P., Maurya, R., Rajendran, S. M., Arya, K. R., & Srivastava, A. K., (2008). Mild antihyperglycaemic activity in *Eclipta alba, Berberis aristata, Betula utilis, Cedrus deodara, Myristica fragrans* and *Terminalia chebula*. *Indian J. Sci. Technol., 1*(5), 1–6.

Ahmed, A. S., Ahmed, Q., Saxena, A. K., & Jamal, P., (2017). Evaluation of *in vitro* antidiabetic and antioxidant characterizations of *Elettaria cardamomum* (L.) Maton (Zingiberaceae), *Piper cubeba* L. f. (Piperaceae), and *Plumeria Rubra* L. (Apocynaceae). *Pak J. Pharm. Sci., 30*(1), 113–126.

Ahmed, S. J., Faraj, R. A., & Al-Shaikh, M. N., (2013). Study effects of cumin and DNA profile in diabetic rats. *IOSR Journal of Pharmacy and Biological Sciences (IOSR-JPBS), 5*(1), 14–16.

Akhlaghi, F., Rajaei, Z., Hadjzadeh, M. A. R., Iranshahi, M., & Alizadeh, M., (2012). Antihyperglycemic Effect of Asafoetida (*Ferula assafoetida* Oleo-Gum-Resin) in Streptozotocin-induced diabetic rats. *World Applied Sciences Journal, 17*(2), 157–162.

Al-Ani, I. M., Santosa, R. I., Yankuzo, M. H., Saxena, A. K., & Alazzawi, K. S., (2017). The antidiabetic activity of curry leaves "*Murraya Koenigii*" on the glucose levels, kidneys, and Islets of Langerhans of rats with streptozotocin-induced diabetes. *Makara J. Health Res.*, *21*(2), 54–60.

Al-Qudah, M. M. A., Haddad, M. A., & EL-Qudah, J. M. F., (2016). The effects of the aqueous ginger extract on pancreas histology and on blood glucose in normal and alloxan monohydrate-induced diabetic rats. *Biomed Res.*, *27*(2), 350–356.

American Diabetes Association, (2014). Diagnosis and classification of diabetes mellitus. *Diabetes Care 37*(1), S81–S90.

Anitha, T., Balakumar, C., Ilango, K. B., Jose, C. B., & Vetrivel, D., (2014). Antidiabetic activity of the aqueous extracts of *Foeniculum vulgare* on streptozotocin-induced diabetic rats. *Int. J. Adv. Pharm. Biol. Chem.*, *3*(2), 487–494.

Arul, J. M., Parameswari, C. S., & Vincent, S., (2016). Antidiabetic, hypolipidemic, and histopathological analysis of zingerone in streptozotocin-induced diabetic rats. *Asian J. Pharm. Clin. Res.*, *9*(3), 220–224.

Arulmozhi, D. K., Kurian, R., Veeranjaneyulu, A., & Bodhankar, S. L., (2007). Antidiabetic and antihyperlipidemic effects of *Myristica fragrans*. In animal models. *Pharmaceutical Biology, 45*(1), 64–68.

Arulselvan, P., Senthikumar, G. P., Sathish, K. D., & Subramanian, S., (2006). Anti-diabetic effect of *Murraya koenigii* leaves on streptozotocin-induced diabetic rats. *Pharmazie, 61*(10), 874–877.

Asha, B., Krishnamurthy, K. H., & Siddappa, D., (2011). Evaluation of anti hyperglycaemic activity of *Zingiber officinale* (Ginger) in albino rats. *J. Chem. Pharm. Res., 3*(1), 452–456.

Atal, S., Agrawal, R. P., Vyas, S., Phadnis, P., & Rai, N., (2012). Evaluation of the effect of piperine *per se* on blood glucose Level in alloxan-induced diabetic mice. *Acta Pol. Pharm.: Drug Research, 69*(5), 965–969.

Augusti, K. T., & Sheela, C. G., (1996). Antiperoxide effect of S-allyl cysteine sulfoxide, an insulin secretagogue, in diabetic rats. *Experientia, 52*, 115–120.

Bhopal, C. H. M. V., Madhavulu, B., Kudagi, B. L., Rama, M. P., Umar, S., & Sreeharsha, H. M., (2015). Anti-hyperglycemic activity of *Murraya koenigii* in comparison with pioglitazone in experimental small animal models. *Int. J. Med. Appl. Sci., 4*(1), 166–172.

Bisht, S., & Sisodia, S. S., (2011). Assessment of antidiabetic potential of *Cinnamomum tamala* leave extract in streptozotocin-induced diabetic rats. *Indian J. Pharmacol., 43*(5), 582–585.

Bokaeian, M., Nakhaee, A., Moodi, B., Farhangi, A., & Akbarzadeh, A., (2010). Effects of garlic extract treatment in normal and streptozotocin diabetic rats infected with *Candida albicans*. *Indian J. Clin. Biochem., 25*(2), 182–187.

Chakraborty, U., & Das, H., (2010). Antidiabetic and antioxidant activities of *Cinnamomum tamala* leaf extracts in STZ-treated diabetic rats. *Global J. Biotech. and Biochem., 5*(1), 12–18.

Chattopadhyay, I., Biswas, K., Bandyopadhyay, U., & Banerjee, K. R., (2004). Turmeric and curcumin: Biological actions and medicinal applications. *Current Science*, *87*(1), 44–53.

Dhandapani, S., Subramanian, V. R., Rajagopal, S., & Namasivayam, N., (2002). Hypolipidemic effect of *Cuminum Cyminum* L. on alloxan-induced diabetic rats. *Pharmacol. Re., 46*(3), 251–255.

Eidi, A., Eidi, M., & Esmaeili, E., (2006). Antidiabetic effect of garlic (*Allium sativum* L.) in normal and streptozotocin-induced diabetic rats. *Phytomedicine, 13*(9/10), 624–629.

Eidi, A., Eidi, M., Haeri, R. A., & Basati, F., (2010). Hypoglycemic effect of ethanolic extract of *carum carvi* l. seeds in normal and streptozotocin-induced diabetic rats. *J. Med Plants,* 9(35), 106–113.

El-Bahr, S. M., & Al-Azraqi, A. A., (2014). Effects of dietary supplementation of turmeric (*Curcuma longa*) and black cumin seed (*Nigella sativa*) in streptozotocin-induced diabetic rats. *Int. J. Biochem. Res. and Rev.,* 4(6), 481–492.

El-Masry, A. A., (2012). Potential therapeutic effect of *curcuma longa* on streptozotocin-induced diabetic rats. *Global Adv. Res. J. Med. Med Sci.,* 1(4), 091–098.

El-Soud, N. A., El-Laithy, N., El-Saeed, G., Wahby, M., Khalil, M., Morsy, F., & Shaffie, N., (2011). Antidiabetic activities of *Foeniculum vulgare* Mill. The essential oil in streptozotocin-induced diabetic rats. *Maced. J. Med. Sci.,* 4(2), 139–146.

Emam, M. A., (2012). Comparative evaluation of antidiabetic activity of *Rosmarinus officinalis* L. and *Chamomile Recutita* in streptozotocin-induced diabetic rats. *Agriculture and Biology Journal of North America,* 3(6), 247–252.

Gallagher, A. M., Flatt, P. R., Duffy, G., & Abdel-Wahab, Y. H. A., (2003). The effects of traditional antidiabetic plants on *in vitro* glucose diffusion. *Nut Res.,* 23, 413–424.

Hamza, R. G., & Elshahat, A. N., (2011). Hypoglycaemic effect of γ-Irradiated clove extract on alloxan-induced diabetic rats. *Egypt. J. Rad. Sci. Applic.,* 24(2), 345–358.

Han, K. L., Choi, J. S., Lee, J. Y., Song, J., Joe, M. K., Jung, M. H., & Hwang, J. K., (2008). Therapeutic potential of peroxisome proliferator-activated receptor-alpha/gamma dual agonist with the alleviation of endoplasmic reticulum stress for the treatment of diabetes. *Diabetes,* 57, 737–745.

Huang, T. H. W., Peng, G., Kota, B. P., Li, G. Q., Yamahara, J., Roufogalis, B. D., & Li, Y., (2005). The anti-diabetic action of *Punica granatum* flower extract: Activation of PPAR-c and identification of an active component. *Toxicol. Appl. Pharmacol.,* 207, 160–169.

Jabbar, A. A. A. S., & Basim, A. K. A. S., (2013). Antihyperglycaemic and pancreatic regenerative effect of an n-butanol extract of celery (*Apium graveolens*) seed in STZ-induced diabetic male rats. *Suez Canal Veterinary Medicine Journal,* 18(1), 71–85.

Jabbar, A. A. A. S., Mohsen, N. A. A., & Ahmed, K. I., (2012). Antioxidant activity of n-butanol extract of celery (*Apium graveolens*) seed in Streptozotocin-induced diabetic male rats. *Res. Pharm. Biotechnol.,* 4(2), 24–29.

Jafri, S. A., Abass, S., & Qasim, M., (2011). Hypoglycemic effect of ginger (*Zingiber officinale)* in alloxan-induced diabetic rats (*Rattusnorvagicus*). *Pak. Vet. J.,* 31(2), 160–162.

Jung, E. H., Kim, S. R., Hwang, I. K., & Ha, T. Y., (2007). Hypoglycemic effects of a phenolic acid fraction of rice bran and ferulic acid in C57BL/KsJ-db/db mice. *J. Agric. Food Chem.,* 55, 9800–9804.

Kaleem, M., Sheema, S. H., & Bano, B., (2005). Protective effects of *piper nigrum* and *Vinca rosea* in alloxan-induced diabetic rats. *Indian J. Physiol. Pharmacol.,* 49(1), 65–71.

Kamble, S., & Rambhimaiah, S., (2013). Antidiabetic activity of aqueous extract of *Cinnamomum cassia* in alloxan-induced diabetic rats. *Biomed. Pharmacol. J.,* 6(1), 83–88.

Kesari, A. N., Gupta, R. K., & Watal, G., (2005). Hypoglycemic effects of *Murraya koenigii* on normal and alloxan diabetic rabbits. *J. Ethnopharmacol.,* 11, 223–231.

Khalil, O. A., Ramadan, K. S., Danial, E. N., Alnahdi, H. S., & Ayaz, N. O., (2012). Antidiabetic activity of *Rosmarinus officinalis* and its relationship with the antioxidant property. *African J. Pharm. Pharmacol.,* 6(14), 1031–1036.

Koga, K., Shibata, H., Yoshino, K., & Nomoto, K., (2006). Effects of 50% ethanol extract from rosemary (*Rosmarinus officinalis*) on α-glucosidase inhibitory activity and the elevation of plasma glucose level in rats, and its active compound. *J. Food Sci.,* 71(7), S507–S512.

Kumanan, R., Manimaran, S., Saleemulla, K., Dhanabal, S. P., & Nanjan, M. J., (2010). Screening of bark of *Cinnamomum tamala* (Lauraceae) by using α-amylase inhibition assay for anti-diabetic activity. *Int. J. Pharm. Biomed Res.*, *1*(2), 69–72.

Kumar, G. S., Shetty, A. K., Sambaiah, K., & Salimath, P. V., (2005). Antidiabetic property of fenugreek seed mucilage and spent turmeric in streptozotocin-induced diabetic rats. *Nut Res.*, *25*, 1021–1028.

Kumar, J. S., Manjunath, S., Mariguddi, D. D., Kalashetty, P. G., Dass, P., & Manjunath, C., (2013). Anti-diabetic effects of turmeric in alloxan induced diabetic rats. *J. Evol. Med. Dent. Sci.*, *2*(11), 1669–1679.

Kumar, S., Sharma, S., & Suman, J., (2011). In vivo anti-hyperglycemic and antioxidant potential of *Piper longum* fruit. *J. Pharm. Res., 4*(2), 471–474.

Labban, L., Mustafa, U. E. S., & Ibrahim, Y. M., (2014). The effects of rosemary (*Rosmarinus officinalis*) leaves powder on glucose level, lipid profile, and lipid peroxidation. *Int. J. Clin. Med., 5*, 297–304.

Lal, A. A. S., Kumar, T., Murthy, P. B., & Pillai, K. S., (2004). Hypolipidemic effect of *Coriandrum Sativum* L. in trition induced hyperlipidemic rats. *Indian J. Exp. Biol.*, *42*, 909–912.

Lee, H. S., (2005). Cuminaldehyde: Aldose reductase and alpha-glucosidase inhibitor derived from *Cuminum cyminum* L. seeds. *J. Agric. Food Chem.*, *53*(7), 2446–2450.

Lekshmi, P. C., Ranjith, A., Nisha, V. M., Menon, A. N., & Raghu, K. G., (2014). In vitro antidiabetic and inhibitory potential of turmeric (*Curcuma longa* L) rhizome against cellular and LDL oxidation and angiotensin-converting enzyme. *J. Food Sci. Technol.*, *51*(12), 3910–3917.

Lemhadri, A., (2004). The anti-hyperglycaemic activity of the aqueous extract of *Origanum vulgare* growing wild in Tafilalet region. *Journal of Ethnopharmacology, 92*, 251–256.

Maha, B., (2013). Antidiabetic potential of turmeric with/without fermented milk enriched with probiotics in diabetic rats. *Am. J. Biomed Life Sci.*, *1*(1), 01–07.

Maiti, R., Das, U. K., & Ghosh, D., (2005). Attenuation of hyperglycemia and hyperlipidemia in streptozotocin-induced diabetic rats by aqueous extract of seeds of *Tamarindus indica*. *Biol. Pharm. Bull.*, *28*, 1172–1176.

Maiti, R., Misra, D. S., & Ghosh, D., (2014). Hypoglycemic and hypolipidemic effect of seed hydromethanolic extract of *Tamarindus indica* L. on streptozotocin-induced diabetes mellitus in rats. *Am. J. Phytochemical. and Clin. Ther.*, *2*(12), 1416–1429.

Mamun, M. A., Hasan, N., Shirin, F., Belal, M. H., Khan, M. A. J., Tasnin, M. N., Islam, M. D., et al., (2017). Antihyperglycemic and antihyperlipidemic activity of ethanol extract of garlic *(Allium sativum)* in streptozotocin-induced diabetic mice. *Int. J. Med. Health Res.*, *3*(2), 63–66.

Manoharan, S., Chellammal, A., Linsa, M. A., Vasudevan, K., Balakrishnan, S., & Ranezab, A. P., (2009). Antidiabetic efficacy of *Tamarindus indica* seeds in alloxan-induced diabetic rats. *Electron J. Pharmacol Ther.*, *2*, 13–18.

Manoharan, S., Silvan, S., Vasudevan, K., & Balakrishnan, S., (2007). Antihyperglycaemic and anti-lipid peroxidative effects of *Piper longum* dried fruits in alloxan-induced diabetic rats. *J. Biol. Sci.*, *6*(1), 161–168.

Mathers, C. D., & Loncar, D., (2006). Projections of global mortality and burden of disease from 2002 to 2030. *PLoS Medicine, 3*(11), 2011–2030.

Mazhar, J., & Mazumder, A., (2013). Evaluation of antidiabetic activity of methanolic leaf extract of *Coriandrum sativum* in alloxan-induced diabetic rats. *Res. J. Pharm. Biol. Chem. Sci.*, *4*(3), 500–507.

Mishra, N., (2013). Haematological and hypoglycemic potential *Anethum graveolens* seeds extract in normal and diabetic Swiss albino mice. *Vet World*, 6(8), 502–507.

Mittal, M., Gupta, N., Parashar, P., Mehra, V., & Khatri, M., (2014). Phytochemical evaluation and pharmacological activity of *Syzygium aromaticum*: A comprehensive review. *Int. J. Pharm. Pharm. Sci.*, 6(8), 67–72.

Mohamed, A. M., EL-Sharkawy, F. Z., Ahmed, S. A. A., Aziz, W. M., & Badwary, O. A., (2009). Glycemic control and therapeutic effect of *Nigella sativa* and *Curcuma longa* on rats with streptozotocin-induced diabetic hepatopathy. *J. Pharmacol. Toxicol.*, 4(2), 45–57.

Murugan, P., & Pari, L., (2007a). Influence of tetrahydrocurcumin on erythrocyte membrane-bound enzymes and antioxidant status in experimental type 2 diabetic rats. *J. Ethnopharmacol.*, 113, 479–486.

Nabi, S. A., Ali, M. S., Natava, R., Tilak, T. K., & Rao, C. A., (2012). Antidiabetic and antioxidant activities of *Piper longum* root aqueous extract in STZ induced diabetic rats. *J. Pharm. Chem.*, 6(3), 30–35.

Nabi, S. A., Kasetti, R. B., Sirasanagandla, S., Tilak, T. K., Kumar, M. V. J., & Rao, C. A., (2013). The antidiabetic and antihyperlipidemic activity of *Piper longum* root aqueous extract in STZ induced diabetic rats. *BMC Complement Altern. Med.*, 13(1), 1–9.

Nahar, L., Nasrin, F., Zahan, R., Haque, A., Haque, E., & Mosaddik, A., (2014). Comparative study of the antidiabetic activity of *Cajanus cajan* and *Tamarindus industry* in alloxan-induced diabetic mice with a reference to *in vitro* antioxidant activity. *Pharmacognosy Research*, 6(2), 180–187.

Nair, S. S., Kavrekar, V., & Mishra, A., (2013). In vitro studies on alpha-amylase and alpha-glucosidase inhibitory activities of selected plant extracts. *Euro. J. Exp. Bio.*, 3(1), 128–132.

Naiz, K., Gull, S., & Zia, M. A., (2013). Antihyperglycemic/hypoglycemic effect of *celery seeds* (Ajwain/Ajmod) in Streptozotocin-induced diabetic rats. *Journal of Rawalpindi Medicinal College*, 17(1), 134–137.

Nampoothiri, S. V., Lekshmi, P. C., Venugopalan, V. V., & Menon, A. N., (2012). Antidiabetic and antioxidant potentials of spent turmeric oleoresin, aby-product from curcumin production industry. *Asian Pac. J. Trop. Dis.*, S169–S172.

Naquvi, K. J., Ali, M., & Ahmad, J., (2012). Antidiabetic activity of aqueous extract of *Coriandrum sativum* l. Fruits in streptozotocin-induced rats. *Int. J. Pharm. Pharm. Sci.*, 4(1), 239–240.

Naveen, S., & Khanum, F., (2012). Anti-diabetic, anti-oxidant, anti-dyslipidemic and hepatoprotective properties of coriander seed extract in streptozotocin-induced diabetic rats. *J. Herbal Med. Toxicol.*, 6(2), 61–67.

Nema, A. M., & Omimah, A. N., (2013). The antihyperglycaemic effect of the aqueous extract of *Origanum vulgare* leaves in streptozotocin-induced diabetic rats. *Jordan Journal of Biological Sciences*, 6(1), 31–38.

Nishiyama, T., Mae, T., Kishida, H., Tsukagawa, M., Mimaki, Y., Kuroda, M., Sashida, Y., Takahashi, K., Kawada, T., Nakagawa, K., & Kitahara, M., (2005). Curcuminoids and sesquiterpenoids in turmeric (*Curcuma longa* L.) suppress an increase in blood glucose level in type 2 diabetic KK-Ay mice. *J. Agric. Food Chem.*, 53(4), 959–63.

Ohnishi, M., Matuo, T., Tsuno, T., Hosoda, A., Nomura, E., Taniguch, H., Sasaki, S., & Morishita, H., (2004). Antioxidant activity and hypoglycemic effect of ferulic acid in STZ-induced diabetic mice and KK-Ay mice. *Biofactors.*, 21(1–4), 315–319.

Olatunde, A., Joel, E. B., Tijjani, H., Obidola, S. M., & Luka, C. D., (2014). Anti-diabetic activity of aqueous extract of *Curcuma longa* (Linn) rhizome in normal and alloxan-induced diabetic rats. *Researcher*, 6(7), 58–65.

Onyesife, C. O., Ogugua, V. N., & Anaduaka, E. G., (2014). Hypoglycemic potentials of ethanol leave extract of black pepper (*Piper Nigrum*) on alloxan-induced diabetic rats. *Annals of Biological Research*, 5(6), 26–31.

Oseni, O. A., & Idowu, A. S. K., (2014). The inhibitory activity of aqueous extracts of Horseradish *Moringa oleifera* (Lam) and Nutmeg *Myristica fragrans* (Houtt) on oxidative stress in alloxan induced diabetic-male Wistar albino rats. *Am. J. Biochem. Mol. Biol.*, 4(2), 64–75.

Oshaghi, E. A., Tavilani, H., Khodadadi, I., & Goodarzi, M. T., (2015). Dill tablet: A potential antioxidant and anti-diabetic medicine. *Asian Pac. J. Trop. Biomed.*, 5(9), 720–727.

Otunola, G. A., & Afolayan, A. J., (2015). Antidiabetic effect of combined spices of *Allium sativum, Zingiber officinale*, and *Capsicum frutescens* in alloxan-induced diabetic rats. *Frontiers in Life Science*, 8(4), 314–323.

Oyindamola, E. A., Bolanle, A. O., & Abass, O. O., (2017). Effect of nutmeg (*Myristica fragrans*) on oxidative stress in alloxan-induced diabetic in Wistar albino rats. *World J. Pharm. Pharm. Sci.*, 6(4), 1901–1908.

Palanisamy, P., Srinath, K. R., Kumar, D. Y., & Chowdary, P. C., (2011). Evaluation of antioxidant and anti-diabetic activities of *Cinnamomum tamala* Linn leaves in streptozotocin-induced diabetic rats. *Int. Res. J. Pharm.*, 2(12), 157–162.

Parle, M., & Khanna, D., (2011). Clove: A champion spice. *Int. J. Res. Ayurveda and Pharmacy*, 2(1), 47–54.

Patil, S. B., Takalikar, S. S., Joglekar, M. M., Haldavnekar, V. S., & Arvindekar, A. U., (2013). Insulinotropic and β-cell protective action of cuminaldehyde, cuminol and an inhibitor isolated from *Cuminum cyminum* in streptozotocin-induced diabetic rats. *Br. J. Nutr.*, 110(8), 1434–1443.

Qin, B., Panickar, K. S., & Anderson, R. A., (2010). Cinnamon: Potential role in the prevention of insulin resistance, metabolic syndrome, and type 2 diabetes. *J. Diabetes Sci. Technol.*, 4(3), 685–693.

Ramchander, T., Rajkumar, D., Sravanprasad, M., Goli, V., & Dhanalakshmi, C. H. A., (2012). Antidiabetic activity of aqueous methanolic extracts of leaf of *Tamarindus indica*. *International Journal of Pharmacognosy and Phytochemical Research*, 4(1), 5–7.

Rao, G., Bhat, S., Rao, G. S., & Bhat, G. P., (2013). Antidiabetic and antioxidant efficacy of a powdered mixture of *Curcuma longa* and *Emblica officinalis* in diabetic rats in comparison with glyburide. *Web med Central Diabetes*, 4(2), 1–13.

Salehi, P., Asghari, B., Esmaeili, M. A., Dehghan, H., & Ghazi, I., (2013). α- Glucosidase and α- amylase inhibitory effect and antioxidant activity of ten plants extracts traditionally used in Iran for diabetes. *J. Med. Plants Res.*, 7(6), 257–266.

Sarfraz, M., Khaliq, T., Khan, J. A., & Aslam, B., (2017). Effect of aqueous extract of black pepper and Ajwa seed on liver enzymes in alloxan-induced diabetic Wister albino rats. *Saudi Pharmaceutical Journal*, 25, 449–452.

Seo, K. I., Choi, M. S., Jung, U. J., Kim, H. J., Yeo, J., Jeon, S. M., & Lee, M. K., (2008). Effect of curcumin supplementation on blood glucose, plasma insulin, and glucose homeostasis related enzyme activities in diabetic db/db mice. *Mol. Nutr. Food Res.*, 52(9), 995–1004.

Shadli, S., Alam, M., Haque, A., Rokeya, B., & Ali, L., (2014). Antihyperglycemic effect of *Zingiber officinale* roscoe bark in streptozotocin-induced type 2 diabetic model rats. *Int. J. Pharm. Pharm. Sci.*, 6(1), 711–716.

Shobha, R. I., Rajeshwari, C. U., & Andallu, B., (2013). Anti-peroxidative and anti-diabetic activities of aniseeds (*Pimpinella anisum* l.) and Identification of bioactive compounds. *Am. J. Phytomed. Clin. Ther.*, 1(5), 516–527.

Shobha, R., & Andallu, B., (2018). Antioxidant, anti-diabetic, and hypolipidemic effects of aniseeds (*Pimpinella anisum* L.): *In vitro* and *in vivo* studies. *J. Complement Med. Alt. Healthcare, 5*(2), 555–656.

Shukla, P., & Rangari, V., (2015). Enhancement of anti-diabetic activity of 4-hydroxyisoleucine in combination with natural bioavailability enhancers. *Int. J. Pharm. Pharm. Sci., 7*(4), 302–306.

Shukri, R., Mohamed, S., & Mustapha, N. M., (2010). Cloves protect the heart, liver, and lens of diabetic rats. *Food Chemistry, 122*(4), 1116–1121.

Singh, R. P., Gangadharappa, H. V., & Mruthunjaya, K., (2017). Cuminum cyminum: A popular spice: An updated review. *Pharmacogn. J., 9*(3), 292–301.

Somani, R. S., & Singhai, A. K., (2008). Hypoglycaemic and antidiabetic activities of seeds of *myristica fragrans* in normoglycaemic and alloxan-induced diabetic rats. *Asian J. Exp. Sci., 22*(1), 95–102.

Soni, R., Mehta, N. M., & Srivastava, D. N., (2013). Effect of ethanolic extract of *Cinnamomum tamala* leaves on wound healing in STZ induced diabetes in rats. *Asian J. Pharm. Clin. Res., 6*(4), 39–42.

Sreelatha, S., & Inbavalli, R., (2012). Antioxidant, antihyperglycemic, and antihyperlipidemic effects of *Coriandrum sativum* leaf and stem in alloxan-induced diabetic rats. *J. Food Sci., 77*(7), T119–23.

Sreenivasan, V., Kandasamy, C. S., Kumar, M. G., Prabhu, K. G., Arulraj, P., Johnson, J. S., Chander, U., & Venkatanarayanan, R., (2015). Review on different natural herbals associated with the anti-diabetic activity. *World Journal of Pharmacy and Pharmaceutical Sciences, 4*(8), 581–595.

Sultana, S., Khan, M. I., Rahman, H., Nurunnabi, A. S. M., & Afroz, R. D., (2014). Effects of ginger juice on blood glucose in alloxan-induced diabetes mellitus in rats. *J. Dhaka Med Coll., 23*(1), 14–17.

Sushruta, K., Satyanarayana, S., Srinivas, N., & Sekhar, J. R., (2006). Evaluation of the blood-glucose reducing effects of aqueous extracts of the selected umbelliferous fruits used in culinary practices. *Tropical Journal of Pharmaceutical Research, 5*(2), 613–617.

Tailang, M., Gupta, B. K., & Sharma, A., (2008). Antidiabetic activity of alcoholic extract of *Cinnamonum zeylanicum* leaves in alloxan induced diabetic rats. *People's Journal of Scientific Research, 1*, 9–11.

Thomson, M., Zainab, M. A. A., Khaled, K. A. Q., Lemia, H. S., & Muslim, A., (2007). Anti-diabetic and hypolipidaemic properties of garlic (*Allium sativum*) in streptozotocin-induced diabetic rats. *Int. J. Diabetes and Metabolism, 15*, 108–115.

Vanschoonbeek, K., Thomassen, B. J., Senden, J. M., Wodzig, W. K., & Van Loon, L. J., (2006). Cinnamon supplementation does not improve glycemic control in postmenopausal type 2 diabetes patients. *J. Nutr., 136*(4), 977–980.

Vavaiya, R. B., Patel, A., & Manek, R. A., (2012). Anti-diabetic activity of *Amomum subulatum* Roxb. Fruit constituents. *International Journal of Pharmaceutical Innovations, 2*(5), 50–63.

Veermanic, C., Pushpavali, G., & Pugalendi, K. V., (2008). Antihyperglycemic of *Cardispermum halicacabum* Linn. Leaf extract on STZ induced diabetic rats. *J. Appl. Biomed., 6*, 19–26.

Verspohl, E. J., Bauer, K., & Neddermann, E., (2005). Antidiabetic effect of *Cinnamomum cassia* and *Cinnamomum zeylanicum in vivo* and *in vitro*. *Phytother. Res., 19*(3), 203–206.

Winn-Deen, E. S., David, H., Sigler, G., & Chavez, R., (1988). Development of a direct assay for alpha-amylase. *Clin. Chem., 34*(10), 2005–2008.

World Health Organization, (2016). *Global Report on Diabetes*. WHO Press, World Health Organization, Geneva, Switzerland.

Yanardag, R., Bolkent, S., Tabakoglu-Oguz, A., & Ozsoy-Saçan, O., (2003). Effects of *Petroselinum crispum* extract on pancreatic b cells and blood glucose of streptozotocin-induced diabetic rats. *Biol. Pharm. Bull.*, *2*(8), 1206–1210.

Yarizade, A., Kumle, H. H., & Niazi, A., (2017). *In vitro* antidiabetic effects of Ferulaassa-foetida extracts through dipeptidyl peptidase IV and α-glucosidase inhibitory activity. *Asian J. Pharm. Clin. Res.*, *10*(5), 357–360.

Yogalakshmi, K., & Vaidehi, J., (2015). Impact of *Pimenta dioica* leaf extract on certain blood parameters in STZ induced diabetic wistar rats. *Int. J. of Modern Research and Reviews*, *3*(8), 739–743.

Younas, J., & Hussain, F., (2014). *In vitro* Antidiabetic evaluation of *Allium sativum* L. *Int. J. Chem. Biochemi. Sci.*, *5,* 22–25.

Zohrameena, S., Mujahid, M., Bagga, P., Khalid, M., Noorul, H., Nesar, A., & Saba, P., (2017). Medicinal uses and pharmacological activity of *Tamarindus indica. World J. Pharm. Sci.*, *5*(2), 121–133.

CHAPTER 25

Aloe Species: Chemical Composition and Therapeutic Uses in Diabetic Treatment

AMIT KUMAR SINGH,[1] RAMESH KUMAR,[1] B. UMARANI,[2] and ABHAY K. PANDEY[1]

Department of Biochemistry, University of Allahabad, Allahabad – 211002, Uttar Pradesh, India, E-mail: akpandey23@redifmail.com (A. K. Pandey)

2P.G. and Research Department of Zoology, Nehru Memorial College (Autonomous), Puthanampatti – 621 007, Tiruchirappalli, Tamil Nadu, India

ABSTRACT

Herbal medicine is being used around the globe and plays an essential role in several developed and developing countries as primary health care. The genus Aloe has about 400 species while some of them (*A. Vera, A. greatheadi, Aloe ferox,* and *Aloe arborescens*) are commercially used in homeopathic and Allopathic medicines for its remarkable pharmaceutical properties. These plants have potent natural sources which have health-promoting effect for mammals including human beings. Phytochemistry of the species of aloe plants reported more than 200 biologically active secondary metabolites with the ability of various biological activities like antidiabetic, antioxidant antidiabetic, hepatoprotective, and renoprotective activities. This review highlighted the diabetic properties of different species of genus *Aloe*.

25.1 INTRODUCTION

Aloe is a genus of family Asphodelaceae which consists of moreover 500 flowering succulent plants species (Stevens, 2001). The term aloe means alloeh (shining bitter substance), derived from the Arabic word (Tyler et al., 1976). The gel and dried leaf exudates of aloe species have been used as medicine since the ancient civilizations of the Greeks, Mediterranean people, and Egyptians (Trease and Evans, 1976). *Aloe* species have a wide variety of traditional usage and are also now used in modernized medicine in and around the world. Leaf exudates of aloe species have important sources of laxative drugs and are also used in the cosmetics properties including skincare creams, shampoos, and shaving creams (Leung, 1977) and also used in the treatment of skin diseases. The bitter agent of aloe exudates used in alcoholic beverages. The leaves and roots of *Aloe* species reported that contains many secondary metabolites belonging to alkaloids, flavonoids, glycoproteins, anthrones, pyrones, chromones, coumarins, naphthalenes, and anthraquinones (Dagne, 1996). This review described the medicinal usage of different species of Aloe especially in the treatment of diabetes.

25.2 DIABETIC IMPORTANCE AND SECONDARY METABOLITES FROM *ALOE* SPECIES

25.2.1 ALOE VERA

A. Vera is an herb with the source of a lot of secondary metabolites. This plant revealed the presence of numerous biologically active phytochemical substances (Chauhan et al., 2016). These substances are minerals, vitamins, enzymes, phenolic compounds, anthraquinones, lignin, sugars, sterols, salicylic acid amino acids, and saponins (Sharrif et al., 2011). Among these, anthraquinones are the best active ingredient with high medical values. *A. Vera* gel contains polysaccharides like acemannan which is actively worked on parenchymatous tissue. This plant contributed significantly towards number of biological activities because of their secondary metabolites (Xiong, 2002; Yang et al., 2000). According to Dagne et al. (2000) *A. Vera*, plant extract possessed multiple biological activities due to the synergistic action several phytocompounds. Through *in vitro* and *in vivo* studies, Aloe has been reported that gel of Aloe Vera lowering blood glucose level and inhibit expression of glucose transporter protein (GLUT4) (Kumar et al. (2011). Devaraj et al. (2008) reported that *A. Vera* gel complex reduced body fat mass, body weight and insulin resistance in pre-diabetes obese and early nontreated diabetic patients. A study conducted by Tanaka et al. (2006) reported that aloeemodin-8-O-glycoside from *A. Vera* gel reduced blood glucose levels of diabetic mice. According to Jain et al. (2010), *Aloe vera* gel has cardioprotective, antioxidant, and antidiabetic properties. *A. Vera* also has exhibit potent activity on separated rat pancreatic islets by increasing mitochondrial activity, and insulin levels while reducing the production of reactive oxygen species Rahimifard et al. (2013).

25.2.2 ALOE FEROX

Aloe ferox (Cape aloe), is widely distributed in South Asia, South Africa, Lesotho, Western, and Eastern Cape. It is a perennial shrub with reaching 2–3 m height, and has thick succulent leaves with brown spines on the margin. From ancient times, the leaf exudates of *A. ferox* have been used as medicine for the treatment of various diseases. *Aloe ferox* contains many bioactive volatile compounds including 1,3,6-octatriene, 2-heptanol, 2,4-decadien-1-ol, 3-cyclohexene-1-acetaldehyde, and -4-dimethyl, 1,3-cyclopentadiene, 1-methyl-2-heptanol), benzene, 1-methyl-2-(2-propenyl)-(3.78%), 1,4-cyclohexadiene, 1-methyl (2,5-dihydrotoluene), 3-carene and theaspirane. These compounds are reported to cure diabetes and its relative complications (Magwa et al., 2006). According to Loots et al. (2011) *A. ferox,* significantly increased insulin level and favorable recovered in ALP and HDL-C levels.

25.2.3 ALOE GREATHEADII

Botes et al. (2008) reported that the ethanol extract of *A. greatheadii* possess many groups of secondary metabolites including alkaloids, aldehydes, ketones, polyphenols, fatty acids, sterols, indoles, alkanes, pyrimidines, organic acids, dicarboxylic acids, and alcohols. These compounds are well known to possess the antidiabetic activity and cure related complications. Loots et al. (2011) observed an increased level of insulin and HDL and decreased level glucose and liver alkaline phosphatase (ALP) in *A. greatheadii* treated diabetic animal model. Further, *A. greatheadii* contains variety of antioxidant phytochemicals and thus it has a potent antidiabetic activity by lowering blood glucose level (Botes et al., 2008). Earlier, Loots et al. (2007) stated that *A. greatheadii* possess potential health benefits, including treating diabetes and antioxidant properties due to their high amount of flavonoid, indoles, and polyphenol contents.

25.2.4 ALOE ARBORESCENS

Aloe arborescens is one of the main varieties of Aloe used commercially around the world. It is cultivated in Brazil, Uruguay, and South Africa and some Asian countries. Beppu et al. (1993) reported that powder of *A. arborescens* activate beta cells in pancreas and significantly reduced the blood glucose level in mice. However, no scientific reports are available on the phytochemical profile of this plant. Thus, it is needed to conduct thorough phytochemical and clinical studies in *A. arborescens.*

25.2.5 ALOE NYERIENSIS

Aloe nyeriensis is one of the species of genus aloe. It is interesting to note that it is endemic to Kenya. It grows 1.5 meters long and packed with red flowers. It grows mostly on rocky soils. It was reported that the compounds isolated viz., Nataloe-emodin-2-O-Glc and Nataloe-emodin (anthraquinones), Nataloin (anthtrone), Aloenin, Aloenin aglycone (pyrones) and -2‖-*p*-coumaroyl ester (Aloenin) from *A. nyeriensis* possessed significant antidiabetic activity (Conner et al., 1987).

25.2.6 ALOE EXCELSA

Aloe excelsa, is an arborescent aloe. It is also known as the Zimbabwe Aloe which grows 5–6 meters large and reaches tree. It is single-stemmed and all but the lowest part of the trunk is swathed in the remains of dead leaves (Gundidza, 2001). According to Coopoosamy and Magwa (2006), two phytochemicals viz., 1,8-dihydroxy-3-hydromethyl-9,10-antracenedione and 10-*C*-b-D-glucopyranosyl-1,8-dihydroxymethyl-9-anthracenone isolated from leaf extract of *A. excelsa* have a potent antidiabetic activity. The leaf extract of *A. excelsa* possessed significant activity on reducing blood glucose levels in diabetic rats (Gundidza et al., 2005).

25.3 CONCLUSION

The Aloe species exhibits many medicinal uses including hypoglycemic and antidiabetic activities. Thus, it is a multipurpose medicinal agent and more controlled studies are required to prove the effectiveness of *Aloe* species under various conditions to develop diabetic medicine and the results may help to develop new antidiabetic drugs.

ACKNOWLEDGMENTS

AKS and RK acknowledge financial support from CSIR New Delhi in the form of a Junior Research Fellowship. All the authors also acknowledge UGC-SAP and DST-FIST facilities of the Biochemistry Department, University of Allahabad, Allahabad, India.

KEYWORDS

- *Aloe excelsa*
- *Aloe* species
- **anthraquinones**
- **antidiabetic activity**
- **bioactive constituents**
- **dicarboxylic acids**

REFERENCES

Beppu, H., Yohichi, N., & Keisuke, F., (1993). Hypoglycaemic and antidiabetic effects in mice of *Aloe arborescens* Miller var. *natalensis* Berger. *Phytotherapy Research, 7,* 37–42.

Botes, L., Van der, W. F., & Du Toit, L., (2008). Phytochemical contents and antioxidant capacities of two *Aloe greatheadii* var. davyana extracts. *Molecules, 13*(9), 2169–80.

Chouhan, A., Karma, A., Artani, N., & Parihar, D., (2016). Overview on cancer: Role of medicinal plants in its treatment. *World Journal of Pharm. and Pharmaceutical Sci., 5,* 185–207.

Conner, J. M., Gray, A. I., Reynolds, T., & Waterman, P. G., (1987). Anthraquinone, anthrone and phenylpyrone components of *Aloe nyeriensis* var. kedongensis leaf exudates. *Phytochemistry, 26*(11), 2995–2997.

Coopoosamy, R. M., & Magwa, M. L., (2006). Antibacterial activity of aloe emodin and Aloin a isolated from *Aloe excels*. *African Journal of Biotechnology, 5*(11), 1092–1094.

Council of Europe, (1981). *Flavoring Substances and Natural Resources of Flavorings* (pp. 6, 9.). Maisonneuve, Moulins-les-Metz, France.

Dagne, E., (1996). Review of the chemistry of Aloe of Africa. *Bull. Chem. Soc. Ethiop., 10*(1), 89–103.

Dagne, E., Bisrat, D., Viljoen, A., & Van, B. E., (2000). Chemistry of *Aloe* species. *Curr. Org. Chem., 4,* 1055–1078.

Devaraj, S., Jialal, R., Jialal, I., & Rockwood, R., (2008). A pilot randomized placebo-controlled trial of 2 *Aloe Vera* supplements in patients with pre-diabetes/metabolic syndrome. *Planta. Med., 74,* SL77.

Gundidza, H., (2001). *Aloe excelsa. Personal Communication.* Harare.

Gundidza, M., Masuku, S., Humphrey, G., & Magwa, M. L., (2005). Anti-diabetic activity of Aloe excels. *Cent. Afr. J. Med., 51*(11/12), 115–120.

Jain, N., Vijayaraghavan, R., Pant, S. C., Lomash, V., & Ali, M., (2010). *Aloe Vera* gel alleviates cardiotoxicity in streptozocin-induced diabetes in rats. *J. Pharm. Pharmacol., 62,* 115–123.

Kumar, R., Sharma, B., Tomar, N. R., Roy, P., Gupta, A. K., & Kumar, A., (2011). *In vivo* evaluation of hypoglycemic activity of Aloe spp. and identification of its mode of action on GLUT-4 gene expression *in vitro. Appl. Biochem. Biotechnol., 164,* 1246–1256.

Leung, A. Y., (1977). *Aloe Vera* in cosmetics. *Excelsa, 8,* 65–68.

Lisa, B., Francois, H., Van Der, W., & Du Toit, L., (2008). Phytochemical contents and antioxidant capacities of two *Aloe greatheadii* var. *davyana* extracts. *Molecules, 13,* 2169–2180.

Loots, D. T., Marlien, P., Shahidul, I. Md., & Lisa, B., (2011). Antidiabetic effect of *Aloe ferox* and *Aloe greatheadii* var. Davyana leaf gel extracts in a low dose streptozotocin diabetic rat model. *S. Afr. J. Science, 107*(7/8).

Loots, D. T., Van Der Westhuizen, F. H., & Botes, L., (2007). *Aloe ferox* leaf gel phytochemical content, antioxidant capacity, and possible health benefits. *J. Agric. Food Chem., 55,* 6891–6896.

Magwa, M. L., Gundidza, M., Coopoosamy, R. M., & Mayekiso, B., (2006). Chemical composition of volatile constituents from the leaves of *Aloe ferox. African Journal of Biotechnology, 5*(18), 1652–1654.

Rahimifard, M., Navaei-Nigjeh, M., & Mahroui, N., (2013). Improvement in the function of isolated rat pancreatic islets through reduction of oxidative stress using traditional Iranian medicine. *Cell J., 16,* 147–163.

Sharrif, M. M., & Verma, S. K., (2011). *Aloe Vera* their chemicals composition and applications: A review. *Int. J. Biol. Med. Res., 2*(1), 466–471.

Stevens, P. F. (2001). Angiosperm Phylogeny Website. Version 14 http://www.mobot.org/MOBOT/research/APweb/ (accessed on 20 February 2020).

Tanaka, M., Misawa, E., & Ito, Y., (2006). Identification of five phytosterols from Aloe Vera gel as anti-diabetic compounds. *Biol. Pharm. Bull., 29,* 1418–1422.

Trease, G. E., & Evans, W. C., (1976). *Pharmacognosy* (12th edn., p. 404). Bailliere Tindall, London.

Tyler, V. E., Brady, L. R., & Robbers, J. E., (1976). *Pharmacognosy* (7th edn.) Lee and Febiger: Philadelphia.

Xiong, Y. Q., (2002). *Aloe* (pp. 53–54). Agricultural Press, Beijing, China.

Yang, Q. Y., & Ma, B. S., (2000). *Aloe*, the best health food in 21st century. Shandong. *Food Ferment, 3,* 40–42.

Index

α

α-amylase inhibition, 323, 324
α-glucosidase, 307, 311, 316, 320, 323, 324
   inhibition, 307, 311, 316, 320, 323

β

β-carotene, 201, 211
β-cells, 239, 240, 270, 288, 289
β-glucoronidase, 98
β-glucosidase, 202

A

abdominal
   cavity, 38
   fullness, 306
abnormal cells, 4, 141
acanthaceae, 107, 130, 252
acidic phase, 254
acridine orange (AO), 119, 121, 122, 124, 130, 131, 135–137, 153, 156, 162, 163
activating transcription factor 3 (ATF3), 187
active
   compounds, 34, 46, 59, 188, 199, 201, 211, 333
   ingredient, 53, 96, 100, 199, 211, 344
   kinase conformation, 23
adenocarcinoma, 49, 51, 82, 153, 204, 206
adenosine tri-phosphate (ATP), 22–25, 172
Adhatoda zeylanica, 250, 252
adipose tissue, 227, 244
Aegle marmelous, 177–185
Aka coralliphaga, 48
Akebia trifoliate, 48
albino rat model, 275, 276, 278, 283, 284
alkaline
   phase, 254
   phosphatase (ALP), 15, 207, 345
alkaloids, 46, 47, 61, 105, 108, 110, 157, 158, 169, 238, 266, 293, 296, 344, 345
Allium sativum, 106, 142, 143, 145, 307, 327, 333, 335
alloxan, 224, 275–280, 307–310, 319, 329, 333

Aloe species, 344, 346, 347
   arborescens, 343, 346
   excelsa, 346, 347
   ferox, 343, 345
   nyeriensis, 346
   vera, 107, 344
alpha
   fetoprotein (AFP), 16
   glucosidase activities, 310
Amaryllidaceae, 142, 143, 145
amentoflavone, 195
American Diabetes Association (ADA), 276, 306
Amino acids, 22, 24, 172, 290, 333, 344
Amomum
   subulatum, 311
   villosum, 186
   xanthioides, 177, 179, 184, 186
analysis of variance (ANOVA), 237, 242, 244, 254
Andrographis
   alata, 250, 252
   paniculata, 107, 129, 130, 250, 252
angiogenesis, 60, 61, 69, 86, 88, 148, 160, 169, 186
anthocyanidin pigment, 108
anthocyanins, 67, 110
anthraquinones, 107, 344, 346, 347
anti-aging properties, 154, 168
anti-angiogenic
   effect, 67
   property, 49
anti-apoptotic proteins, 67
antiarrhythmic effect, 89
anticancer, 30, 37, 43–53, 59–63, 65–72, 79, 83–85, 87, 96–99, 104–107, 109–111, 117, 118, 125, 130, 135, 141–150, 153, 154, 157, 158, 160, 161, 164, 169, 177–188, 191, 192, 194, 196, 199–201, 206, 208–212, 306, 333
   activity, 47, 79, 98, 99, 105, 107, 111, 118, 130, 153–155, 157, 158, 160, 161, 164, 184, 185, 191, 192, 196, 200

cell lines procurement, 155
cell viability assay, 156
evaluation, 193
agents, 45, 46, 65, 110
compounds, 45
drug, 44, 45, 142, 143, 164, 177, 191
  activity, 109
  discovery, 141–143, 150
  properties, 30
effects, 186
properties, 49
spices, 144, 145
  Allium sativum, 145
  Capsicum annuum, 146
  Carum carvi, 146
  Cinnamomum zeylanicum, 147
  Coriandrum sativum, 147
  Crocus sativus, 147
  Cuminum cyminum, 148
  Curcuma longa, 148
  Elettaria cardamomum, 148
  Foeniculum vulgare, 148
  Myristica fragrans, 149
  Piper nigrum, 149
  Syzygium aromaticum, 149
  Trigonella foenum-graecum, 150
anticancerogenic
  agents, 70
  properties, 66
anti-cancerous
  agents, 111
  effect, 125, 126
  property, 207
anticarcinogenic, 100, 177, 178, 180–182, 185–188
  activities, 99
  effects, 82
  properties, 148
anti-diabetic, 96, 219, 220, 224, 226–229, 233, 234, 236–238, 245, 246, 249, 253, 260, 261, 263, 270–272, 276–278, 280, 283, 284, 290, 291, 294–296, 302, 305–310, 316, 333–335, 343, 345–347
  activity, 238, 239, 245, 246, 253, 258, 260, 261, 263, 265, 270–272, 276–278, 280, 283, 284, 291, 294, 295, 305, 307, 309, 310, 316, 333, 334, 345–347
  agents, 270, 308
  effect, 226, 284

potency, 309
potentials, 307
properties, 226, 296, 302, 305, 308–310, 333, 334, 345
antiglycative activities, 49
anti-hyperglycemic
  activity, 276
  potential, 224
  properties, 227
antihyperlipidemic
  activity, 236, 242, 246
  properties, 201
anti-inflammatory
  agent, 227
  potential, 146
antilipidmtic properties, 219
anti-malignant effect, 169
anti-metastatic effect, 52
antimicrobial activity, 333, 335
anti-mycobacterial activity, 48
antineoplastic
  agent, 122
  drugs, 47
anti-nephrotoxic activity, 306
antioxidant, 70, 106, 157, 211
  activity, 66, 69, 70, 237, 267, 306, 310, 333
  effects, 98
  properties, 146, 154, 211, 267, 345
anti-proliferation, 194, 196, 205
  activity, 48, 66, 69, 137, 138, 193
  compounds, 117
  effect, 67, 107, 129
anti-tumor
  active components, 43
  activity, 83, 87, 96–98, 101, 179, 185
  agents, 44, 45
  drugs, 44–47
  effects, 53, 67, 83, 84, 95, 135, 147, 186
  leads, 45
  properties, 83, 86
anti-tumorigenic property, 210
apigenin, 49, 64, 96
Apocynaceae, 79, 81, 108
apoptosis, 49, 52, 61–63, 67, 69, 71, 84–88, 95, 96, 100, 106, 110, 111, 117–120, 122–124, 126, 130–138, 147, 156, 160, 162–164, 186, 187, 204–207, 209, 210
apoptosome complex, 205

Index

apoptotic
   activities, 99
   cells, 121, 130, 131, 136, 163
   morphological changes, 130, 133–136
   morphology, 130, 206
   traits, 124
aristolochic acids, 5
aromatic
   qualities, 70
   rings, 65, 66
arrhythmias, 80
ascorbic acid, 187, 188
asymptomatic patients, 3, 12
atherosclerosis, 66, 245
athymic nude mice cancer, 210
ATP type I–VI inhibitors, 24, 25

B

B-cell, 15, 16, 290
Benedict test, 158
berberine, 49
beta-ionone, 180
betulinic acid, 49
biflavones, 195
bioactive
   component, 309, 310
   compounds, 46, 95, 100, 101, 155, 159, 168, 169, 172, 174, 177, 178, 185–187, 199–201, 203, 204, 207, 211, 212, 220, 235, 289, 293, 295
   identification, 235
   constituents, 307, 335, 347
   secondary metabolites, 46
   terpenoids, 226
bioassays, 35, 37
biochemical
   analysis, 253, 284
   constituent, 310, 335
   estimation, 266, 286
   parameters, 204, 237, 254
   properties, 30
biochemistry, 5, 130
bio-components, 333, 334
biological
   gradient, 10
   marker, 3, 10
biomarkers, 3, 6, 8, 10–12, 16, 17
   alpha-fetoprotein (AFP), 16

calcitonin, 16
carcinoembryonic antigen (CEA), 13, 14
fecal occult blood (FOB), 9, 14
human chorionic gonadotropin (HCG), 17
lactate dehydrogenase, 17
papanicolaou test, 12
prostate-specific antigen (PSA)
   screening, 13
urinary homovanillic, 16
vanillylmandelic acid (VMA), 16
Biophytum sensitivum, 191, 192, 196, 275–277, 279, 280
biopsy, 6–8, 13, 17
biosensing, 266
biosynthesis, 263, 272
biosynthetic pathways, 47
bladder habits, 4
blastoma, 4
blood
   brain-barrier (BBB), 31, 32, 36, 88, 90, 120, 124, 125
   glucose, 221, 224, 233, 236, 239–242, 253, 258, 259, 263, 265, 270, 271, 275, 277–280, 283–287, 289, 290, 306, 308–316, 318–331, 333, 344–346
body mass index, 5
boiling point, 202
brain-derived neurotrophic factor, 88
Bridelia retusa, 279
Brownian motion, 35
buccal method, 38
butyric acid, 185

C

caffeic acid, 52, 148, 180, 185–187, 334
Camptotheca acuminata, 48
camptothecin, 83
cancer
   cell pathways, 68
   definition, 60
   drug
      bioactive compounds, 46
      development stages, 36
      properties, 40
   early detection, 5
      bowel cancer, 9
      breast cancer, 6
      cervical cancer, 8

early diagnosis, 6
ovarian cancer, 8
prostate cancer, 7
screening, 9, 12
stem cells, 62
therapy, 85
treatment, 60
canthaxanthin, 211
capsaicinoids, 146
Capsicum annum, 142, 143
carbohydrates, 14, 158, 169, 201, 237, 240, 242, 275, 276, 283, 284, 290, 293, 296
carbon
skeletons, 69
tetrachloride, 206
carbonic acid, 228
methyl phenyl ester, 226, 227
carcinogen detoxification, 71
carcinogenesis, 64, 65, 68, 88, 206, 208
carcinogenic agents, 208
carcinogenicity, 207, 209
carcinoma, 4, 13, 14, 16, 49, 51, 52, 63, 83, 96, 97, 100, 107, 205–207, 209
carderolytes, 83
cardiac
arrhythmia, 84
glycosides, 80, 82, 85–87, 89
toxicity, 60
cardiotonic glycosides, 85
carotenes, 70, 149
carotenoids, 70, 202, 208, 211
terpenoids, 70, 72
Carum carvi, 99, 142, 143, 146, 317
caryophyllene, 109
catalytic therapy, 200
catechins, 67
catechol moiety, 105
catecholamines, 16
Catharanthus sroseus, 108
cationic dye, 131
cell
adhesion, 13
apoptosis, 136
culture conditions, 119
death, 4, 46, 47, 49, 86, 111, 118, 120, 194, 205, 206, 210
division, 4, 44, 49, 60, 103, 205

line, 50, 51, 64, 108, 117–119, 122, 137, 162, 164, 191, 192, 204, 205, 207, 208, 210
assays, 37
conditions, 119
membrane, 70
cleavage, 163
proliferation, 52, 61, 69, 71, 85, 87, 118, 147, 169, 199, 207, 209
propagation, 46
signaling, 87, 99, 110
surface adhesion molecules, 109
cellular
barriers, 35
lipid membranes, 87
membrane, 69
metabolic activities, 21
proliferations, 299
signaling pathways, 49
topoisomerase, 147, 205
Centaur Immuno Assay Instrument, 236, 254, 266, 286
cerebrospinal fluid, 12
cervical
intraepithelial neoplasia (CIN), 12
screening, 12
smear, 12
chalcones, 67
chemical
carcinogens, 5
compounds, 98, 155, 219, 227
chemiluminescence, 236, 254, 266, 286
immune assay (CLIA), 236, 254, 266, 286
chemiluminescent technology, 223
chemokines, 109
chemo-preventive
activity, 63
agents, 208
effects, 99
potentials, 98
properties, 208
chemoprotective agent, 108
chemotherapeutic
agent, 108
compounds, 137
drug resistance, 47
chemotherapy, 30, 45, 60, 61, 65, 82, 95, 103, 104, 108, 111, 135, 154, 178, 192, 200

Index

chemotypes, 47
chiral counterparts, 35
chitosan (CS), 249–252, 254–261
  amino groups, 254
  nanoparticles (CNPs), 249–261
  preparation, 250
chloroform, 154–157, 159, 161–164, 194, 277, 285–287, 324
  extract, 157, 160, 161, 163
chlorogenic acid, 187
chmnotherapy, 43
cholangiocarcinoma cells, 50, 53
chromatin, 121, 130, 131, 137, 163
chromatographic analysis, 89
chronic
  cancers, 147
  diseases, 46, 61, 95, 284, 296
  hepatitis, 207
  inflammatory abnormalities, 203
  lymphocytic leukemia (CLL), 125
  myelogenous leukemia (CML), 22, 209
cinnamic acid, 66
Cinnamomum zeylanicum, 142, 143, 147, 323
cirrhosis, 15, 207
cis-jasmone (CJ), 117, 118, 120, 123–125
citric acid, 181
Citrus
  limon, 177, 178, 184, 186, 188
  sinensis, 177–185
clinical trials, 11, 30, 38–40, 44, 65, 79, 82, 117, 149, 211, 221
coherence, 11
colchicine, 50
colon cancer, 8, 14, 50–52, 62, 64, 71, 96, 99–101, 106, 108, 149, 186
colonoscopy, 8, 14
colorectal cancer, 49, 52, 146, 187, 205
combretastatin A-4, 48
Combretum caffrum, 48
Committee for the Permission and Control of Supervision on Experimental Animals (CPCSEA), 221, 236, 253
Coriandrum sativum, 142, 143, 147, 326, 333
crocetin, 96, 199–203, 206–212
crocin, 147, 199, 202, 204–212
Crocus sativus, 142, 143, 147, 200, 212
crystalline materials, 255

C-terminal lobe, 22, 23
Cuminum cyminum, 142, 143, 148, 313, 334
cupressuflavone, 195
Curcuma longa, 48, 62, 64, 68, 96, 99, 142, 143, 148, 179–184, 186–188, 200, 309, 321
curcumin, 48, 50, 62, 64, 65, 68, 69, 96, 99, 109, 148, 181, 187, 200, 309, 310, 322
curcuminoids, 66, 68, 333
cyanidin glycosides, 50
Cyathea nilgiriensis, 233–241, 243, 245, 246, 293–298, 301, 302
cyclin dependant kinases (CDK), 49, 161
cyclocurcumin, 68
cyclooxygenase, 87, 148, 157, 206
  COX-2, 87, 109, 148
cysteine, 25, 26, 106, 333
cystoscopy, 13
cytokines, 69, 109, 147, 206
cytology screening methods, 9
cytoplasm, 131, 132, 136, 163, 288
cytoplasmic organization, 156
cytoprotective proteins, 208
cytostatic activity, 86
cytotoxic
  activity, 45, 64, 66, 96, 98, 99, 108, 126, 135, 138, 162, 164, 191, 209
  effect, 82, 106, 120, 123, 125, 135, 153, 161, 191, 194, 196
cytotoxicity, 85, 86, 107, 109, 118, 120, 122, 123, 125, 130, 135, 156, 160, 161, 185, 193, 194, 209

D

data-mining, 178
deacetylation, 249, 251, 256
deoxyribonucleic acid (DNA), 8, 12, 47, 49, 60, 70, 71, 83, 104, 119–122, 124, 129–131, 136, 137, 146, 147, 204–207, 209–211, 308
deproteinization, 251
dermatology, 37
detoxification
  enzymes, 208
  systems, 208
diabetes mellitus, 219, 234, 240, 246, 259, 265, 270, 271, 275, 280, 283–285, 289, 291, 293, 294, 301, 305, 306

diabetic
    albino rat models, 234
    compounds, 294
    control group, 270, 271, 286
    obese mice, 224
    rats, 224, 233, 236, 239–241, 243–245,
        253, 254, 258, 259, 270, 275, 277–280,
        283, 287–290, 307–310, 314, 319, 322,
        323, 329, 346
    targets, 294
diallyl disulfide (DADS), 146
diarrhea, 98, 104, 204, 306
DICARBOXYLIC ACID, 129–137, 202,
    226, 345, 347
diethylnitrosamine (DEN), 206, 207
digital rectal examination (DRE), 7, 13
digitoxin, 85
digoxin, 80
    molecules, 84
dimethyl sulfoxide (DMSO), 120, 156, 192,
    193, 278
diterpenoids, 70
diverticulosis, 14
docking, 167–169, 171, 172, 174, 175,
    293–296, 298–300, 302
    analysis, 295
    studies, 295, 298
double contrast barium enema, 14
doxorubicin, 60, 83
drug
    bioavailability, 32, 35
    carrier, 249, 250
    development, 11, 17, 26, 29, 30, 36–38,
        44, 45
    discovery
        natural products, 45
    molecule, 35
    properties
        lipophilicity, 32
        other properties, 36
        physicochemical properties, 34
        PK, 34
        structural properties, 31, 32
ductal carcinoma *in situ* (DCIS), 49

E

Ehrlich ascites carcinoma (EAC), 185, 208
eicosanoic acid, 335

electrophoresis, 210
electrostatic interactions, 295
Elettaria cardamomum, 142, 143, 148
ellagic acid, 66, 67, 200
ellagitannins, 66
embryonic tissue, 4
encapsulation efficiency, 252
encephalopathy, 276
endometriosis, 8
endophytic fungi, 44
endoscopy, 6, 14
enzyme-linked immunosorbent assay
    (ELISA), 305, 307, 311, 313, 320, 322,
    328–330
epidermal growth factor receptor (EGFR),
    22, 51, 52, 210
epigallocatechin-3-gallate (EGCG), 49, 67
epithelial
    cancer cells, 106
    cells, 4, 80
    mesenchymal trans-differentiation
        (EMT), 207
esophageal cancer, 15, 62
essential oils, 105, 109, 186
estrogen receptors, 210
ethanolic fraction, 209
ethidium bromide (EB), 130, 131, 135–137,
    153, 156, 162, 163
eukaryotic genes, 22
extrinsic pathways, 205

F

fatty acids, 149, 201, 345
fecal occult blood (FOB), 9, 14
    test (FOBT), 14
ferulic acid, 181, 185, 187, 309, 333
fetal lung fibroblasts, 204
fibroblast growth factor-2 (FGF-2), 84–88
field emission scanning electron microscope
    (FE-SEM), 253
flatulence, 98, 306
flavanoids, 149, 157
flavanones, 67
flavone, 108, 157, 159, 167–169, 171–175,
    309
flavonoids, 46, 49, 61, 62, 65–67, 80,
    105–107, 136, 153, 154, 157, 158, 161,
    164, 167–169, 174, 195, 201, 226, 227,

237, 238, 240, 245, 266, 289, 293, 296, 305, 309, 310, 333, 334, 344
flavonols, 67
Florence fennel, 148
fluorescence
  microphotograph, 133–135
  microscope, 121, 131
fluorescent microscope, 137, 156, 157
Foeniculum vulgare, 142, 143, 148, 315
folk medicine, 110
Food and Drug Administration (FDA), 11, 13, 21, 22, 25, 82
formylpeptide (FMLP), 88
Fourier transform infrared spectroscopy (FTIR)
  analysis, 249, 253, 256, 265, 267
  spectroscopy, 253
free
  fatty acid, 244
  radicals, 70, 105, 136, 146, 206, 211, 242
furocoumarins, 105

G

gallic acid, 147
gallotanins, 237
gamma-seminoprotein, 13
gas chromatography-mass spectrometry (GC-MS), 138, 153, 155, 157, 159, 164, 167–170, 174, 175, 223, 226–229, 237–239, 293, 295–298
  analysis, 155
    components identification, 155
gastric cancer, 14
gastrointestinal (GI), 14, 15, 33, 36, 60, 186, 220
gelation, 249, 251, 254
genistein, 50, 97, 99, 158
genomics, 12
genotoxicity, 209
germ cell tumors, 16, 17
glioblastoma, 83
glucokinase (GK), 240, 290
glucometer, 314, 317, 319, 321, 322, 324, 326–330
gluconeogenesis, 227, 240, 289, 290
glucose oxidase
  method (GOD), 222, 223, 236, 254, 305, 307, 314, 315, 317, 319, 320, 323–327

peroxidase, 236, 254, 305, 314, 317, 324, 325
glucosides, 49
glutathione (GSH), 206, 207, 209, 310
  peroxidase (GPx), 206, 207
  S-transferase (GST), 206, 207
glycoalbumin, 290, 291
glycogen, 241, 242, 260, 275, 277–280, 289, 334
glycogenesis, 289
glycogenolysis, 239, 289
glyconeogenesis, 239, 289
glycoprotein, 13–16, 309, 344
glycosides, 67, 79, 80, 82–87, 89, 93, 107, 158, 169, 203, 237, 238, 293, 296
glycosylated hemoglobin, 236, 253, 254, 266, 277, 286
glycosylphosphatidylinositol (GPI), 13
gold nanoparticles (AuNPs), 270, 271
growth factors, 69, 109
Gymnema sylvestre, 249, 250, 252

H

halofuginone, 50
Hashimoto's thyroiditis, 89
heat shock protein 90 (Hsp 90), 167–169, 171–175
HeLa cells, 98, 206, 211
helicases, 205
Helicobacter pylori, 5
hematological parameters, 204
hematopoietic cancer, 85
hemoglobin, 236, 240, 289, 308
hemorrhoids, 14
hepatic cancer, 207
hepatocellular carcinoma (HCC), 15, 16, 50, 191, 192, 194
hepatocytes, 71
hepatoprotective activities, 343
herbal
  medicine, 95, 196, 234, 276, 343
  plant, 192, 196, 234, 275, 291, 294, 305
hespertin, 51
heterocycle, 65, 67
heterocyclic system, 24
heterogeneous metabolic disorder, 284
heterogenic group, 66

high density lipoproteins (HDL), 233, 237, 244, 245, 345
histoarchitecture, 288
histone deacetylase (HDAC), 48
homeostasis, 223, 227, 234, 241, 290
    model assessment (HOMA), 223, 225
        insulin resistance, 223, 225
homovanillic acid (HVA), 16
hormone
    regulation, 210
    therapy, 8
human
    breast tumor cell line, 50
    chorionic gonadotropin (hCG), 17
    intestinal absorbance (HIA), 120, 124, 125
    papillomavirus (HPV), 5, 8, 12, 44
hydroalcoholic leaf extract, 105
hydrocarbons, 70
hydrogen
    bond, 172, 299
        acceptors (HBA), 120, 124, 125, 171, 172
        donors (HBD), 120, 124, 125, 171, 172
    bonding, 35, 69, 256
hydrophobicity, 171, 298
hydroxybenzoic acid, 66
hydroxycinnamic acids, 66, 110
hydroxyl groups, 65, 66, 69, 105
hypercholesterolemia, 243, 244
hyperchromic nucleus, 288
hyperglycemia, 221, 239, 240, 270, 275, 276, 289, 290, 306
hyperglycemic effect, 219, 224
hyperinsulinemia, 219, 220, 224, 229
hyperlipidemia, 227, 242, 243
hyperphosphorylation, 62
hyperplasia, 87
hypertension, 306
hypertriglyceridemia, 244
hypoglycemic, 227, 233, 276, 278–280, 284, 296, 306
    activity, 280, 294, 307
    agents, 234, 294
    conditions, 224
hypolipidemic activity, 245
hypothyroidism, 17

I

ibrutinib inhibitors, 26
immune modulators, 65, 72
in silico
    docking, 171
    models, 298, 302
    pharmacokinetic evaluation, 120
in vitro
    analysis, 334
    conditions, 130
incubation, 85, 120, 131, 132
inflammation, 46, 48, 66, 87, 88, 100, 109, 111, 147
inflammatory
    agents, 109
    bowel disease, 14
    response, 136, 206, 207
inhibition
    assay, 305, 307, 311, 314, 316, 318, 320, 323, 324
    concentration (IC), 34, 48, 87, 100, 117, 119, 120, 123–125, 156, 157, 161, 194, 309
inhibitors, 21–26, 48, 161, 299
inhibitory activity, 100, 309
inorganic acids, 5
inotropic properties, 80
insomnia, 37
insulin, 67, 219–221, 223–225, 227, 233, 234, 239–241, 243, 259, 270, 271, 275, 278, 279, 283, 284, 286, 287, 289, 290, 294, 299, 306, 308–315, 317–320, 329–331, 334, 344, 345
    like growth factor 2 (IGF-II), 67
inter-cellular junction tightness, 36
interleukin-1 (IL-1), 88, 206
intestinal
    absorption, 33, 34
    tumor cells, 99
intoxication, 82, 89
intracellular signals, 205
intranasal method, 38
intraperitoneal method (IP), 38
intravenous (IV), 25, 35, 38, 82, 236, 241, 242, 244, 253, 259, 260, 271, 278, 285, 287
intrinsic pathway, 100, 205
irinotecan, 48, 65
isoflavones, 67

Index

J

jasmonates, 117, 118, 120, 123–126
   antiproliferative activity measurement, 120
   cytotoxic effect, 123
   pharmacokinetic evaluation, 124
jasmonic acid (JA), 117–125
   apoptotic traits study, 120

K

kaempferol, 51, 64, 97, 100
kallikrein-3 (KLK3), 13
Kaposi sarcoma-associated herpesvirus (KSHV), 5
kinase
   activation, 24
   activity, 24, 209, 299
   domain, 22, 25
   drug discovery, 22
   families, 26
   functioning, 22
   inhibitors, 21, 23, 25
   proteins, 22
   structure, 22
   targets, 26

L

lactate dehydrogenase LDHA, 17, 206
Lamarckian genetic algorithm, 295
laxative drugs, 344
Lee index, 222
leguminosae, 168
leukemia, 4, 5, 16, 22, 44, 49–52, 80, 85, 86, 108, 118, 149, 204, 208
ligand molecule, 172
lipid, 69, 71, 242
   peroxidation (LPO), 206–209, 242, 310, 318
   sensors, 226
lipolysis, 227, 242, 244
lipophilicity, 29, 31–33, 40
liver
   cancer, 4, 50, 63, 147, 191–194, 196, 207
   glycogen, 237, 240, 241, 253, 254, 259, 275, 277–280, 289
low-density lipid (LDL), 233, 237, 244, 245
lung carcinogenesis, 62
luteolin, 161, 195, 309, 333, 334
lycopene, 64, 72, 97, 100, 211
lymphatic systems, 60
lymphoma, 4, 5, 15, 16, 30, 49, 50, 52, 85, 118, 208
lypoxygenase, 157
lysophosphatidic acid, 8

M

malondialdehyde (MDA), 64, 105, 206, 207, 210
mammography, 6, 7, 9
mass
   spectra, 155
   spectrometer, 223
   spectrometry, 226
   spectrum, 168, 235
medicinal plants, 53, 61, 79, 96, 97, 101, 103–105, 111, 118, 153, 188, 192, 194, 220, 276, 280, 294
   active compounds, 98
      carum carvi seeds, 99
      cuminum cyminum seeds (ZEERA), 98
      curcumin from turmeric, 99
      fesitin from berries, 99
      genistein from soybean, 99
      gingerol from gingers, 100
      kaempferol from beverages, 100
      lycopene from tomato, 100
      nigella sativa seeds (black cumin), 98
   anti-cancerous agents, 105
      Aegle marmelos, 105
      Allium cepa, 106
      Allium sativum, 106
      Aloe vera, 107
      Andrographis paniculata, 107
      Catharanthus roseus, 108
      Cedrus deodara, 108
      Curcuma longa, 109
      Vitis vinifera, 110
      Withania somnifera, 110
      Zingiber officinale, 110
melanoma
   cancers, 50
   cells, 86
membrane
   fluidity, 88
   transporter expression, 36

menopause, 7
mesothelin, 8
metabolic
   disorders, 177, 227
   enzymes, 33, 34
metalloenzymes, 136
metastasis, 14, 61, 63, 64, 69, 88, 97, 104, 109, 148, 186
metastasize, 44, 199
methoxy groups, 69
methyl
   ester, 105, 157, 160
   jasmonate (MJ), 118, 120, 123–125
   thiazol tetrazolium (MTT), 83, 120, 123, 125, 153, 156, 161, 162, 185, 193, 194
methylcholanthrene (MCA), 206
Michigan cancer foundation-7 (MCF-7), 50, 153–156, 161–164, 210
microorganisms, 47, 129
microtubules, 49, 147
microvascular complications, 289
microviscosity, 88
mitochondria, 86
mitochondrial
   activity, 345
   death pathway, 186
   membrane, 205
   reduction, 83
molecular
   design suit (MDS), 120, 124
   docking, 167–169, 171, 175, 293, 294, 298
   formula (MF), 155, 158–160, 168, 171
   linkages, 254
   mechanism, 199, 204
   profiling, 3, 11
   properties, 171
   scaffolds, 171
   signaling, 200
   structure (MS), 34, 89, 96, 97, 138, 153, 157–160, 164, 167–170, 174, 175, 223, 226–228, 235, 237–239, 293, 295–298
   targets, 21, 26, 45
   weight (MW), 22, 31, 33, 35, 80, 120, 125, 155, 158–160, 168, 171, 202, 203, 223
monoclonal antibodies, 16, 47
mononuclear cell (MNCs), 86
monosodium glutamate (MSG), 219–222, 224, 225, 227, 229
monoterpenoids, 70
mutagenesis, 87
mutant tumors, 22
myeloperoxidase activity, 206
myopathy, 276
Myristica fragrans, 142, 143, 149, 329

N

nanomaterials, 264, 272
nanomedicine, 250
nanoparticles (NPs), 249–261, 263, 264, 266–270, 272
nanoscale, 263, 269
nanotechnology, 264
National
   Cancer Institute (NCI), 105, 147, 187
   Center for Cell Science (NCCS), 130, 155, 193
   Institute of Standard and Technology (NIST), 155, 168, 223, 235
natural products, 45–47, 53, 65, 69, 82, 117, 161, 199, 208, 284
naturotherapy, 111
nausea, 104, 204
n-decanoic acid, 237, 238
necrosis cell, 288
necrotic cells, 157, 163
negative predictive value, 10
neonatal kidneys, 204
neoplasia, 9, 161
neovascularization, 88
nephropathy, 234, 276, 306
nephrotoxicity, 97
Nerium oleander, 79–83, 85, 89, 90
   extracts, 82
   therapeutic applications, 82
nerve growth factor (NGF), 88
neural expression, 88
neuroblastoma, 16, 51, 130
neuroectodermal tumors, 107
neurologic dysfunction, 60
neuropathy, 234, 306
neuroprotective
   activity, 90
   effect, 227
neutral drug molecules, 34
n-hexacosane, 333
Nigella sativa, 98, 200

Index

noncompetitive inhibitors, 25
non-flavonoids, 66
non-insulin dependent diabetic, 224
non-malignant tumors, 44
non-polar lipid environment, 32
nontoxic biopolymer, 249
non-volatile compounds, 200
novel antitumor medicine, 45
N-terminal lobe, 22, 23
nuclear
  factor
    erythroid 2-related factor 2 (Nrf2), 206
    kappa-light-chain-enhancer from activated B chain (NF-κB), 67, 80, 86, 87, 100, 148
  fragmentation, 136
  morphology, 136, 156
nutraceuticals, 143

O

oleander cultivation, 81
oleandrin, 79–90
  anti-cancer properties, 84
  chemistry, 80
    Nerium oleander L., 81
  cytotoxic effects, 84
    anti-tumor promoting effects, 86
    oleandrin and cancer, 84
  effects, 87
  poisoning, 89
    toxic dose, 89
  tumor cell growth-inhibitory effect, 87
    neuroprotective activity oleandrin, 88
    oleandrin and proliferation, 87
oleanolic acid, 48, 333
oleic acid, 183
oncogenesis, 208
oncogenic transformation, 61
opalescent suspension, 251, 252, 254
Opisthorchis viverrini, 5
oral
  cancer, 52
  doses, 38
  hypoglycemic agents, 306
organic compounds, 46, 222
organosulfur compounds, 59, 70–72
ornithine decarboxylase (ODC), 87
osteopontin, 8

osteosarcoma (OS), 51, 52, 88
ovarian
  cancers, 106, 149
  germ cell cancers, 49
oxalidaceae, 192, 276
oxidative stress, 71, 240, 242, 289, 310

P

pancreatic
  cancer cells, 86, 210
  capillaries, 288
  cells, 83, 87
  islets, 221, 345
Papanicolaou (Pap), 9, 12
parathyroid hormone (PTH), 17
parenchymatous tissue, 344
parkinsonism, 203
pathogenic kinases, 21
pathogens, 65, 95
peak area (PA), 87, 155, 158–160, 223
peptic ulcers, 14
peripheral neuropathy, 104, 276
permeability, 31–36, 40, 124, 125
peroxisome proliferatoractivated receptors (PPARs), 226
Petroselinum crispum, 310, 318
pharmaceutical areas, 334
pharmacodynamics (PD), 86
pharmacokinetics (PK), 30, 34–36, 86
  environments, 33
  evaluation, 126
  parameters, 120, 124, 126
  properties, 118
pharmacological activities, 100, 143, 203
pharmacology, 29, 40
phenolic
  acids, 66, 67, 237
  compounds, 66, 105, 110, 147, 245, 266, 334, 344
phenols, 153, 157, 158, 169, 237, 238, 293, 296, 309, 310, 333, 334
phenylpropanoid pathway, 68
pheochromocytoma, 16
phosphate group, 69
phosphoenolpyruvate carboxykinase (PEPCK), 308
phosphoprotein, 15
phosphorylation, 21, 84, 87

photodynamic therapy, 200
photomicrography, 193
photosynthesis, 70
physicochemical properties, 30, 36
   permeability, 35
   solubility, 34
phytochemical, 49, 59–61, 65, 68, 70, 107, 109, 110, 143, 150, 155, 177–179, 186–188, 226, 235, 245, 266, 267, 280, 296, 345, 346
   analysis, 223, 296, 334
   anticancer, 150
   components, 223, 226
   compounds, 154, 161, 294
   database, 177
   data-mining, 178
   drugs, 144
   leads, 144
   profile, 346
   screening, 155, 158, 168, 169, 235, 237, 238, 246, 295, 296, 302, 334
   substances, 344
phytochemistry, 138, 164, 175, 178, 343
phytocomponents, 227
phytocompounds, 157, 169, 177, 178, 185–187, 192, 228, 235, 344
phytoconstituents, 98, 205, 211
phytosterols, 59, 61, 63, 71, 72, 157, 158, 169, 237
picrocrocin, 199–202, 211, 212
Piper
   longum, 308, 312, 334
   nigrum, 142, 143, 149, 311
Pisum sativum, 168, 175
plasma, 16, 33, 85, 124, 125, 307, 309, 312–314, 317, 329, 330
   protein, 124, 125
      binding (PPB), 120, 124, 125
plasmon vibrations, 266, 267
plasticizer, 130–132, 134–137
plausibility, 11
pleiotropic effects, 61
pluripotent cells, 4
podophyllotoxin, 49, 52, 61
polar
   drug molecules, 36
   substituents, 35
   surface area (PSA), 32, 125
polyanions, 254

polycationic substance, 250
polyherbal
   drug, 250, 252–254, 256, 260, 261
   mixture, 252
polymannans, 107
polymerases, 205
polymeric carriers, 47
polymorphonuclear cells (PMNs), 206
polypeptide hormone, 284, 294
polyphenolic
   components, 310
   structure, 68
polyphenols, 59, 62, 65–67, 106, 136, 147, 157, 237, 289, 305, 310, 333, 345
polyphones, 65
polyphosphoric groups, 256
polysaccharide, 46, 250
Pombe cells, 119
positive predictive value, 10
positron emission tomography, 6
post-initiation periods, 209
potent anti-carcinogenic compound, 187
potential
   anti-diabetic properties, 306, 307
   catalytic sites, 172
   health benefits, 345
preclinical activities, 37
   dosing route pre-determination, 38
   formulation, 37
   preclinical tests, 37
preliminary phytochemical screening, 157, 235
premalignant lesions, 12, 17
pro-apoptotic properties, 210
prodrugs, 47
progenesis, 36
pro-inflammatory cytokines, 206
proliferation, 3, 53, 69, 83, 86, 88, 105, 130, 142, 171, 208–210
prostate
   cancer, 7, 8, 13, 49, 51, 52, 84, 85, 100, 206
   cells, 52, 63, 84, 85, 88, 100, 101
   specific antigen (PSA), 7
protein
   data bank (PDB), 169, 295
   domain structure, 22
   kinases, 22, 25, 26
   tyrosine kinases, 22
proteolytic enzymes, 136
proteomics, 12

Index 361

pseudo-molecular ion, 89
Pterocarpus marsupium, 219, 220, 222, 229

Q

quadruple double focusing analyzer, 155
quercetin, 51, 52, 62, 97, 183
   oligosaccharides, 106

R

radioimmunoassay kit, 307, 317–319
radiotherapy, 59, 60, 111, 178, 200
Rattus norvegicus, 265, 276, 280, 285
reactive oxygen species, 129, 345
relative light units (RLUs), 223
renal
   cell carcinoma, 51, 117, 126
   tumors, 306
renoprotective activities, 343
resveratrol, 51, 62, 97, 110
retention time (RT), 155, 158–160, 170, 237, 296
retinoblastoma cancer cells (RB355), 111
retinopathy, 203, 234, 276, 306
Rhamnaceae, 264, 284
rheumatoid arthritis, 15
rotatable bonds (RotB), 32, 124, 125

S

saffron, 96, 147, 199–212
   anticancer activity, 207
      breast and cervical cancer, 210
      hepatic cancer, 207
      leukemia, 208
      lung cancer, 209
      pancreatic cancer, 210
      skin cancer, 208
   current biomedical status, 203
safranal, 199–202, 204, 209, 211, 212
salicylic acid, 344
saponins, 46, 106, 107, 158, 237, 245, 296, 302, 305, 309, 334, 344
scanning electron microscopy (SEM), 249, 253, 256, 258, 261, 263, 265, 269
Schistosoma haematobium, 5
secondary metabolites, 45, 46, 66, 68, 80, 177, 192, 195, 235, 263, 284, 343–345
   Aloe arborescens, 346
   Aloe excelsa, 346
   Aloe ferox, 345
   Aloe greatheadii, 345
   Aloe nyeriensis, 346
   Aloe vera, 344
semi-autoanalyzer, 236, 286
serum
   blood glucose, 236, 253, 254, 261, 277
   insulin, 223, 224, 236, 241, 253, 254, 263, 270, 271, 277, 278, 280, 283, 286, 289, 290
sesquiterpenoids, 70, 333, 335
shikimic-acid, 184
sigmoidoscopy, 14
silver nanoparticles (AgNPs), 263–272
small-molecule kinase inhibitors (SMKIs), 22
solubility, 29, 33–36, 40, 249
soxhlet apparatus, 154, 192, 235, 295
squamous intraepithelial lesion (SIL), 12
standard
   deviation (SD), 157, 161, 162, 237, 314
   drug glibenclamide, 270, 272, 286, 287
   error of mean (S.E), 132–134
   mass spectra, 155
statistical
   analysis, 224, 246
   package for the social sciences (SPSS), 156, 157, 237, 254
steroidal lactones, 110
steroids, 237, 296
sterols, 71, 344, 345
stigma, 147, 201, 202, 204
stilbenes, 66, 68
Streprocaulon tomentosum, 83
streptomycin, 119, 155, 193
streptozotocin (STZ), 224, 233, 234, 236, 239–241, 244, 245, 253, 258, 259, 263, 265, 270, 271, 278, 283–291, 307–310, 333
structural-activity relationships (SARs), 29
Student-Newman Keuls (SNK), 237, 242, 244, 254, 260
subcutaneous (SC), 38, 219, 221
supercoils, 205
superoxide
   dismutase (SOD), 206
   radicals, 154
suppository doses, 38
surfactants, 35
surrogate endpoint, 11, 12, 17
synthetic
   anti-cancer drugs, 53
   drugs, 46, 283, 284

Syzygium aromaticum, 142, 143, 149, 318

T

Tagetes erecta, 153, 154, 164
tannins, 110, 157, 158, 169, 238, 289, 296, 305, 309, 310, 333, 334
tartaric acid, 147
taxifolin, 108
T-cell neoplasms, 16
telomerase activity, 147, 205
teratogenicity, 37
terpenoids, 46, 59, 69, 70, 153, 157, 158, 226, 227, 237, 280, 296
tetradecanoylphorbol-13-acetate (TPA), 87, 88
therapeutic
　activities, 85
　agents, 103, 111, 161, 210
　compounds, 22, 98
　drugs, 84
　effects, 62, 67, 82, 87
　index, 82, 89
　potential, 59, 126, 200
　properties, 66, 80, 98, 100
　systems, 45
therapeutical potentials, 79
therapy optimization study, 40
thermodynamic solubility, 35
threonine kinases, 22, 26
thrombocytopenia, 104
thymoquinone, 96, 98, 100, 101, 200
thyrocalcitonin, 16
thyroid cancer cell, 62
topoisomerase I (Topo I), 47–49, 148
toxic dose, 39, 84, 90
transitional cell carcinoma (TCC), 52, 208
trans-membrane proteins, 36
transmission electron microscope (TEM), 263, 265, 269
triglyceride (TG), 237, 244, 245, 308
Trigonella foenum-graecum, 142, 143, 150
tripolyphosphate (TPP), 250–252, 254, 256
triterpenoids, 64, 70, 169, 237, 238, 245, 296
trophoblastic diseases, 17
tumor
　cell, 3, 4, 14, 36, 50, 66, 82–86, 95, 96, 98–100, 171
　growth-inhibitory effects, 79
　necrosis factor (TNF), 85, 88, 106, 147, 149, 205, 206
　-α (TNF-α), 106, 206
　proliferation, 14, 84
tumorigenesis, 14, 62, 68, 88, 171
tumorigenic micro-organisms, 49
type 2 diabetes, 220, 221, 227, 229

U

ultrasound, 6, 13
ultraviolet (UV), 5, 263, 267
University Grant Commissions (UGC), 164, 174, 211, 212, 261, 272, 280, 291, 347

V

vanillic acid, 184, 187
vanillylmandelic acid (VMA), 16
vascular endothelial growth factor, 147
Ventilago maderaspatana, 263, 264, 272, 283, 284, 291
vernacular names, 148–150
very low density lipoprotein (VLDL), 233, 244, 245
vinblastine, 61, 108, 200
vincristine (VCR), 48, 61, 65, 108, 118, 200
visual inspection with acetic acid (VIA), 9

W

World Health Organization (WHO), 4–6, 95, 105, 220, 276, 294, 306

X

xenograft prostate tumors, 100
x-ray diffraction (XRD), 249, 255, 256, 261
　analysis, 255, 261
　study (XRD study), 252

Z

zeaxanthin, 202, 211
Zingiber officinale, 110, 179–184, 187, 200, 320, 333
Zingiberaceae (ginger family), 109, 142, 143, 148, 177–179, 184, 185, 188, 311, 320, 321